Lecture Notes in Mathematics

1642

Editors:
A. Dold, Heidelberg
F. Takens, Groningen

T0213440

Springer
Berlin
Heidelberg
New York
Barcelona
Budapest
Hong Kong
London
Milan
Paris
Santa Clara
Singapore
Tokyo

Michael Puschnigg

Asymptotic Cyclic Cohomology

Springer

Author

Michael Puschnigg
Mathematics Institute
University of Heidelberg
Im Neuenheimer Feld 288
D-69120 Heidelberg, Germany
e-mail: puschnig@mathi.uni-heidelberg.de

Cataloging-in-Publication Data applied for

Die Deutsche Bibliothek – CIP-Einheitsaufnahme

Puschnigg, Michael:
Asymptotic cyclic cohomology / Michael Puschnigg. – Berlin; Heidelberg; New
York; Barcelona; Budapest; Hong Kong; London; Milan; Paris; Santa Clara;
Singapore; Tokyo: Springer, 1996
  (Lecture notes in mathematics; 1642)
  ISBN 3-540-61986-0

NE: GT

Mathematics Subject Classification (1991): 19D55, 18G60, 19K35, 19K56

ISSN 0075-8434
ISBN 3-540-61986-0 Springer-Verlag Berlin Heidelberg New York

Typesetting: Camera-ready TeX output by the author
SPIN: 10520141   46/3142-543210 - Printed on acid-free paper

Introduction

This work is a contribution to the study of topological K-Theory and cyclic cohomology of complete normed algebras. The aim is the construction of a cohomology theory, defined by a natural chain complex, on the category of Banach algebras which

a) is the target of a Chern character from topological K-theory (resp. bivariant K-theory).

b) has nice functorial properties which faithfully reflect the properties of topological K-theory.

c) is closely related to cyclic cohomology but avoids the usual pathologies of cyclic cohomology for operator algebras.

d) is accessible to computation in sufficiently many cases.

The final goal is to establish a Grothendieck-Riemann-Roch theorem for the constructed Chern character which for commutative C^*-algebras reduces to the classical Grothendieck-Riemann-Roch formula.

In his "Noncommutative Geometry" Alain Connes has developed the framework for a large number of far reaching generalisations of the index theorems of Atiyah and Singer. To motivate the problem addressed in this book and to put it in the right context we recall some basic principles of index theory and noncommutative geometry.

The classical index theorem for an elliptic differential operator D on a compact manifold M identifies the Fredholm index of this operator with the direct image of the symbol class of the operator under the Gysin map in topological K-Theory:

$$Ind_a(D) = \pi!(\sigma(D))$$

$$\pi! : K^*(T^*M) \to K^*(pt.) \simeq \mathbb{Z}$$

In more general situations where one considers not necessarily compact manifolds (for example operators on the universal cover of a compact manifold which are invariant under deck transformations, operators on a compact manifold differentiating only along the leaves of a foliation and being elliptic on the leaves, or elliptic operators of bounded geometry on an open manifold of bounded geometry) the considered elliptic operators are not Fredholm operators anymore. Nevertheless it is still possible to associate an index invariant with them which now has to be interpreted as an element of the operator K-group of some C^*-algebra. Moreover, Kasparov and Connes proved a number of very general index theorems of the form:

$$Ind_a(D) = \pi!(\sigma(D)) \in K_0(C^* - algebra)$$

The C^*-algebras occuring in this way can be of quite general type and their K-groups usually cannot be identified with the K-groups of some topological space as in the classical cases.

As far as applications are concerned, the classical index theorem, formulated and proved in the context of topological K-theory, gains its full power only after being translated into a cohomological index formula with the help of a differentiable Grothendieck-Riemann-Roch Theorem. This theorem claims that for any K-oriented map $f : X \to Y$ of smooth compact manifolds the diagram

$$
\begin{array}{ccc}
K^*(X) & \xrightarrow{\;f!\;} & K^*(Y) \\
ch \downarrow & & \downarrow ch \\
H_{dR}^*(X) & \xrightarrow[f_*(-\cup Td(f))]{} & H_{dR}^*(Y)
\end{array}
$$

commutes. Here

$$
ch : K^* \to H_{dR}^*
$$

denotes the Chern character which is given by a universal characteristic class that identifies complexified topological K-theory of a manifold with its de Rham cohomology:

$$
ch : K^*(M) \otimes_{\mathbb{Z}} \mathbb{C} \xrightarrow{\;\simeq\;} H_{dR}^*(M).
$$

Under this translation the direct image in K-theory can be identified with an explicit pushforward map in cohomology. Together, the index and Grothendieck-Riemann-Roch theorem yield a formula expressing the Fredholm index of an elliptic operator D as integral over the manifold of a universal characteristic class associated to the symbol of D:

$$
Ind_a(D) = \int_M \text{characteristic class}(\sigma(D))
$$

To obtain index formulas from the generalized index theorems above it is necessary to develop a Grothendieck-Riemann-Roch formalism in the context of operator K-theory. This means that one looks for a (co)homology theory on the category of C^*-, Banach-, resp. abstract algebras, which is defined by a natural chain complex and carries enough additional structure to provide a commutative diagram

$$
\begin{array}{ccc}
K_*(A) & \xrightarrow{\;f!\;} & K_*(B) \\
ch \downarrow & & \downarrow ch \\
H_*(A) & \xrightarrow[?]{} & H_*(B)
\end{array}
$$

On the subcategory of algebras of smooth (resp. continuous) functions on compact manifolds it should correspond to the classical Grothendieck-Riemann-Roch theorem.

So the Grothendieck-Riemann-Roch problem consists of three parts:

1. Define a (co)homology theory for Banach- (C^*-) algebras which generalizes the deRham (co)homology of manifolds.

2. Construct a Chern-character from K-theory to this noncommutative deRham-(co)homology.

3. Find a cohomological pushforward map and establish a suitable Grothendieck-Riemann-Roch theorem.

After having formulated this program, Alain Connes also made the first real breakthrough concerning a solution of the problem. In his foundational paper "Noncommutative Differential Geometry" [CO] he introduced a generalization of de Rham theory in the noncommutative setting, cyclic (co)homology HC_* (resp. HC^*), which can be calculated as the (co)homology of a functorial chain complex vanishing in negative dimensions , and he constructed an algebraically defined Chern character

$$ch : K_* \to HC_* .$$

The dual Chern character pairing

$$ch : K_* \otimes HC^* \to \mathbb{C}$$

generalizes the pairing between idempotent matrices and traces in degree zero and the pairing between invertible matrices and closed one-currents on the given algebra in degree one.

Cyclic cohomology proved to be a very powerful tool in many areas of K-theory, as the large number of well known applications shows. The project of constructing characteristic classes for operator K-theory however soon faced serious difficulties. Whereas the $\mathbb{Z}/2\mathbb{Z}$-periodic version

$$HP^* := \lim_{\to_k} HC^{*+2k}$$

of cyclic cohomology of the algebra of smooth functions on a manifold coincides with the deRham homology of the manifold,

$$HP^*(\mathcal{C}^\infty(M)) \simeq H_*^{dR}(M) ,$$

the periodic cyclic cohomology of its enveloping C^*-algebra of continuous functions equals the space of Borel measures on M in even degree and vanishes in odd degree.

$$HP^*(C(M)) \simeq \begin{cases} C(M)' & * = 0 \\ 0 & * = 1 \end{cases}$$

Thus while the Chern character pairing between reduced K-theory and reduced periodic cyclic cohomology yields a perfect pairing for the Fréchet algebra $\mathcal{C}^\infty(M)$, it vanishes for its enveloping C^*-algebra $C(M)$. (Note that both algebras can be considered as equivalent as far as K-theory is concerned). This example shows how cyclic cohomology and K-theory can behave quite differently in certain situations and that the Chern character from K-theory to cyclic homology can be far from being an isomorphism.

Actually the pathological behaviour of the Chern character pairing for (stable) C^*-algebras has nothing to do with the particular structure of cyclic cohomology but is a consequence of the continuity of the Chern character as the following argument shows:

Let C_* be any cyclic theory, i.e. a functor from Banach algebras to chain complexes equipped with a Chern character $ch : K_* A \to h(C_* A)$ associating a cycle to each idempotent (resp. invertible) matrix over A. Let φ be an even cocycle for this theory (the argument for odd cocycles is similar). This cocycle yields a map (still denoted by the same letter)

$$\varphi : \{ e \in A, \ e^2 = e \} \to \mathbb{C}$$

which provides the pairing of the cohomology class of φ with $K_0(A)$.

Suppose that the Chern character pairing satisfies the following conditions:
(They hold for the Chern character pairings with continuous periodic cyclic cohomology HP^* and with entire cyclic cohomology HC_ϵ^*.)

1) $\varphi(e)$ depends only on the homotopy class of e.

2) $\varphi(e) = \varphi(e') + \varphi(e'')$ if $[e] = [e'] + [e'']$ in $K_0(A)$.

3) $|\varphi(e)| \leq F(\| e \|)$ for some function F on the real half-line.

Then if A happens to be a stable C^*-algebra, the pairing $K_* A \otimes h(C^* A) \to \mathbb{C}$ equals zero:

In fact one observes that the image of the map φ, viewed as a subset of $\mathbb{C}$, is closed under addition because A is stable and condition 2) holds. On the other hand this image is bounded by conditions 1) and 3), as any idempotent in a C^*-algebra is homotopic to a projector (selfadjoint idempotent) and nonzero projectors in C^*-algebras have norm 1. So the image of φ is a bounded subset of $\mathbb{C}$ closed under addition and thus zero.

This fact is quite annoying because the generalized index theorem and the hypothetical Grothendieck-Riemann-Roch are theorems about C^*-algebras and do not hold for more general Banach or Fréchet algebras (bivariant K-theory is well behaved only for C^*-algebras). Moreover, it is just the study of the K-theory and the cohomology of C^*-algebras which is at the heart of the most important applications: in the index-theoretic approach to the Novikov-conjecture on higher signatures of manifolds, for example, one has to analyse the K-theory and cyclic cohomology of the group-C^*-algebra $C_{red}^*(\Gamma)$ of the fundamental group of the manifold under consideration. Finally another difficulty in establishing a Grothendieck-Riemann-Roch formula is that the pushforward maps of operator K-theory have no counterpart in cyclic homology.

Connes and Moscovici defined in [CM] a modified version of cyclic cohomology, called asymptotic cyclic cohomology, and pointed out that this theory should provide a nontrivial cohomology theory on the category of C^*-algebras. Our work can be viewed as attempt to realize this program. This also explains the title of the book. The initial setup of asymptotic cyclic cohomology in [CM] had to be modified in several ways and the theory we are going to develop is however not equivalent to the one originally defined by Connes and Moscovici.

Our aim is to develop a cyclic theory, called asymptotic cyclic cohomology after [CM], which is the target of a Chern character that appropriately reflects the structure and the typical properties of operator K-theory. The theory will generalize ordinary and entire cyclic cohomology providing thus a framework for the explicit construction of (geometric) cocycles and the calculation of their pairing with concrete elements of K-groups. Finally we establish a generalized Grothendieck-Riemann-Roch theorem for the Chern character from operator K-theory to stable asymptotic homology. This will be achieved by the construction of a bivariant Chern character on Kasparovs bivariant K-theory with values in bivariant stable asymptotic cyclic cohomology.

The above argument for the vanishing of the Chern character pairing gives a first hint how one has to modify cyclic cohomology to get a theory with the desired properties. Cochains should consist of densely defined and unbounded rather than of bounded functionals or, as Connes-Moscovici propose in [CM], continuous families of unbounded cochains with larger and larger domains of definition.

To realize our goal we however start from a quite different line of thought. Our point of departure is on one hand the work of Connes, Gromov and Moscovici [CGM] on almost flat bundles and of Connes and Higson [CH] on asymptotic morphisms and bivariant K-theory, and on the other hand the work of Cuntz and Quillen [CQ] on cyclic cohomology and universal algebras.

In [CH] Connes and Higson made the important observation, that K-theory becomes in a very natural way a functor on a much bigger category than the ordinary category of Banach (C^*-algebras), namely on the category with the same objects but with the larger class of so called "asymptotic morphisms" as maps. Especially they showed that every pushforward map in K-theory associated to a generalized index theorem is induced from an explicitely constructible asymptotic morphism of the C^*-algebras involved.

A (linear) asymptotic morphism of Banach algebras is a bounded, continuous family $(f_t, t > 0)$ of continuous (linear) maps $f_t : A \to B$ such that

$$\lim_{t \to \infty} f_t(aa') - f_t(a)f_t(a') = 0 \ \forall a, a' \in A$$

The deviation from multiplicativity

$$\omega(a, a') := f_t(aa') - f_t(a)f_t(a')$$

is called the curvature of f_t at (a, a').

The interest in this notion originates (among other things) from the fact, that the E-theoretic K-groups, which are a modification of Kasparov's KK-groups, can be described as groups of asymptotic morphisms.

A cohomology theory that is the target of a good Chern character on operator K-theory should certainly have the same functorial properties as K-theory itself. Cyclic (co)homology however is by no means a functor on the asymptotic category. Therefore it is no surprise that the Chern character in cyclic homology fails to be an isomorphism in general.

On the other hand Connes, Gromov and Moscovici showed in [CGM], that the pullback of a trace τ on an algebra B under a linear map $f : A \to B$ may be interpreted as an even cocycle in the cyclic bicomplex of A:

$$f^* \tau = \sum_{n=0}^{\infty} \varphi^{2n} .$$

Moreover its components (φ^{2n}) decay exponentially fast

$$|\varphi^{2n}(a^0, \ldots, a^{2n})| \leq C^{-n}$$

when evaluated on tensors with entries $a^0, \ldots, a^{2n}$ belonging to a fixed finite subset Σ of A. The constant C depends on the deviation of f from being multiplicative on Σ.

Cochains with this growth behaviour occur already in the calculations of localized analytic indices of Connes and Moscovici [CM], where the authors point out that a cyclic theory for C^*-algebras should be based on such cocycles.

Relating this to the approach to cyclic cohomology via traces on universal algebras by Cuntz and Quillen [CQ] suggests that it might be possible to pull back arbitrary cochains in the cyclic bicomplex under linear maps and that in fact every even(odd)-dimensional cocycle in the cyclic bicomplex could be obtained as the pullback of a trace (resp. a closed one-current) under a linear map.

Thus one might hope to reinterpret cyclic cohomology as being given by a chain complex that behaves functorially under linear maps and to obtain an asymptotic cyclic theory as the envelope under linear asymptotic morphisms of the ordinary cyclic theory. Cochains in this theory should be characterized by natural growth (resp. continuity) conditions as in the example above. In fact any cyclic theory which is functorial under asymptotic morphisms would possess the pushforward maps necessary to formulate a GRR theorem.

So our starting point for the construction of asymptotic cyclic cohomology will be to take ordinary cyclic theory and to extend it to a functor on the linear asymptotic category $\mathcal{C}$. (We restrict ourselves to linear asymptotic morphisms. It would have been possible to dispense with this restriction but only at the cost of making the formulas much more complicated without providing a wider range of applications.) This means the following. First we choose a natural chain complex C^* calculating cyclic cohomology, i.e. a functor

$$C^* : \text{Algebras} \to \text{Chain Complexes}$$

such that

$$H^*(C^*) \simeq HC^* .$$

Then we consider pairs (C_α^*, Φ) consisting of

a)
a functor

$$C_\alpha^* : \mathcal{C} \to \text{Chain Complexes} ,$$

such that the corresponding homology groups define a homotopy functor

$$HC_\alpha^* := H^*(C_\alpha^*) : Homot\,\mathcal{C} \to \mathbb{C} - \text{Vector Spaces}$$

b)
a morphism of functors

$$\Phi : C^* \to C_\alpha^*|_{Algebras}$$

on the category of algebras inducing a natural transformation

$$HC^* \to HC_\alpha^*$$

from ordinary to asymptotic cyclic cohomology.

Among all such pairs we look for a minimal one, i.e. a pair satisfying the obvious universal property. By an argument due to J.Cuntz any such cohomology theory will be Bott-periodic, so that C_α^* (and C^*) should in fact be $\mathbb{Z}/2\mathbb{Z}$-graded complexes.

In [CO] Connes introduced a natural $\mathbb{Z}/2\mathbb{Z}$-graded complex, the (b,B)-bicomplex CC_* of a unital algebra. An equivalent (but not identical) complex Ω_*^{PdR}, the periodic de Rham complex, has been constructed later on by Cuntz and Quillen [CQ]. These are both complexes of modules of formal differential forms over the given algebra and carry a natural filtration (Hodge filtration), derived from the degree filtration on differential forms. The quotient complexes with respect to the Hodge filtration successively compute the cyclic homology groups HC_* and the completed complexes $\widehat{\Omega}_*^{PdR}$ (with respect to the Hodge filtration) calculate the periodic cyclic homology HP_* of Connes. The periodic de Rham complex provides in our opinion the best choice for the complex C^* above and it is therefore Ω_*^{PdR} that will be extended to a functor on the linear asymptotic category.

The universal problem above can be solved provided that the forgetful functor

$$\text{Banach-algebras} \to \mathcal{C}$$

has a right adjoint $R_\mathcal{C}$. An explicit solution would then be given by

$$\widehat{\Omega}_*^{PdR,\alpha} := \widehat{\Omega}_*^{PdR} \circ R_\mathcal{C}$$

If one forgets the topology for the moment and looks at the problem at a purely algebraic level, there is indeed an adjoint, provided by a canonical quotient of the full tensor algebra:

$$RA := TA/(1_A - 1_\mathbb{C})$$

This would lead to

$$\widehat{\Omega}_*^{PdR,\alpha}(A) = \widehat{\Omega}_*^{PdR}(RA)$$

The algebras RA are of Hochschild cohomological dimension one, which makes it possible to calculate their periodic cyclic homology via the quotient complex of the periodic de Rham complex by the second step of the Hodge filtration, the so called X-complex of Cuntz-Quillen:

$$\widehat{\Omega}_*^{PdR}(RA) \xrightarrow{qis} X_*(RA)$$

where the X-complex is given by

$$X_*(A): \quad \to \quad A \xrightarrow{d} \Omega^1 A/[\Omega^1 A, A] \xrightarrow{b} A \quad \to$$

In fact, Cuntz and Quillen [CQ] showed that cyclic homology can be developed starting from the X-complex of tensor algebras (resp. quasifree algebras). Moreover one obtains in this way a very natural and advantageous viewpoint of the basic features of the theory.

A basic observation is that the tensor algebras RA are canonically filtered by powers of the ideal

$$0 \to IA \to RA \xrightarrow{mult} A \to 0$$

So although the algebra RA depends only on the underlying vector space of A, the I-adic filtration on RA makes it possible to recover the multiplicative structure of A. Remarkably, the X-complex of RA with its I-adic filtration turns out to be quasiisomorphic, as filtered complex, to the periodic de-Rham complex of A with its Hodge filtration. So whereas the complex $X_*(RA)$ is easy to manipulate algebraically it also contains all information encoded in the periodic de Rham complex of A with its Hodge filtration. Especially one recovers the periodic cyclic homology of A as the homology of the X-complex of the (algebraic) I-adic completion of RA:

$$\widehat{\Omega}_*^{PdR}(A) \xleftarrow{qis} X_*(\widehat{RA})$$

$$HP_*(A) = H_*(X_*(\widehat{RA}))$$

In fact, the I-adic completion $\widehat{RA}$ of RA is still of cohomological dimension one although quite far from being free.

The description of periodic cyclic (co)homology using the X-complex of tensor algebras exhibits the functoriality of the (uncompleted) cyclic complexes with respect to linear maps which is crucial for us but somewhat hidden if one uses Connes original cyclic (b, B)-bicomplex.

Moreover the Cuntz-Quillen approach enables one to construct product operations and homotopy operators for cyclic theories on the level of chain complexes by a uniform procedure. One tries to guess the right formulas for the periodic de Rham complex on differential forms of degree zero and one modulo error terms of higher degree. For free algebras, which are of Hochschild cohomological dimension one, the second step of the Hodge filtration is contractible, so that it becomes possible to get rid of the error terms in this case. This yields by passing to the quasiisomorphic quotient complexes a map of X-complexes of free algebras. For free algebras of the form RA one finally recovers by taking the associated graded complexes with respect to the I-adic filtration the whole periodic de Rham complex of the initial algebra A, this time with a globally defined chain map reducing to the initial formula on forms of low degree. As homotopic initial maps on forms of low degree provide homotopic global chain maps in the end, the effect of the constructed chain maps on homology is determined by their effect on ordinary cyclic homology of degree zero and one, respectively.

There is a "Cartesian square" of functors

$$\begin{array}{ccc} \text{Algebras} & \xrightarrow{\text{forget}} & \text{Algebras, linear maps} \\ {\scriptstyle R,I-adic filt}\downarrow & & \downarrow {\scriptstyle R} \\ \text{Filtered Alg.} & \xrightarrow{\text{forget}} & \text{Algebras} \end{array}$$

on the level of morphism sets.

This shows that the I-adic filtrations on the complexes $X_*(RA)$ are never preserved by a homomorphism of tensor algebras which is induced by a linear morphism that is not multiplicative. Therefore not the degree, but only the parity of an ordinary cyclic cycle is preserved under pushforward by a linear morphism. In fact any even (odd) cocycle (in the $\mathbb{Z}$-graded setting) occurs as the linear pullback of a trace (closed one-current). This explains again why only a $\mathbb{Z}/2\mathbb{Z}$-graded theory can be defined on the linear asymptotic category.

Concerning the original aim of making cyclic cohomology functorial under linear asymptotic morphisms our goal can be described (in terms of the Cuntz-Quillen approach) as follows.

Consider the diagram

$$\begin{array}{ccccccc} \text{Morphisms:} & \text{linear} & & \epsilon\text{-mult.} & & \text{mult.} & \\ \text{Algebras:} & A & = & A & = & A & \\ & \downarrow & & \downarrow & & \downarrow & \\ \text{Algebras:} & RA & \subset & \mathcal{R}A =? & \subset & \widehat{RA} & \\ & \downarrow & & \downarrow & & \downarrow & \\ \text{Chain complexes:} & X_*(RA) & \subset & X_*(\mathcal{R}A) & \subset & X_*(\widehat{RA}) & \end{array}$$

In the right column the Cuntz-Quillen procedure for obtaining the cyclic complex of A is described. The universal way to extend this construction to the category of algebras with linear maps as morphisms is given in the left column: one replaces the given algebra by its tensor algebra and constructs the cyclic complex of the latter algebra. The tensor algebra already being free one can directly pass to its X-complex. The complex $X_*(RA)$ cannot be interesting homologically however. It has to be contractible because every linear map is linearly homotopic to zero. Being interested in a nontrivial homology theory which is functorial under asymptotic morphisms, i.e. a functor on a "category of ϵ-multiplicative linear maps" we have to look for an intermediate theory. One has to find a topological completion of the tensor algebra RA which is not contractible but functorial under ϵ-multiplicative maps. If it is moreover of cohomological dimension one one can again take its X-complex to arrive at a reasonable theory (middle column). Such a completion is constructed as follows.

Let $f : A \to B$ be an almost multiplicative linear map of Banach algebras. Then the induced homomorphism $Rf : RA \to RB$ of tensor algebras will not preserve I-adic filtrations but the norms of the occuring "error terms" will decay exponentially fast with their I-adic valuation. This suggests the following construction: Fix a multiplicatively closed subset K of A and consider tensors over A with entries in K. Expand a given element of this subalgebra of RA in a standard basis with respect to the I-adic filtration. A weighted L^1-norm for the coefficients of such an expansion is then introduced allowing the coefficients to grow exponentially to the basis $N > 1$ with respect to the I-adic valuation. Denote the corresponding completion by $RA_{(K,N)}$. It is a Fréchet algebra and possesses the following crucial property: If $f : A \to B$ is linear with curvature uniformly bounded on $K \subset A$ by a sufficiently small constant then Rf induces a continuous homomorphism $Rf : RA_{(K,N)} \to RB_{(K',N')}$ for suitable $K' \subset B, N' > 1$. Usually f will be a linear asymptotic morphism. As the curvature of an asymptotic morphism is uniformly bounded only over compact sets, the multiplicatively closed subsets $K \subset A$ used for the construction above will always be assumed to be compact. It turns out that the algebras $RA_{(K,N)}$ are also of cohomological dimension one.

The Fréchet algebras $RA_{(K,N)}$ form an inductive system with formal inductive limit $\mathcal{R}A$. This limit could be called the topological I-adic completion of RA. It should be viewed as virtual infinitesimal thickening of A as the kernel of the projection $\pi : \mathcal{R}A \to A$ is formally topologically nilpotent (i.e. the spectrum of its elements equals zero).

We define the analytic X-complex X^ϵ_* of a Banach algebra to be the reduced X-complex of the topological I-adic completion of the tensor algebra of its unitalization. The cohomological analytic X-complex is closely related to the entire cyclic bicomplex of Connes. It turns out to be convenient to introduce also a bivariant analytic X-complex $X^*_\epsilon(-,-)$ of a pair of algebras as the Hom-complex of the associated analytic X-complexes. The bivariant analytic X-complex is a bifunctor on the category of Banach algebras and its cohomology groups are smooth homotopy bifunctors. There exists an obvious composition product

$$X^*_\epsilon(A, B) \otimes X^*_\epsilon(B, C) \to X^*_\epsilon(A, C) .$$

The fundamental functoriality of the locally convex algebras $RA_{(K,N)}$ under almost multiplicative linear maps implies that every linear asymptotic morphism

$$f_t : A \to B, t > 0$$

induces a continuous homomorphism of formal inductive limit algebras

$$\mathcal{R}f : \mathcal{R}A \to \mathcal{R}B \otimes_\pi \mathcal{O}_\infty(\mathcal{R}^\infty_+) .$$

Here $\mathcal{O}_\infty(\mathcal{R}^\infty_+)$ is the algebra of germs around ∞ of smooth functions on the asymptotic parameter space $\mathcal{R}^\infty_+$. This leads one to define the (cohomological) asymptotic X-complex $X^*_\alpha(A)$ of a Banach algebra A as the cohomological X-complex of the formal topological I-adic completion $\mathcal{R}A$ with coefficients in the formal inductive limit algebra $\mathcal{O}_\infty(\mathcal{R}_+^\infty)$. The bivariant asymptotic X-complex $X_\alpha^*(A, B)$ of the pair (A, B) is introduced as the complex of germs at ∞ of homomorphisms between the X-complexes of the formal topological I-adic completions $\mathcal{R}A$ and $\mathcal{R}B$ (See chapter 6). By construction any linear asymptotic morphism defines an even cocycle in the bivariant asymptotic X-complex. The composition product carries over to the asymptotic setting and turns $X_\alpha^*(-, -)$ into a bifunctor on the linear asymptotic category. Moreover bivariant asymptotic cohomology becomes a (continuous) asymptotic homotopy bifunctor.

So much for the motivation and definition of the asymptotic cyclic theory. We have to be more precise at one point however. Asymptotic morphisms do not consist of a single, but of whole families of linear maps, and one has to keep track of the chain homotopies provided by evaluation at different "parameter values" in such families.

We do this by working throughout in the category of differential graded algebras and differential graded chain complexes. The asymptotic X-complex of the universal enveloping differential graded algebra of the given algebra is large enough to contain the higher homotopy information needed. One obtains then Cartan homotopy formulas for the "change of asymptotic parameters".

There are natural maps

$$CC^* \to X_\alpha^*, \qquad CC_\epsilon^* \to X_\alpha^*$$

in the derived category yielding natural transformations

$$HP^* \to HC_\alpha^*, \qquad HC_\epsilon^* \to HC_\alpha^*$$

on cohomology.

For the algebra of complex numbers the maps on cohomology above are isomorphisms. More generally, analytic and asymptotic homology coincide:

$$HC_\alpha^*(\mathbb{C}, A) \simeq HC_\epsilon^*(\mathbb{C}, A) \ .$$

The corresponding cohomology groups are in general quite different however.

The well known pairings between cyclic theories and K-theory extend to a pairing $K_* \otimes HC_\alpha^* \to \mathbb{C}$. It is uniquely determined by its naturality with respect to asymptotic morphisms and by demanding that it restricts to the classical pairing between idempotents and traces (resp. invertible elements and closed one-currents) on the ordinary cyclic complex. As for a given value of the asymptotic parameter a cocycle is given by a sequence of densely defined multilinear functionals on the underlying algebra A, the pairing can be defined for this choice of parameter only for special representatives of a finite number of classes in K_*A. Taking a family of parameter values which approaches ∞ in the asymptotic parameter space allows to define the pairing on larger and larger subsets of K_*A which finally exhaust the whole K-group and yield the pairing globally. This behaviour explains why the argument at the beginning of the introduction showing the pathological nature of the Chern chracter pairing for the classical cyclic theories on stable C^*-algebras does not apply to the asymptotic theory. Indeed there is a large class of stable C^*-algebras for which the pairing of K-theory with asymptotic cohomology is nondegenerate.

The most striking new phenomenon of asymptotic cyclic theory is that inclusions of holomorphically closed subalgebras become cohomology equivalences in many cases. This often allows one to construct asymptotic cocycles on C^*-algebras by lifting well known cyclic cocycles from a suitable dense subalgebra.

Since these subalgebras are not Banach algebras anymore, we develop the theory for the slightly larger class of admissible Fréchet algebras, i.e. Fréchet algebras possessing an analogue of the open unit ball of Banach algebras. These algebras seem to provide the natural framework for our theory.

The descent principle to holomorphically closed dense subalgebras can be used to show that asymptotic cyclic cohomology is stably Morita invariant: for any C^*-algebra A the inclusion

$$A \hookrightarrow A \otimes_{C^*} \mathcal{K}(\mathcal{H})$$

induces an asymptotic (co)homology equivalence.

In particular

$$HC_\alpha^*(\mathcal{K}(\mathcal{H})) = \begin{cases} \mathbb{C} & * = 0 \\ 0 & * = 1 \end{cases}$$

in sharp contrast to the cyclic theories known so far.

In order to go further it is necessary to develop product operations. By the principles explained above we are able to construct a chain map

$$\times : X_* R(A \otimes B) \to X_* RA \widehat{\otimes} X_* RB$$

which is associative up to homotopy and yields exterior products

$$X_{\epsilon,\alpha}^*(A) \widehat{\otimes} X_{\epsilon,\alpha}^*(B) \to X_{\epsilon,\alpha}^*(A \otimes_\pi B)$$

both for analytic and asymptotic cohomology.

It behaves naturally with respect to asymptotic morphisms. Moreover, the pairing of K-theory with analytic (resp. asymptotic) cohomology is compatible with exterior products. To be precise, the compatibility of the products in K-theory resp. the cyclic theories holds only up to a factor $2\pi i$ if the involved classes are of odd dimension: the cyclic theories are a priori $\mathbb{Z}/2\mathbb{Z}$-graded, whereas the product of odd classes in K-theory has to be defined using Bott periodicity, which causes the "period" factor $2\pi i$. This makes me believe that the exterior product on cohomology coincides up to normalization constants with Connes's product. I have not investigated this point however.

The attempt to define an exterior product of bivariant X-complexes was only partially successful up to now. The main difficulty lies in the construction of a homotopy inverse of the exterior product map for the ordinary X-complexes above. (See [P], where meanwhile a natural homotopy inverse has been constructed.) At least it is possible to establish a particular consequence of a bivariant product operation, namely the existence of a slant product

$$K_*(A) \otimes HC^*_{\epsilon,\alpha}(A \otimes_\pi B) \to HC^*_{\epsilon,\alpha}(B)$$

It is constructed in such a manner that any idempotent (or invertible) matrix over A gives rise to an explicit map $X^*_{\epsilon,\alpha}(A \otimes_\pi B) \to X^*_{\epsilon,\alpha}(B)$ of chain complexes. Its homotopy class depends only on the K-theory class of the given matrix. The slant product behaves naturally with respect to asymptotic morphisms and is compatible with the exterior product. It represents a convenient tool to prove the split injectivity of the exterior product with cohomology classes in the image of the Chern character. As an application we show that the exterior (resp. slant) product with the fundamental class of the circle yields an isomorphism

$$HC^*_\alpha(S\mathbb{C}, S\mathbb{C}) \simeq HC^*_\alpha(\mathbb{C}, \mathbb{C})$$

of the bivariant asymptotic cohomology of $\mathbb{C}$ and its suspension $S\mathbb{C} = C_0(I\!\!R)$. Extending this argument from $\mathbb{C}$ to more general admissible Fréchet algebras A by taking the exterior product with the bivariant cohomology class of the identity on A unfortunately fails: the exterior product is only defined for unital algebras and unitalization does not commute with taking tensor products (the suspension of an algebra is nonunital). In fact it seems to me to be a difficult question, whether an admissible Fréchet algebra is equivalent in asymptotic cohomology to its double suspension (this could be called a cohomological Bott periodicity theorem). In fact such a periodicity theorem would be highly desirable because it necessarily has to hold for any theory with reasonable excision properties.

At this point the E-theoretic description of Bott periodicity [CH] fortunately saves us as it realizes the bivariant Bott- resp. Dirac elements inducing the K-theoretic periodicity maps stably by (nonlinear) asymptotic morphisms. This allows to prove a stable version of cohomological periodicity: there are natural asymptotic cohomology equivalences

$$\alpha_{SA} \in HC^1_\alpha(S^2A, SA), \qquad \beta_{SA} \in HC^1_\alpha(SA, S^2A),$$

inverse to each other under the composition product.

Suspending an algebra therefore only produces a shift of its stable asymptotic cohomology groups $HC^*_\alpha(S-, S-)$, so that stable asymptotic cohomology becomes in fact a bifunctor on the stable linear asymptotic homotopy category. This opens the way to derive exactness and excision properties of stable asymptotic cohomology which make these groups quite accessible in many situations. By adapting a well known argument from stable homotopy theory, it can be shown that the long cofibre (Puppe) sequence associated to a homomorphism $f : A \to B$ of admissible Fréchet algebras induces six term exact séquences on (bivariant) stable asymptotic cohomology relating the stable cohomology groups of A and B to those of the mapping cone C_f of f. A short exact sequence

$$0 \to J \to A \xrightarrow{p} B \to 0$$

of admissible Fréchet algebras gives rise to six term exact cohomology sequences if and only if stable excision holds. This means that the inclusion of the kernel J into the cofibre C_p of the quotient map p induces a stable asymptotic (co)homology equivalence. Following an argument of Connes and Higson we show that stable excision holds for any epimorphism of separable C^*-algebras that admits a bounded linear section. This is the only place where we have to restrict ourselves to a particular class of admissible Fréchet algebras, as we need the existence of a bounded, positive, quasicentral approximate unit in the kernel J of p.

With all this machinery developed it becomes possible to extend the Chern character to the bivariant setting, i.e. to construct a transformation of bifunctors:

$$ch : KK^*(-,-) \to HC_\alpha^*(S-, S-)$$

from Kasparov's KK-theory to stable bivariant asymptotic cohomology. In principle it is given by the "composition" (see [CH])

$$KK^* \to E^* \text{ " } \to \text{ " } HC_{\alpha,st}^* \,,$$

where the "arrow" on the right hand side maps an asymptotic morphism to the corresponding bivariant asymptotic cocycle. As the asymptotic morphisms of E-theory are nonlinear however, one has to be careful in the actual construction of the bivariant Chern character. In particular, one obtains a Chern character on K-homology defined for arbitrary Fredholm modules and generalizing the constructions known so far. The Kasparov product on bivariant K-theory corresponds to the composition product on asymptotic cohomology, which is precisely the

Grothendieck-Riemann-Roch Theorem:

The diagram

$$\begin{array}{ccc}
KK^*(A,B) \otimes KK^*(B,C) & \xrightarrow{\ \otimes\ } & KK^*(A,C) \\
{\scriptstyle ch \otimes ch} \Big\downarrow & & \Big\downarrow {\scriptstyle ch} \\
HC_\alpha^*(SA,SB) \otimes HC_\alpha^*(SB,SC) & \xrightarrow{\frac{1}{2\pi i^{**}} \otimes} & HC_\alpha^*(SA,SC)
\end{array}$$

commutes. For $A = \mathbb{C}$ this yields a Grothendieck-Riemann-Roch formula as asked for in the beginning.

(The factor $2\pi i$ occurs for the same reason as in the comparison theorem of the ordinary Chern character with products). Consequently the Chern character of a KK-equivalence yields a stable asymptotic (co)homology equivalence. The bivariant Chern character becomes an isomorphism between complexified KK-theory and stable bivariant asymptotic cohomology on a class of separable C^*-algebras containing $\mathbb{C}$ and being closed under extensions with completely positive lifting and

KK-equivalences. This shows that with asymptotic cyclic cohomology we have come much closer to the "right" target of a Chern character in operator K-theory.

Finally we present two explicit calculations of asymptotic cyclic cohomology groups in concrete examples. In the first the functorial and excision-properties of the theory are used to determine the stable asymptotic homology of separable, commutative C^*-algebras. If A is a separable, commutative C^*-algebra with associated locally compact space X, then

$$HC_*^\alpha(S^2A) \simeq \bigoplus_{n=-\infty}^{\infty} H_c^{*+2n}(X, \mathbb{C})$$

where on the right hand side sheaf cohomology with compact supports is understood. In the second example we outline and illustrate a procedure to calculate asymptotic homology by standard methods of homological algebra.

Besides the calculation of asymptotic cohomology groups in concrete examples the most obvious questions not studied in this paper are the determination of more explicit versions of the Grothendieck-Riemann-Roch theorem and their application to generalized index problems. We plan to treat these topics elsewhere.

The plan of this book is as follows:

In chapter 1 we introduce the linear asymptotic homotopy category and the notion of an admissible Fréchet algebra. It is shown that these algebras behave like Banach algebras as far as spectral properties, holomorphic functional calculus and K-theory are concerned. Finally we demonstrate that K-theory becomes a homotopy functor under asymptotic morphisms.

Chapter 2 begins with a heuristic motivation of the definition of the periodic de Rham complex of an algebra, starting from the desired formal properties of the pairing of K-theory with de Rham homology. Then the (co)homological and bivariant X-complexes are introduced and studied in the ordinary as well as the differential graded setting.

Chapter 3 presents the idea of extending functors from categories of algebras to larger linear categories and develops the approach to cyclic cohomology of Cuntz and Quillen, adapted to the differential graded case.

Except for the definition of an admissible Fréchet algebra the first three chapters collect material due to Connes, Connes-Higson, and Cuntz-Quillen. This was done on one hand for the convenience of the reader and on the other hand to document the modifications necessary for our needs.

In chapter 4 Cartan homotopy formulas are derived by the method explained in the summary above. They are used to show that derivations on an algebra act trivially on its cohomology, as well as for controlling the change of asymptotic parameters in the differential graded case. Finally some comparison theorems between ordinary, bivariant and differential graded X-complexes are presented. All these results could have been shown by a short abstract argument given at the end of the chapter. We have however decided to go the longer way of giving complete and explicit constructions on the level of chain complexes whenever possible. This was done to make continuity properties readily accessible and especially to enable one to do explicit calculations in concrete examples.

We begin to discuss the analytical aspects of the theory in chapter 5. The topological I-adic completion of the tensor algebra over an admissible Fréchet algebra is studied. Then the definition of the analytic X-complex is given and elementary properties of analytic cyclic (co)homology are derived.

The asymptotic X-complex and asymptotic cyclic cohomology are treated analogously in chapter 6. The demonstrations are more involved however. A notable difference is that whereas analytic cohomology is a homotopy functor only with respect to smooth homotopies, in the asymptotic case even continuous homotopies may be allowed. Explicit formulas for the pairing between K-theory and asymptotic cohomology are given at the end of the section. They are a bit more general than the well known Chern character formulas of Connes and Cuntz-Quillen. This will be of use when we analyse the compatibility of the pairing with products.

With the derivation lemma in chapter 7, we get the theory off the ground. The criteria for the inclusion of a dense subalgebra to be an asymptotic (co)homology equivalence apply in two cases: for the inclusion of the subalgebra of smooth elements with respect to the action of a one parameter automorphism group and for the inclusion of the domain of a positive, unbounded trace on a separable C^*-algebra. Some examples are discussed which will reappear frequently in the remaining chapters.

The construction of product operations is given in chapter 8. The exterior product $HC^*_{\epsilon,\alpha}(A)\widehat{\otimes}HC^*_{\epsilon,\alpha}(B) \to HC^*_{\epsilon,\alpha}(A \otimes_\pi B)$ is defined on the level of chain complexes. The utility of this product depends heavily on the derivation lemma, which enables one to lift the product of two cohomology classes to topological tensor products other than the projective one. This makes it possible for example to prove the stable Morita invariance of the asymptotic cohomology of C^*-algebras. After this application the compatibility of the Chern character with exterior products is shown. This finally justifies our choice of constants in the definition of the exterior product. The chapter ends with the construction of the slant product.

Chapter 9 is devoted to exact sequences of bivariant asymptotic cohomology groups. It begins with the stable periodicity theorem and the calculation of the coefficient groups of stable asymptotic cohomology. The proofs are quite tedious as one has to descend several times to different "smooth" subalgebras in order to avoid nonlinearity of the involved asymptotic morphisms. From the stable periodicity theorem the excision theorems for mapping cones of morphisms of admissible Fréchet algebras and for separable C^*-algebras are derived.

In chapter 10 we discuss the bivariant Chern character from KK-theory to stable asymptotic cohomology and prove the generalized Grothendieck-Riemann-Roch theorem.

In the final chapter 11 the stable asymptotic homology of separable, commutative C^*-algebras is computed and a general scheme for the calculation of asymptotic cyclic (co)homology groups is outlined.

For a more detailed overview consult the introductions of the various chapters.

Apart from the last chapter the text coincides with the authors 1994 doctoral thesis at the Universität Heidelberg.

Acknowledgments:

First of all, I want to thank my advisor Professor Joachim Cuntz most heartily. His constant support and his patience made it possible for me to carry out this project over the last three years. Discussions with him and his advice were of great help both mathematically and psychologically, especially when I arrived at a point, where it became clear that large parts of the theory had to be redeveloped from a modified point of view, which happened more than once. I am especially indebted to Professor Ryszard Nest with whom I had very fruitful discussions on the subject during a stay in Copenhagen and in Heidelberg. To him I owe the idea to work with the "simplicial asymptotic parameter space" to get a rigid definition of the composition of asymptotic morphisms. Finally I want to express my deep gratitude to Professor Alain Connes. His courses and seminars in Paris and the enlightning discussions with him were a rich source of motivation and ideas for my study of noncommutative geometry and cyclic cohomology. I am glad that he was willing to be a referee of this thesis.

Contents

Chapter 1: The asymptotic homotopy category

In this chapter the asymptotic homotopy category of Connes-Higson [CH] is recalled. It is the natural domain of definition of the topological K-functor for C^*-algebras and will play a central role in the book. There are a few differences between the presentation in [CH] and our presentation though.

Whereas nonlinear asymptotic morphisms play a crucial role in the work of Connes and Higson we will consider only linear ones. On the other hand the definition of the asymptotic category will be modified such that the composition of two asymptotic morphisms (and not only its homotopy class) can be constructed explicitly. This is done by replacing the positive real halfline $I\!R_+$ as asymptotic parameter space by a cosimplicial space $\mathcal{R}^\infty_+$ given in codimension n by $I\!R^n_+ \cup \infty$ (with a nonstandard topology around ∞).

As objects of the asymptotic homotopy category we introduce the class of admissible Fréchet algebras. A Fréchet algebra is called admissible if it possesses an analogue of the open unit ball of Banach algebras. An open neighbourhood of zero will be called an "open unit ball" or "small" if the multiplicative closure of any compact subset of this "unit ball" is relatively compact in the given Fréchet algebra. The most basic examples of admissible Fréchet algebras are Banach algebras, for which the open unit ball in the norm sense is an "open unit ball" in our sense. In fact, admissible Fréchet algebras share a number of formal properties with Banach algebras: the spectra of its elements are compact and nonempty and holomorphic functional calculus is valid for admissible Fréchet algebras. The advantage of this class of algebras compared to Banach algebras is that it is stable under "passage to holomorphically closed dense subalgebras".

This is why admissible Fréchet algebras are taken as the class of algebras considered in this paper. The chapter ends with the definition and the study of elementary properties of topological K-theory for the linear asymptotic homotopy category of admissible Fréchet algebras.

1-1 Asymptotic parameters

In this chapter a category will be considered whose morphisms consist not of single maps but in fact of a whole family of maps of a certain kind. If morphisms f, g are given by families of maps parameterized by a single space X, (to take the simplest case) then the possible compositions of n morphisms will be parameterized by the space X^n. The natural candidate for the total parameter space of morphisms will then be the union $\mathcal{X} = \bigcup_n X^n$. This is now carried out in the case we will be concerned with.

Let $\mathbb{R}_+ := [0,\infty[$ denote the closed real halfline. The disjoint union of the algebras $C(\mathbb{R}_+^n)$ $(C^\infty(\mathbb{R}_+^n))$ of continuous (continuous and smooth on the interior) complex valued functions are denoted by

$$C(\mathbb{R}_+^\infty) := \bigcup_n C(\mathbb{R}_+^n) \quad C^\infty(\mathbb{R}_+^\infty) := \bigcup_n C^\infty(\mathbb{R}_+^n)$$

The union of the subalgebras of bounded functions are denoted similarly by $C_b(\mathbb{R}_+^\infty)$. A map from an algebra A to $C(\mathbb{R}_+^\infty)$ is by definition a map from A to $C(\mathbb{R}_+^n)$ for some n.

Two maps $f, f' : A \to C(\mathbb{R}_+^\infty)$ given by $f : A \to C(\mathbb{R}_+^n), f' : A \to C(\mathbb{R}_+^m)$ are identified if there exists an order preserving inclusion $i : \{1,\ldots,m\} \to \{1,\ldots,n\}$ such that the triangle

$$\begin{array}{ccc} & & C(\mathbb{R}_+^n) \\ & f \nearrow & \uparrow i^* \\ A & \xrightarrow[f']{} & C(\mathbb{R}_+^m) \end{array}$$

commutes. The maps f occurring in that way are called degenerate.

The algebras $C(\mathbb{R}_+^n)$ are augmented by

$$C(\mathbb{R}_+^n) \to \mathbb{C}; \ f \to f(0,\ldots,0)$$

There is a canonical map $C(\mathbb{R}_+^\infty) \otimes C(\mathbb{R}_+^\infty) \to C(\mathbb{R}_+^\infty)$ given by

$$C(\mathbb{R}_+^n) \otimes C(\mathbb{R}_+^m) \ \to \ C(\mathbb{R}_+^{n+m})$$

$$f \otimes g \ \to \ \pi_0^* f \pi_1^* g$$

It is compatible with the identifications made above.

Remark:

In the terminology of algebraic topology $\mathbb{R}_+^\infty$ is a cosimplicial space and $C(\mathbb{R}_+^\infty)$ is a simplicial algebra. A nondegenerate map into $C(\mathbb{R}_+^\infty)$ in our terminology corresponds to a map into a nondegenerate simplex of $C(\mathbb{R}_+^\infty)$ and the identifications made mean that degenerate simplices are identified with their corresponding nondegenerate simplex. The reason for not using the standard terminology is that the asymptotic parameter space $\mathcal{R}_+^\infty$ obtained by introducing a new topology on $\mathbb{R}_+^\infty \cup \{\infty\}$ will not be a cosimplicial space anymore (there will be no coface maps).

In order that asymptotic morphisms (to be defined later) compose well it is necessary to change the topology of the parameter space $\mathcal{R}_+^\infty$:

Definition 1.1:

For $n \in \mathbb{N}$ the topological space $\mathcal{R}_+^n$ is defined to be the space with underlying set $\mathbb{R}_+^n \cup \{\infty\}$ and the topology defined by the standard topology on $\mathbb{R}_+^n$ and the fundamental system of neighbourhoods of ∞ given by

$$U_{f_1,\ldots,f_n} := \{(x_1,\ldots,x_n) \in \mathbb{R}_+^n; x_1 > f_1(0), x_2 > f_2(x_1),\ldots,x_n > f_n(x_{n-1})\} \cup \{\infty\}$$

where f_i are positive, strictly monotone increasing, convex, unbounded, realvalued functions. $\mathcal{R}_+^\infty := \bigcup_n \mathcal{R}_+^n$ is called the **asymptotic parameter space**.

$\square$

The projection map

$$\mathbb{R}_+^{n+m} \to \mathbb{R}_+^n \times \mathbb{R}_+^m$$

extends to a continuous map

$$\mathcal{R}_+^{n+m} \to \mathcal{R}_+^n \times \mathcal{R}_+^m$$

and yields therefore a canonical homomorphism

$$C_0(\mathcal{R}_+^n) \otimes C_0(\mathcal{R}_+^m) \to C_0(\mathcal{R}_+^{n+m})$$

One observes

Lemma 1.2:

a) Under the projection map

$$\pi_n: \quad U_{f_1,\ldots,f_n} \cap \mathbb{R}_+^n \quad \to \quad \mathbb{R}_+$$

$$(x_1,\ldots,x_n) \quad \to \quad x_n$$

the inverse images of relatively compact sets are relatively compact.

b) If $f : A \to C(\mathbb{R}_+^n)$ is degenerate with associated nondegenerate map $f' : A \to C(\mathbb{R}_+^m)$ then $f \in C_0(\mathcal{R}_+^n)$ iff $f' \in C_0(\mathcal{R}_+^m)$

c) The fundamental open neighbourhoods of ∞ are convex.

$\square$

1-2 Asymptotic morphisms [CH]

Following Connes and Higson we introduce in this paragraph the linear asymptotic homotopy category.

Definition 1.3:

Let A, B be associative $\mathbb{C}$-algebras and $\varrho : A \to B$ a $\mathbb{C}$-linear map. The **curvature** ω_ϱ of ϱ at $(x, y) \in A \times A$ is defined to be

$$\omega_\varrho(x, y) := \varrho(xy) - \varrho(x)\varrho(y)$$

$\square$

The curvature of a linear map satisfies the

1.4 Bianchi identity

$$\varrho(a)\omega(b, c) - \omega(a, b)\varrho(c) = \omega(ab, c) - \omega(a, bc)$$

If A, B, f are unital the curvature descends to a map

$$\omega : A/\mathbb{C}.1 \otimes A/\mathbb{C}.1 \to B$$

Definition 1.5:(Connes-Higson)

Let A, B be (augmented) complex Fréchet algebras. **A (smooth) asymptotic morphism**

$$\varrho : A \to B$$

is a continuous linear map

$$\varrho_t : A \to C_b(\mathbb{R}_+^\infty, B) \quad (\text{ resp. } \varrho_t : A \to \mathcal{C}^\infty(\mathbb{R}_+^\infty, B) \cap C_b(\mathbb{R}_+^\infty, B))$$

satisfying

i) If A is augmented, then ϱ_t is unital and compatible with augmentation maps

ii)

$$\omega_{\varrho_t}(x, y) \in C_0(\mathcal{R}_+^\infty, B) \quad \forall (x, y) \in A \times A$$

$\square$

Lemma 1.6:

Let $\varrho : A \to B$ be an asymptotic morphism. Then

$$\lim_{t \to \infty} \omega_t(x, y) = 0$$

uniformly on compact subsets of $A \times A$.

Proof:

As $\omega_t(x, y)$ is bilinear and continuous on $A \times A$ it suffices to remark that any compact subset of a Fréchet space is contained in the closed convex hull of a nullsequence:

Lemma 1.7:

Let $\mathcal{A} \subset A$ be a dense subspace of a Fréchet space A. Choose a strictly monotone decreasing sequence of positive real numbers

$$(\lambda_n) \; ; \; \lambda_n > 0; \; \sum_{n=0}^{\infty} \lambda_n = 1$$

Let $K \subset A$ be compact. Then there exists a countable set $\mathcal{B} \subset A$ with single accumulation point 0 such that

$$K \subset \{ \sum_{i=0}^{\infty} \lambda_i \, b_i \mid b_i \in \mathcal{B} \}$$

Proof:

As $\mathcal{A}$ is dense the balls (in some metric d on A)

$$\mathcal{M}_j := \bigcup_{x \in \mathcal{A}} B(x, \frac{1}{2} \lambda_j^2)$$

form an open covering of A and therefore of K for all $j \in \mathbb{N}$.
Choose $(x_1^j, \ldots, x_{n_j}^j) \in \mathcal{A}$ such that

$$K \subset \bigcup_{k=1}^{n_j} B(x_k^j, \frac{1}{2} \lambda_j^2) \quad \text{and} \quad K \cap B(x_k^j, \frac{1}{2} \lambda_j^2) \neq \emptyset \; \forall j, k$$

One has then

$$x_k^j \in B(x_{k'}^{j-1}, \lambda_{j-1}^2)$$

for some k'.
Put

$$\mathcal{B} := \{ \frac{x_1^0}{\lambda_0}, \ldots, \frac{x_{n_0}^0}{\lambda_0} \} \cup \{ \frac{x_k^j - x_{k'}^{j-1}}{\lambda_{j-1}} \; ; j \in \mathbb{N}; \; d(x_k^j, x_{k'}^{j-1}) \leq \lambda_{j-1}^2 \}$$

$\mathcal{B}$ is a subset of $\mathcal{A}$ which has the only accumulation point 0 and is therefore also compact: for any given defining seminorm on A there are only finitely many elements of $\mathcal{B}$ which in the given seminorm are larger than any given $\epsilon > 0$.

Let now $y \in K$. Choose for any $j \in I\!N$ an integer $k_j(y) \in \{1, \ldots, n_j\}$ such that

$$d(y, x^j_{k_j(y)}) \leq \frac{1}{2}\lambda_j^2$$

Then

$$y = \lambda_0 \frac{x^0_{k_0(y)}}{\lambda_0} + \sum_{j=1}^{\infty} \lambda_{j-1}\left(\frac{x^j_{k_j(y)} - x^{j-1}_{k_{j-1}(y)}}{\lambda_{j-1}}\right)$$

and

$$d(x^j_{k_j(y)}, x^{j-1}_{k_{j-1}(y)}) \leq d(x^j_{k_j(y)}, y) + d(y, x^{j-1}_{k_{j-1}(y)}) \leq \frac{1}{2}\lambda_j^2 + \frac{1}{2}\lambda_{j-1}^2 \leq \lambda_{j-1}^2$$

Therefore

$$\frac{x^j_{k_j(y)} - x^{j-1}_{k_{j-1}(y)}}{\lambda_{j-1}} \in \mathcal{B}$$

and the lemma is proved.

$\square$

Every homomorphism of (augmented) Fréchet algebras defines an asymptotic morphism in an evident way. For nontrivial examples see [CH]. (Note however that in their paper asymptotic morphisms maybe nonlinear in general).

Composition of asymptotic morphisms

Let $\varrho : A \to B, \varrho' : B \to C$ be (smooth) asymptotic morphisms represented by

$$\varrho_t : A \to C_b(I\!R^n_+, B), \varrho'_t : B \to C_b(I\!R^m_+, C)$$

Define $\varrho' \circ \varrho : A \to C$ as the composition

$$(\varrho' \circ \varrho)_t : \quad A \xrightarrow{\varrho} C_b(I\!R^n_+, B) \xrightarrow{C_b(\varrho')} C_b(I\!R^n_+, C_b(I\!R^m_+, C)) \simeq C_b(I\!R^{n+m}_+, C)$$

$$A \to \mathcal{C}^{\infty}(I\!R^n_+, B) \to \mathcal{C}^{\infty}(I\!R^n_+, \mathcal{C}^{\infty}(I\!R^m_+, C)) \simeq \mathcal{C}^{\infty}(I\!R^{n+m}_+, C)$$

Proposition 1.8:

(Augmented) Fréchet algebras and asymptotic morphisms form a category under the composition defined above.

Proof:

Let $\varrho : A \to B, \varrho' : B \to C$ be as above. It is clear that $(\varrho' \circ \varrho)_t : A \to C_b(I\!R^{n+m}_+)$ is bounded and augmentation preserving if ϱ and ϱ' are. For the curvature of $\varrho' \circ \varrho$ one finds:

$$\omega_{\varrho' \circ \varrho}(x, y) = \varrho' \varrho(xy) - \varrho' \varrho(x)\varrho' \varrho(y) =$$

$$= \varrho'(\varrho(xy) - \varrho(x)\varrho(y)) + \varrho'(\varrho(x)\varrho(y)) - \varrho' \varrho(x)\varrho' \varrho(y) = \varrho'(\omega_\varrho(x, y)) + \omega'_\varrho(\varrho(x), \varrho(y))$$

Curvature estimates:

Let $\| \; \|_\beta, \| \; \|_\gamma$, be seminorms on B, C such that there exists $\| \varrho' \| \in \mathbb{R}$ with

$$\| \varrho'(b)(x) \|_\gamma \leq \| \varrho' \| \| b \|_\beta \quad \forall b \in B, x \in \mathbb{R}_+^\infty$$

Choose $U_{f_1,\ldots,f_n} \subset \mathcal{R}_+^n$ such that

$$\| \omega_\varrho(x,y) \|_\beta < \frac{\epsilon}{2 \| \varrho' \|}$$

on $U_{f_1,\ldots,f_n}$.

For $j \in \mathbb{N}$ define

$$K_j := \{ \varrho_t(x)(U_{f_1,\ldots,f_n} \cap \pi_n^{-1}([0,j+1]), \varrho_t(y)(U_{f_1,\ldots,f_n} \cap \pi_n^{-1}([0,j+1])\} \subset B$$

These sets are relatively compact by Lemma 1.2. By the uniformness of curvature estimates (Lemma 1.6) there are open sets

$$U^j_{h^j_1,\ldots,h^j_m} \subset \mathcal{R}_+^m \quad \text{with} \quad \sup_{x',y' \in K,} \| \omega_{\varrho'}(x',y') \|_\gamma < \tfrac{\epsilon}{2} \quad \text{on } U^j$$

Construct inductively strictly monotone positive functions

$$g_1,\ldots,g_m \in C(\mathbb{R}_+) \lim_{t \to \infty} g_i(t) = +\infty$$

satisfying

$$\begin{aligned}
g_1(t) &> \sup_{0 \leq i \leq j} h^i_1(0) \quad \text{for} && t > j \\
g_2(t) &> \sup_{0 \leq i \leq j} h^i_2(t) \quad \text{for} && t > g_1(j) \\
&\qquad \cdots \\
g_m(t) &> \sup_{0 \leq i \leq j} h^i_m(t) \quad \text{for} \quad t > g_{m-1} \circ \cdots \circ g_1(j)
\end{aligned}$$

Then

$$\| \omega_{\varrho' \circ \varrho}(x,y) \|_\gamma < \epsilon$$

on

$$U_{f_1,\ldots,f_n,g_1,\ldots,g_m} - \{\infty\} \subset \mathbb{R}_+^{n+m}$$

The associativity of the composition is obvious.

$\square$

For two (augmented) Fréchet algebras A, B we denote by $A \otimes_\pi B$ their projective Fréchet tensor product, i.e. the (augmented) algebra $A \otimes B$ completed with respect to the cross seminorms

$$\| m \|_{\alpha,\beta} := \inf_{\sum a_i \otimes b_i = m} \sum_i \| a_i \|_\alpha \| b_i \|_\beta$$

where $\| \; \|_\alpha, \| \; \|_\beta$ ranges over a system of seminorms defining the topology on A, B.

Lemma 1.9:

There exist natural product and cylinder-maps

$$\otimes : Hom_\alpha(A, B) \times Hom_\alpha(A', B') \to Hom_\alpha(A \otimes_\pi A', B \otimes_\pi B')$$

$$Cyl : Hom_\alpha(A, B) \to Hom_\alpha(A[0, 1], B[0, 1])$$

Proof:

Define the product $\varrho \otimes \varrho'$ of $\varrho : A \to B$, $\varrho' : A' \to B'$ as the composition

$$A \otimes_\pi A' \xrightarrow{\varrho_t \otimes \varrho'_t} C_b(I\!\!R_+^n, B) \otimes_\pi C_b(I\!\!R_+^m, B') \to C_b(I\!\!R_+^{n+m}, B \otimes_\pi B')$$

This is clearly a bounded, linear, (augmentation preserving) map. To obtain curvature estimates, we observe that any element (and indeed any compact subset) of $A \otimes_\pi A'$ is contained in the closed convex hull of a nullsequence of elements of the algebraic tensor product $A \otimes B$ (Lemma 1.7), so that by the bilinearity of the curvature it suffices to check the necessary estimates on simple tensors $a \otimes a' \in A \otimes A'$.

The curvature formula

$$\omega_{\varrho \otimes \varrho'}(a_0 \otimes a'_0, a_1 \otimes a'_1) =$$

$$= \omega_\varrho(a_0, a_1) \otimes \varrho'(a'_0 a'_1) + \varrho(a_0 a_1) \otimes \omega_{\varrho'}(a'_0, a'_1) - \omega_\varrho(a_0, a_1) \otimes \omega_{\varrho'}(a'_0, a'_1)$$

shows that if

$$\| \omega_\varrho(a_0, a_1) \|_\alpha < \epsilon$$

on $U_{f_1, \dots, f_n}$ and

$$\| \omega_{\varrho'}(a'_0, a'_1) \|_\beta < \epsilon$$

on $U_{g_1, \dots, g_m}$ for some finite set of seminorms defining the topology on B, B', then

$$\| \omega_{\varrho \otimes \varrho'}(a_0 \otimes a'_0, a_1 \otimes a'_1) \|_{\alpha \otimes \beta} \leq (\| \varrho'(a'_0 a'_1) \|_\beta + \| \varrho(a_0 a_1) \|_\alpha + \epsilon) \epsilon$$

on $U_{f_1, \dots, f_n, g_1, \dots, g_m}$

The construction of the cylinder map is evident: it is the map given by $\varrho \otimes Id : A \otimes C[0, 1] \to B \otimes C[0, 1]$ on the algebraic tensor product. The above proof of the curvature estimates applies for the completion $A[0, 1]$ of $A \otimes C[0, 1]$ as well as for the projective completion $A \otimes_\pi C[0, 1]$.

□

Definition 1.10:

a) Two asymptotic morphisms $\varrho_0, \varrho_1 : A \to B$ are called **homotopic** if there exists an asymptotic morphism

$$\chi : A \to B[0,1]$$

such that

$$\varrho_0 = i_0 \circ \chi, \ \varrho_1 = i_1 \circ \chi$$

where $i_{0,1} : B[0,1] \to B$ are the evaluation maps at $0, 1 \in [0,1]$ $(0, 1 \in \mathbb{R})$.

$\square$

Theorem 1.11:

a) (Augmented) Fréchet algebras form under linear asymptotic morphisms a category, the **linear asymptotic category** $\mathcal{D}$.

b) Similarly, (augmented) Fréchet algebras form a category under smooth asymptotic morphisms, called the **smooth linear asymptotic category** $\mathcal{D}^\infty$.

c) The associated homotopy categories are denoted as the **linear asymptotic homotopy category** $\mathcal{D}_{homot}$, resp. the **smooth linear asymptotic homotopy category** $\mathcal{D}^\infty_{homot}$.

d) The natural functor

$$\mathcal{D}^\infty_{homot} \to \mathcal{D}_{homot}$$

induces an equivalence of categories.

The set of homotopy classes of linear, asymptotic morphisms from A to B will be denoted by $[A, B]_\alpha$.

Proof:

The composition of asymptotic morphisms clearly preserves homotopy, smoothness and
smooth homotopy. (This follows from the existence of the cylinder functor Lemma 1.9).

d) It has to be shown that $\mathcal{D}^\infty_{homot}(A, B) \to \mathcal{D}_{homot}(A, B)$ is bijective for any augmented Fréchet algebras A, B. Let

$$\varrho \in \mathcal{D}(A, B); \varrho_t : A \to C_b(\mathbb{R}^n_+, B)$$

Choose a continuous family ν_t^n of positive smooth functions on $\mathbb{R}^n_+ \times \mathbb{R}^n_+$ with support close to the diagonal $\Delta \subset \mathbb{R}^n_+ \times \mathbb{R}^n_+$ and approximating the delta distribution along the diagonal. Then convolution

$$A \quad \xrightarrow{\varrho_t} \quad C_b(\mathbb{R}^n_+, B) \quad \xrightarrow{(*\nu^n_t)} \quad C_b(\mathbb{R}^{n+1}_+, B)$$
$$f \quad \to \quad \nu^n_t * f$$

defines an asymptotic morphism canonically homotopic to ϱ whose image lies in $C^\infty(\mathbb{R}^{n+1}_+, B)$. This shows the surjectivity of the map under consideration. The injectivity is clear.

□

1-3 Admissible Fréchet algebras

The appropriate category of Fréchet algebras for which asymptotic cohomology will be developed is introduced now. It turns out that the category of Banach algebras is not large enough for our purposes because we want to construct asymptotic cocycles by lifting cocycles from a "smooth" subalgebra to the whole algebra and "smooth" (C^∞) subalgebras of Banach algebras are not Banach algebras anymore. On the contrary admissible Fréchet algebras are stable under passing to smooth subalgebras.

Definition 1.12:

Let A be an algebra and let $K \subset A$ be a nonempty subset. The multiplicative closure $K^\infty \subset A$ of K is defined as

$$K^\infty := \bigcup_{n=1}^\infty K^n = \bigcup_{n=1}^\infty \{\prod_1^n a_i | a_i \in K\}$$

□

Definition and Lemma 1.13:

A Fréchet algebra A is called **admissible** if one of the following equivalent conditions is satisfied:

a) Every nullsequence has an end whose multiplicative closure is relatively compact.

b) Every nullsequence has an end whose multiplicative closure is a nullsequence.

c) To every compact subset $K \subset A$ there exists a neighbourhood U of 0 such that $K \cap U$ has relatively compact multiplicative closure.

d) There exists a neighbourhood U of 0 such that every compact subset of U has relatively compact multiplicative closure.

Open neighbourhoods of 0 satisfying condition d) are called "**small**".

Proof:

a)$\Rightarrow$ b):

Let (a_n) be a nullsequence and choose a strictly increasing sequence (λ_n) of positive, real numbers tending to infinity such that $(\lambda_n a_n)$ is still a nullsequence.

Choose N big enough so that $\lambda_n \geq 2$ for $n \geq N$ and that the multiplicative closure of $\{\lambda_n a_n; n \geq N\}$ is relatively compact. Denote its closure by K. K being compact $\lim_{\lambda \to 0} \lambda x = 0$ uniformly on K. This shows finally that the multiplicative closure of $(a_n) = (\frac{1}{\lambda_n} \lambda_n a_n)$ $n \geq N$ is a nullsequence.

b)$\Rightarrow$ c):

Let $K \subset A$ be compact. If K does not contain 0 one has $K \cap U = \emptyset$ for some neighbourhood U of 0 and the assertion is trivial. So we may suppose $0 \in K$. According to Lemma 1.7. one can choose a nullsequence (b_n) such that any element $y \in K$ can be expressed as

$$y = \sum_{i=0}^{\infty} \mu_i b_i \ \mu_i > 0; \ \sum_{i=0}^{\infty} \mu_i \leq 1$$

for some sequence (μ_i) of positive, real numbers.

By assumption some end of (b_i) has a relatively compact multiplicative closure. By having a closer look at the proof of 1.7. one sees that (in the notations used there) for fixed $j = j_0$ every element $y \in K \cap B(0, \frac{1}{2}\lambda_{j_0}^2)$ may be written as

$$y = \sum_{j=j_0}^{\infty} \lambda_j c_j \ c_j \in \mathcal{B} = \{b_n, n \in I\!N\}$$

where for j_0 large enough the elements $c_j, j \geq j_0$ belong to any given end of the sequence (b_n) and $\sum_{j=j_0}^{\infty} \lambda_j \leq \frac{1}{2}$. The assertion becomes then obvious.

c)$\Rightarrow$ d):

Let (U_k) be a fundamental sequence of neighbourhoods of 0 and suppose that $K_k \subset U_k$ are compact sets whose multiplicative closure is not relatively compact.

As $\bigcup_{k=1}^{\infty} K_k \cup \{0\}$ is compact, this contradicts c).

d)$\Rightarrow$ a): Trivial.

$\square$

Checking whether a given Fréchet algebra is admissible is facilitated by the following

Lemma 1.14:

Let A be a Fréchet algebra and $\mathcal{A} \subset A$ a dense subspace. Then A is admissible if any of the conditions 1.12.a)-d) is satisfied for nullsequences resp. compact sets contained in $\mathcal{A}$.

Proof:

This is a consequence of Proposition 1.7.

□

The class of admissible Fréchet algebras is closed under taking subalgebras and quotients.

Lemma 1.15:

If A is admissible, then so is $\tilde{A}$, the algebra obtained by adjoining a unit.

Proof:

Let $U \subset A$ be "small" and let $D \subset \mathbb{C}$ be the open unit ball. We claim that $\tilde{U} := (\frac{1}{2}U \times \frac{1}{2}D) \subset \tilde{A}$ is "small. Let $\tilde{K} \subset \tilde{U}$ be compact. Then there exists $\lambda > 1$ such that $\pi_1(2\lambda\tilde{K}) \subset U \subset A, \pi_2(2\lambda\tilde{K}) \subset D$ are compact. Let $K \subset A$ be a compact cone with vertex 0 containing the multiplicative closure of $\pi_1(2\lambda\tilde{K})$. Every $\tilde{x} \in \tilde{K}$ can then be written as $\tilde{x} = \frac{1}{2\lambda} + \frac{1}{2\lambda}\mu$ with $x \in K$ and $\mu \in \mathbb{C}, |\mu| \leq 1$. Therefore $\prod_1^n \tilde{x}_i = (2\lambda)^{-n} \sum_{2^n terms} y_j$ where y_j either belongs to $K \subset A$ or $D \subset \mathbb{C}$. If $\| - \|'$ is a seminorm on $\tilde{A}$ corresponding to the seminorm $\| - \|$ on A then $\| \prod_1^n \tilde{x}_i \|' \leq (2\lambda)^{-n} \sum_{2^n terms} \| y_j \| \leq \lambda^{-n}C$ where $C := \sup_{y \in K \times D} \| y \|'$. Thus $\| (\tilde{K})^n \|' \leq \lambda^{-n}$ which proves the claim.

□

Examples:

a) Every Banach algebra is admissible.

b) Let $\delta : A \to A$ be a densely defined, unbounded derivation on an admissible Fréchet algebra and let $\mathcal{A}^\infty$ be the subalgebra of smooth elements with respect to δ with its obvious Fréchet topology. Then $\mathcal{A}^\infty$ is admissible, too. (See 7.4.)

□

Although apparently more general than Banach algebras, admissible Fréchet algebras inherit some basic features from them.

Proposition 1.16:

a) The subset of invertible elements in a unital admissible Fréchet algebra is open and the inversion is continuous.

b) The spectrum of an element of an admissible Fréchet algebra is compact and nonempty.

c) Holomorphic functional calculus is valid for admissible Fréchet algebras.

Proof:

By Lemma 1.15 we may suppose A to be unital in b),c).

a) Let U be a "small" neighbourhood of 0. We claim that $U' := 1 - U$ consists of invertible elements. If $x \in U$ then also $y := \lambda x \in U$ for some $\lambda > 1$. Then $z := 1 - x \in U'$ is invertible with inverse

$$z^{-1} = \sum_{n=0}^{\infty} \lambda^{-n} y^n$$

where the sum converges because the multiplicative closure $\{y^n; n \in I\!N\}$ of y is relatively compact and thus bounded for any seminorm on A.

For the continuity let $(x_n) \in A$, $\lim_{n \to \infty} x_n = 1$ be a sequence of elements of A converging to 1. Then $x_n = 1 - y_n$ where (y_n) is a nullsequence. Choose a monotone increasing, unbounded sequence of positive real numbers $\lambda_n \geq 1$ such that $(\lambda_n y_n)$ remains still a nullsequence. After deleting finitely many elements one may suppose that $\{\lambda_n y_n; n \in I\!N\}$ is contained in a "small" ball. Therefore there exists a compact set $K \subset A$ containing the multiplicative closure of the sequence $(\lambda_n y_n)$. Let $\| - \|$ be a seminorm on A. Then $\sup_{x \in K} \| x \| \leq C$ for some $C > 0$. One finds

$$x_n^{-1} = (1 - y_n)^{-1} = \sum_{k=0}^{\infty} \lambda_n^{-k} (\lambda_n y_n)^k$$

and thus

$$\| x_n^{-1} - 1 \| \leq \sum_{k=1}^{\infty} \lambda_n^{-k} \| (\lambda_n y_n)^k \| \leq \frac{C}{\lambda_n - 1} < \epsilon \text{ for } n >> 0$$

as $(\lambda_n y_n)^k \in K$. This shows the continuity of the inversion.

b) The spectrum of an element is closed by a). Let $x \in A$ and choose $C > 0$ such that $\{\lambda x, |\lambda| < C\}$ is contained in a "small" ball U. Then $Sp(x) \subset \{z \in \mathbb{C}, |z| < \frac{1}{C}\}$ (by a)) is bounded and therefore compact. Let $\| - \|$ be a seminorm on A and denote by A' the Banach space obtained by completing A with respect to $\| - \|$. By a) the composition

$$\mathbb{C} \quad \to \quad A \quad \to A'$$

$$\lambda \quad \to \quad (\lambda - x)^{-1}$$

is an analytic function on $\mathbb{C} - Sp(x)$ with values in the Banach space A' which is bounded near infinity. If $Sp(x)$ were empty, one would conclude with Liouville's theorem that the image of $(\lambda - x)^{-1}$ in A' is independent of λ for any seminorm on A which is impossible.

c) Let A be admissible and $x \in A$. Let f be holomorphic on an open neighbourhood V of $Sp(x) \subset \mathbb{C}$. Choose a closed curve Γ in V not meeting $Sp(x)$ and such that the winding number of Γ with respect to any point in $Sp(x)$ equals 1. The integral

$$f(x) := \frac{1}{2\pi i} \int_{\Gamma} f(\lambda) (\lambda - x)^{-1} d\lambda$$

yields then a well defined element of A by a). The map $\lambda \to f(\lambda)(\lambda - x)^{-1}$ is analytic. If Γ' is a similar curve not intersecting Γ, then

$$\| \int_\Gamma f(\lambda)\,(\lambda - x)^{-1}\,d\lambda - \int_{\Gamma'} f(\lambda)\,(\lambda - x)^{-1}\,d\lambda \,\| = 0$$

for any seminorm on A by the Cauchy integral formula and an argument similar to that in b). Therefore $f(x)$ is independent of the choice of the curve Γ and consequently of the choice of V. If $\mathcal{O}(V)$ denotes the algebra of holomorphic functions on V, the map

$$i_x : \quad \mathcal{O}(V) \quad \to \quad A$$

$$f \quad \to \quad f(x)$$

is in fact a homomorphism of algebras. The linearity is clear and the multiplicativity can be derived from the identity

$$(\lambda - x)^{-1}(\mu - x)^{-1} = \frac{1}{\lambda - \mu}\left((\mu - x)^{-1} - (\lambda - x)^{-1}\right)$$

Let $f, g \in \mathcal{O}(V)$ and choose curves Γ, Γ' as above with Γ' contained in the bounded component of $\mathbb{C} - \Gamma$. Then

$$f(x)g(x) = \left(\frac{1}{2\pi i}\int_\Gamma f(\lambda)\,(\lambda - x)^{-1}\,d\lambda\right)\left(\frac{1}{2\pi i}\int_{\Gamma'} g(\mu)\,(\mu - x)^{-1}\,d\mu\right)$$

$$= (\frac{1}{2\pi i})^2 \int_\Gamma \int_{\Gamma'} f(\lambda)g(\mu)\,(\lambda - x)^{-1}(\mu - x)^{-1}\,d\lambda d\mu$$

$$= \frac{1}{2\pi i}\int_{\Gamma'} g(\mu)\left(\frac{1}{2\pi i}\int_\Gamma f(\lambda)\,(\lambda - \mu)^{-1}\,d\lambda\right)(\mu - x)^{-1}d\mu$$

$$+ \frac{1}{2\pi i}\int_\Gamma f(\lambda)\left(\frac{1}{2\pi i}\int_{\Gamma'} g(\mu)\,(\mu - \lambda)^{-1}\,d\mu\right)(\lambda - x)^{-1}d\lambda$$

$$= \frac{1}{2\pi i}\int_{\Gamma'} g(\mu)f(\mu)\,(\mu - x)^{-1}d\mu = (fg)(x)$$

as the second integral vanishes because λ belongs to the unbounded component of $\mathbb{C} - \Gamma'$.

$\square$

Lemma 1.17:

If A, B are admissible Fréchet algebras, then so is $A \otimes_\pi B$.

Proof:

Let $\{ \| - \|_k, \, k \in I\!N \}$ ($\{ \| - \|_l', \, l \in I\!N \}$) be sequences of seminorms defining the topologies of A, B.

We may suppose that $\| - \|_{k'} \geq \| - \|_k$ ($\| - \|_{l'}' \geq \| - \|_l'$) for $k' \geq k \, (l' \geq l)$.

According to Lemma 1.14 it suffices to check criterion 1.13 a) for nullsequences of elements of the algebraic tensor product $A \otimes B \subset A \otimes_\pi B$.

Let (γ^j) be such a nullsequence. It can be written as

$$\{ \gamma^j = \sum_{k=1}^{n_j} a_k^j \otimes b_k^j \, ; 4 \| \gamma^j \|_{f(j)} \geq \sum_{k=1}^{n_j} \| a_k^j \|_{f(j)} \| b_k^j \|_{f(j)}' \}$$

where f is a monotone increasing unbounded function and

$$\lim_{j \to \infty} \| \gamma^j \|_{f(j)} = 0; \quad \| \gamma^j \|_{f(j)} \neq 0; \quad \| a_k^j \|_{f(j)} \neq 0; \quad \| b_k^j \|_{f(j)}' \neq 0 \quad \forall j, k \in I\!N$$

It follows then that

$$K' := \{ \alpha_k^j := 2 \| \gamma^j \|_{f(j)}^{\frac{1}{2}} \| a_k^j \|_{f(j)}^{-1} a_k^j \} \subset A$$

$$K'' := \{ \beta_k^j := 2 \| \gamma^j \|_{f(j)}^{\frac{1}{2}} \| b_k^j \|_{f(j)}^{-1} b_k^j \} \subset B$$

are nullsequences.

By hypothesis, K' and K'' possess ends

$$K' \cap U' = \{ \alpha_k^j, \, j \geq N_0 \}, \, K'' \cap U'' = \{ \beta_k^j, \, j \geq N_1 \}$$

with relatively compact multiplicative closure. It is then clear by the choices made that

$$\{ \gamma^j; \, j \geq N_0 + N_1 \} \subset A \otimes_\pi B$$

has relatively compact multiplicative closure.

$\square$

1-4 K-theory of admissible Fréchet algebras

Theorem 1.18:

Let $\mathbb{C}[e]$ and $\mathbb{C}[u, u^{-1}]$ be the universal algebras generated by an idempotent and an invertible element, respectively. The abelian groups

$$K_0(-) := Groth \lim_n [\mathbb{C}[e], M_n(-)] \quad K_1(-) := \lim_n [\mathbb{C}[u, u^{-1}], M_n(-)]$$

define a Bott-periodic homotopy functor on the category of admissible Fréchet algebras. (Here Groth(-) denotes the Grothendieck group of an abelian monoid.)

Proof:

The proof of Bott periodicity for Banach algebras carries over to admissible Fréchet algebras because holomorphic functional calculus is valid for them as well. See for example [B].

□

It was a basic observation of Connes and Higson that topological K-theory behaves functorially not only under ordinary but in fact under asymptotic morphisms of (Banach) algebras.

Theorem 1.19:(Connes-Higson)[CH]

a) For any admissible Fréchet algebra the natural maps

$$[\mathbb{C}[e], A] \to [\mathbb{C}[e], A]_\alpha \quad [\mathbb{C}[u, u^{-1}], A] \to [\mathbb{C}[u, u^{-1}], A]_\alpha$$

are bijective.

b)

$$K_0(A) := Groth \lim_n [\mathbb{C}[e], M_n(A)]_\alpha \quad K_1(A) := \lim_n [\mathbb{C}[u, u^{-1}], M_n(A)]_\alpha$$

c) $K_*(-)$ is a covariant functor on the (linear) asymptotic homotopy category of admissible Fréchet algebras.

Proof:

Clearly $b) \Rightarrow c)$. For $a) \Rightarrow b)$ note that A admissible implies $M_n(A) = M_n(\mathbb{C}) \otimes_\pi A$ admissible by 1.17. It remains to show a). For this we proof the

Lemma 1.20: (Cuntz-Quillen)[CQ]

Let A be an admissible Fréchet algebra.

a) Let

$$\varrho: \mathbb{C}[e] \to A$$

be a unital, linear map. Suppose that

$$4\omega_\varrho(e, e) = 4(\varrho(e) - \varrho(e)^2)$$

is "small", i.e. contained in a "small" neighbourhood of zero in A. Then

$$Sp(\varrho(e)) \subset \mathbb{C} - \{\frac{1}{2} + i\mathbb{R}\}$$

and functional calculus with

$$F: \mathbb{C} - \{\frac{1}{2} + i\mathbb{R}\} \to \mathbb{C} \quad F(z) := \begin{cases} 1 & Re(z) > \frac{1}{2} \\ 0 & Re(z) < \frac{1}{2} \end{cases}$$

yields an idempotent

$$\tilde{e} := F(e) \in A$$

It is given by the sum

$$\tilde{e} = \varrho(e) + \sum_{k=1}^{\infty} \binom{2k}{k} (\varrho(e) - \frac{1}{2}) \omega(e, e)^k$$

b) Let

$$\varrho : \mathbb{C}[u, u^{-1}] \to A$$

be a unital, linear map. Suppose that

$$\omega(u, u^{-1}) = 1 - \varrho(u)\varrho(u^{-1}) \text{ and } \omega(u^{-1}, u) = 1 - \varrho(u^{-1})\varrho(u)$$

are "small". Then $\varrho(u)$ is invertible in A and its inverse is given by

$$\varrho(u)^{-1} = \sum_{k=0}^{\infty} \varrho(u^{-1})\omega(u, u^{-1})^k = \sum_{k=0}^{\infty} \omega(u^{-1}, u)^k \varrho(u^{-1})$$

□

Proof:

We follow Cuntz and Quillen:

a) Put

$$v := 2\varrho(e) - 1 \quad \varrho(e) = \frac{1}{2}(1 + v)$$

Then

$$Sp(\varrho(e)) \subset \mathbb{C} - \{\frac{1}{2} + i\mathbb{R}\} \Leftrightarrow Sp(v) \subset \mathbb{C} - \{i\mathbb{R}\}$$

Moreover

$$1 - v^2 = 4\omega(e, e)$$

is "small" so that

$$t(v + i\sqrt{\frac{1-t}{t}})(v - i\sqrt{\frac{1-t}{t}}) = t(v^2 + \frac{1-t}{t}) = 1 - t\,4\omega(e, e)$$

is invertible for $t \in]0, 1]$ which shows the assertion about the spectrum of $\varrho(e)$.

The explicit calculation of $\tilde{e} = F(\varrho(e))$ can be done by using the equality

$$F(z) = \frac{1}{2}(1 + G(2z - 1))$$

with

$$G : \mathbb{C} - \{i\mathbb{R}\} \to \mathbb{C} \quad G(z) = z\,(z^2)^{-\frac{1}{2}} = \begin{cases} +1 & Re(z) > 0 \\ -1 & Re(z) < 0 \end{cases}$$

One obtains

$$\tilde{e} = F(\varrho(e)) = \frac{1}{2}(1 + G(v)) = \frac{1}{2}\left(1 + v\,(1 - 4\omega(e,e))^{-\frac{1}{2}}\right)$$

$$= \frac{1}{2}\left(1 + (2\varrho(e) - 1)\sum_{k=0}^{\infty}\binom{-\frac{1}{2}}{k}(-4\omega(e,e))^k\right)$$

$$= \varrho(e) + (\varrho(e) - \frac{1}{2})\sum_{k=1}^{\infty}(-\frac{1}{2})^k\frac{1.3\ldots 2k-1}{1.2\ldots k}(-4\omega(e,e))^k$$

$$= \varrho(e) + (\varrho(e) - \frac{1}{2})\sum_{k=1}^{\infty}\frac{(2k)!}{(k!)^2}(\omega(e,e))^k = \varrho(e) + \sum_{k=1}^{\infty}\binom{2k}{k}(\varrho(e) - \frac{1}{2})(\omega(e,e))^k$$

which converges as long as $4\omega(e,e)$ is "small" because $\binom{2k}{k} \le 4^k$.

For part b) one finds that the series

$$v := \sum_{k=0}^{\infty}\varrho(u^{-1})\omega(u,u^{-1})^k, \quad w := \sum_{k=0}^{\infty}\omega(u^{-1},u)^k\varrho(u^{-1})$$

converge provided that $\omega(u,u^{-1})$ and $\omega(u^{-1},u)$ are "small". As

$$\varrho(u)v = (1 - \omega(u,u^{-1}))\sum_{k=0}^{\infty}(\omega(u,u^{-1}))^k = 1$$

$$= \sum_{k=0}^{\infty}(\omega(u^{-1},u))^k(1 - \omega(u^{-1},u)) = w\varrho(u)$$

we are done.

$$\square$$

With the help of the lemma it is now easy to demonstrate Theorem 1.19.

Let $[\varrho] \in [\mathbb{C}[e], A]_\alpha$ be represented by

$$\varrho : \mathbb{C}[e] \to \mathcal{C}^\infty(\mathcal{R}_+^m, A)$$

Choose a connected, punctured neighbourhood U of ∞ in $\mathcal{R}_+^m$ such that $4\omega(e,e)$ is "small" on U. Then

$$\begin{aligned}\varrho' : \quad \mathbb{C}[e] \quad &\to \quad \mathcal{C}^\infty(U, A)\\ e \quad &\to \quad F(\varrho(e))\end{aligned}$$

defines a homotopy class $[\varrho'] \in [\mathbb{C}[e], A]$ which is independent of all choices made and provides an inverse to the natural map

$$[\mathbb{C}[e], A] \to [\mathbb{C}[e], A]_\alpha$$

The odd case is similar.

$$\square$$

Chapter 2: Algebraic de Rham complexes

2-1 The periodic de Rham complex [CQ]

A de Rham complex for arbitrary Fréchet- (Banach-, C^*-) algebras A should have the following properties:

a) There should be a pairing

$$h(\Omega^*_{PdR}(A)) \otimes K_*(A) \to \mathbb{C}$$

generalizing the pairings

$$\{\text{Traces on } A\} \otimes K_0(A) \to \mathbb{C}$$

$$\{\text{Closed traces on } \Omega^1 A\} \otimes K_1(A) \to \mathbb{C}$$

b) It should be periodic because any stable homotopy functor on the category of C^*-algebras has to be Bott-periodic (Cuntz).

The first condition asks that traces should be zero-cocycles and closed 1-traces should be 1-cocycles in such a complex.

If one starts in analogy with ordinary de Rham cohomology with the universal enveloping differential bigraded algebra ΩA of a graded algebra A:

$$\Omega A \quad \xrightarrow{\cong} \quad \bigoplus_n A \otimes A/\mathbb{C}^{\otimes n}$$

$$a^0 da^1 \ldots da^n \quad \leftrightarrow \quad a^0 \otimes \bar{a}^1 \otimes \cdots \otimes \bar{a}^n$$

$$d(a^0 da^1 \ldots da^n) := da^0 da^1 \ldots da^n$$

$$|a^0 da^1 \ldots da^n| := |a^0| + |a^1| + \cdots + |a^n|$$

and asks that the periodic (cohomological) de Rham complex should consist of linear functionals on ΩA:

$$\Omega^*_{PdR}(A) \subset Hom(\Omega A, \mathbb{C})$$

then the trace condition on 0,1-cocycles is satisfied if one takes as differential the transpose of the operator

$$b_s : \quad \Omega^n A \quad \to \quad \Omega^{n-1} A$$

$$\omega da \cdot \to \quad (-1)^{deg\omega}[\omega, a]_s$$

([,] denoting the supercommutator) satisfying

$$b_s^2 = 0$$

A trace on $\Omega^1 A$ would in addition be closed if it were annihilated by the transpose of

$$d : \Omega^n A \to \Omega^{n+1} A$$

So a first guess would be to try to form a (now $\mathbb{Z}/2$-graded) complex by putting

$$?^*(A) := Hom(?_*(A), \mathbb{C}) \quad ?_*(A) := (\Omega A, b + d)$$

This however clearly fails as

$$(b + d)^2 = bd + db \neq 0$$

So one has to alter the definition and try

$$??_*(A) := \{\Omega A / (bd + db)\Omega A, b + d\}$$

This is indeed a complex but one sees soon that it cannot give the right cohomology: $h(??_*(A))$ contains $bB(\Omega A)$. (In the notations of [CO]).

A modified attempt however leads to a reasonable theory: one replaces the exterior differential by the operator

$$\begin{aligned} Nd : \quad \Omega^n A \quad &\to \quad \Omega^{n+1} A \\ \omega \quad &\to \quad (n + 1) \, d\omega \end{aligned}$$

and defines

Definition 2.1: (Cuntz,Quillen)

Let A be a graded unital algebra. The **periodic de Rham complex** of A is given by

$$\Omega_*^{PdR}(A) := \{\Omega A / (b(Nd) + (Nd)b)\Omega A, b + Nd\}$$

It is a filtered $\mathbb{Z}/2$-graded chain complex. The Hodge filtration by the subcomplexes $F^n \Omega_*^{PdR}$ generated by differential forms of degree at least n yields a sequence of quotient complexes $\Omega_*^{PdR}(A)/F^n \Omega_*^{PdR}(A)$ that approximate successively the periodic de Rham complex itself. We will be especially interested in the smallest of these quotient complexes, the X-complex:

2-2 X-Complexes [CQ]

Definition 2.2:

a) The **X-complex** of a graded, unital $\mathbb{C}$-algebra is the periodic chain complex

$$X_* A := \Omega_*^{PdR}(A)/F^2 \Omega_*^{PdR}(A)$$

Explicitely, it is given by

$$X_* A : \to A \xrightarrow{d} \Omega_\natural^1 A := \Omega^1 A / [A, \Omega^1 A]_s \xrightarrow{b} A \to$$

$$d(a) := 1 da \quad b(\overline{x dy}) := [x, y]_s$$

The **reduced X-complex** is defined as

$$\overline{X}_*A := X_*A/\mathbb{C}.1$$

b) The **(reduced) cohomological X-complex** is defined by duality:

$$X^*A := Hom_{\mathbb{C}}(X_*A, \mathbb{C}) \quad (\overline{X}^*A := Hom_{\mathbb{C}}(\overline{X}_*A, \mathbb{C}))$$

c) The **(reduced) bivariant X-complex** of the pair (A, B) of unital algebras is the Hom-complex

$$X^*(A, B) := Hom^*_{lin}(X_*A, X_*B) \quad (\overline{X}^*(A, B) := Hom^*_{lin}(\overline{X}_*A, \overline{X}_*B))$$

If $\tilde{A}$ is obtained from A by adjoining a unit then $\overline{X}_*\tilde{A} \simeq X_*A$ if A was already unital. □

The ordinary (cohomological, bivariant) (reduced) X-complex defines a covariant (contravariant, bivariant) functor from the category of unital (augmented) complex algebras to the category of $\mathbb{Z}/2$-graded chain complexes.

One easily verifies in the notation of Connes [CO] that

$$h_0(X^*A) = HC^0(A)/ker S \quad h_1(X^*A) = HC^1(A)$$

for trivially graded algebras (look at the Connes-Gysin sequence
$HH^1(A) \xrightarrow{B} HC^0(A) \xrightarrow{S} HC^2(A)$ for the calculation of $h_0(X_*A)$).

Let A now be a Fréchet algebra. The enveloping differential graded algebra ΩA can be topologized by declaring

$$\Omega A \quad \xleftarrow{\cong} \quad \bigoplus_{k=0}^{\infty} A \otimes \overline{A}^{\otimes k}$$

$$a^0 da^1 \ldots da^k \quad \leftarrow \quad a^0 \otimes \overline{a}^1 \otimes \cdots \otimes \overline{a}^k$$

to be a topological isomorphism, where each summand on the right hand side is given the projective tensor product topology and the whole sum is given the product topology.

We denote by $\Omega^1 A_\natural$ the quotient of $\Omega^1 A$ by the closure of the commutator subspace $[A, \Omega^1 A] \subset \Omega^1 A$. It is a Fréchet space in a natural way. The differentials of the X-complex $d : A \to \Omega^1 A_\natural$ $b : \Omega^1 A_\natural \to A$ are continuous so that the X-complex becomes a complex of Fréchet spaces.

Definition 2.3:

The composition of linear homomorphisms defines a bilinear map of chain complexes

$$X^*(A, B) \otimes X^*(B, C) \to X^*(A, C)$$

The induced map on cohomology is called the **composition product**.

□

Let δ be a (graded) derivation on A. It extends canonically to an action on the complex X_*A by putting

$$\delta(a^0 da^1) := \delta a^0 da^1 + (-1)^{|\delta||a^0|} a^0 d(\delta a^1)$$

and defines a homomorphism of the super-Lie algebra of graded derivations on A to the super-Lie algebra of graded endomorphisms of X_*A.

2-3 Differential graded X-Complexes

Our final aim will be a version of the X-complex that behaves functorially under asymptotic morphisms, i.e. under morphisms that are given by a whole family of maps. In order to get a reasonable theory the homotopy information of such a family has to be encoded into higher homotopy information in the associated X-complexes. This can be taken care of by replacing an algebra by its resolution as a differential graded algebra and by working in a differential graded setting throughout.

Let (A, ∂, N) be a differential graded algebra (∂ denotes the differential and N the number operator multiplying homogeneous elements by their degree). ∂ and N are both derivations on A of degree $+1$ and 0 respectively.

The remark following 2.3 about the action of derivations on X-complexes implies then

Lemma 2.4:

If (A, ∂, N) is a differential graded algebra, then X_*A becomes a $\mathbb{Z}/2$ graded complex of differential graded (DG)-modules. The differentials in the X-complex are morphisms of DG-modules:

$$b, d \in Hom_{DG}(X_iA, X_{i+1}A)$$

The same holds for the topological X-complex and the reduced X-complex provided that $\partial 1 = 0$.

$\square$

Definition 2.5:

Let A be a (trivially graded) unital Fréchet algebra and let $V.$ be a locally convex topological DG-module.

a) The **differential graded X-complex** of A with coefficients in $V.$ is the complex

$$X^*_{DG,V.}(A) := Hom^{cont}_{DG}(X_*(\Omega A), V.)$$

of degree preserving continuous DG-homomorphisms from the X-complex of the differential graded Fréchet algebra ΩA to $V..$

b) The **bivariant differential graded X-complex** of the pair (A, B) with coefficients in V. is the complex

$$X_{DG,V.}^*(A, B) := Hom_{DG,cont}^*(X_*(\Omega A), X_*(\Omega B)\widehat{\otimes}_\pi V.)$$

where $\widehat{\otimes}$ denotes graded tensor products.

□

The composition product extends to a map of complexes

$$X_{DG,V.}^*(A, B) \otimes X_{DG,W.}^*(B, C) \to X_{DG,V.\widehat{\otimes}_\pi W.}^*(A, C)$$

Moreover, there is a natural pairing

$$X_{DG,V.}^*(A, B) \otimes X_{DG,W.}^*(B) \to X_{DG,V.\widehat{\otimes}_\pi W.}^*(A)$$

The basic example of a locally convex DG-module will be the following

Example 2.6:

Let U be a smooth manifold and let $\oplus_k \Gamma(\Lambda^k T^*U)$ be the algebra of smooth exterior differential forms on U. If $V \subset U$ is a relatively compact open submanifold, if $x_1, \dots, x_n$ are coordinates on V, $Y_1, \dots, Y_k \in \Gamma(TU)$ are smooth vector fields and α is a multiindex, then

$$\Gamma(\Lambda^k T^*U) \quad \to \quad \mathbb{R}_+$$

$$\omega \quad \to \quad \sup_{x \in V} |\frac{\partial^\alpha}{\partial x^\alpha}(i_{Y_1} \dots i_{Y_k}\omega)(x)|$$

is a seminorm. The differential graded algebra $\Gamma(\Lambda^*T^*U)$ equipped with the topology defined by all these seminorms will be denoted by $\mathcal{E}(U)$.

□

One has $\mathcal{E}(U)\widehat{\otimes}_\pi\mathcal{E}(U') \simeq \mathcal{E}(U \times U')$ as locally convex DG-modules. In the sequel we will be interested in the simplicial DG-module $V. := \mathcal{E}(\mathbb{R}_+^\infty)$. We denote the associated X-complexes by

$$X_{DG}^*(A) := X_{DG,\mathcal{E}(\mathbb{R}_+^\infty)}^*(A)$$

$$X_{DG}^*(A, B) := X_{DG,\mathcal{E}(\mathbb{R}_+^\infty)}^*(A, B)$$

Under the identification $\mathcal{E}(\mathbb{R}_+^\infty)\widehat{\otimes}_\pi\mathcal{E}(\mathbb{R}_+^\infty) \simeq \mathcal{E}(\mathbb{R}_+^\infty \times \mathbb{R}_+^\infty) \simeq \mathcal{E}(\mathbb{R}_+^\infty)$ the composition product yields maps

$$X_{DG}^*(A, B) \otimes X_{DG}^*(B, C) \to X_{DG}^*(A, C)$$

$$X_{DG}^*(A, B) \otimes X_{DG}^*(B) \to X_{DG}^*(A)$$

It is possible now to replace in our considerations single homomorphisms by whole families of algebra homomorphisms.

Lemma 2.7:

Consider unital Fréchet algebras and form a new category Alg^∞ by defining the set of morphisms to be

$$mor_{Alg^\infty}(A, B) := Hom_{alg}^{cont}(A, \mathcal{C}^\infty(\mathbb{R}_+^\infty, B))$$

Then there is a canonical map

$$mor_{Alg^\infty}(A, B) \to X_{DG}^0(A, B)$$

defining a natural transformation of bifunctors

$$mor_{Alg^\infty} \to X_{DG}^0$$

Especially, the (bivariant) differential graded X-complex becomes a contravariant (bivariant) functor on the category Alg^∞.

Proof:

Recall that $\mathcal{C}^\infty(\mathbb{R}_+^\infty, A) \simeq A \otimes_\pi \mathcal{C}^\infty(\mathbb{R}_+^\infty)$. Let

$$\varrho \in mor_{Alg^\infty}(A, B) = Hom_{alg}(A, B \otimes_\pi \mathcal{C}^\infty(\mathbb{R}_+^\infty))$$

We define $\varrho_* \in X_{DG}^0(A, B)$ to be the composition

$$X_*(\Omega A) \xrightarrow{X_*(\Omega \varrho)} X_*(\Omega(B \otimes_\pi \mathcal{C}^\infty(\mathbb{R}_+^\infty))) \to X_*(\Omega(B) \widehat{\otimes}_\pi \Omega(\mathcal{C}^\infty(\mathbb{R}_+^\infty)))$$

$$X_*(\Omega(B) \widehat{\otimes}_\pi \Omega(\mathcal{C}^\infty(\mathbb{R}_+^\infty))) \to X_*(\Omega(B) \widehat{\otimes}_\pi \mathcal{E}) \xrightarrow{\psi} X_*(\Omega(B)) \widehat{\otimes}_\pi \mathcal{E}$$

The map

$$\psi : X_*(\Omega(B) \widehat{\otimes}_\pi \mathcal{E}(\mathbb{R}_+^\infty)) \to X_*(\Omega(B)) \otimes_\pi \mathcal{E}(\mathbb{R}_+^\infty)$$

is given by

$$a \otimes \omega \quad\to\quad a \otimes \omega \qquad \text{on } X_0$$

$$(a^0 \otimes \omega^0)d(a^1 \otimes \omega^1) \quad\to\quad (-1)^{|\omega^0||a^1|}a^0 da^1 \otimes \omega^0\omega^1 \quad \text{on } X_1$$

One verifies easily that $\varrho_* \in X_{DG}^0(A, B)$ (i.e. it preserves degrees and intertwines the differentials ∂ on ΩA and d_{dR} on $\mathcal{E}(\mathbb{R}_+^\infty)$). Moreover, the identity

$$\varrho_{1*} \circ \varrho_{0*} = (\varrho_1 \circ \varrho_0)_*$$

becomes obvious once the two expressions are written down explicitly.

$\square$

The component of degree zero of the map above is just the family of maps of X-complexes induced by a given family of algebra homomorphisms. The components of higher degree however contain higher homotopy information which will be necessary to obtain homotopy formulas comparing the cohomology classes of members of a family of cocycles for different parameter values.

2-4 The algebraic Chern character [CO]

The interest in cyclic homology comes from the existence of the well known Chern character map

Definition 2.8:

There is a natural transformation

$$ch : K_* \to h(X_*(M_\infty(-)))$$

on the category of admissible Fréchet algebras from topological K-groups to the homology of the X-complex of stable matrices.

It is defined on idempotent (resp.invertible) matrices by

$$[e]; e^2 = e \in A' := M_\infty A \quad \to \quad e \in A'/[A',A'] = h_0(X_*A')$$

$$[u]; uu^{-1} = u^{-1}u = 1 \in A' \quad \to \quad u^{-1}du \in \Omega^1(A')/([A',\Omega^1(A')] + dA') = h_1(X_*A')$$

$\square$

So by duality there are pairings with cyclic cohomology:

Lemma 2.9:

There exists a natural pairing

$$K_*A \otimes h(X^*(M_\infty A)) \to \mathbb{C}$$

of functors on the category of algebras. It extends to a pairing

$$K_*A \otimes h(X^*_{DG}(M_\infty A)) \to \mathbb{C}$$

on the extended category of Fréchet algebras with smooth families of algebra homomorphisms as morphisms (2.7).

Proof:

The latter pairing is clearly given by

$$K_*A \otimes h(X^*_{DG}(M_\infty A)) \xrightarrow{ch \otimes id} h(X_* M_\infty A) \otimes h(X^*_{DG}(M_\infty A)) \to$$

$$\xrightarrow{i \otimes id} h(X_*(\Omega(M_\infty A))) \otimes h(Hom(X_*(\Omega(M_\infty A)), \Omega_{dR}(\mathbb{R}^\infty_+))) \to \Omega_{dR}(\mathbb{R}^\infty_+)$$

It remains to be shown that the image of the pairing which may be a priori any differential form on $\mathbb{R}^\infty_+$ is in fact a constant function which can be identified with a complex number. This follows however from the following

Lemma 2.10:

Let A be a graded algebra, A_0 the subalgebra of elements of degree 0. Then any graded derivation on A annihilates the image of

$$ch : K_* A_0 \to h(X_*(M_\infty A))$$

$\square$

Applying the lemma to $\Omega A, \partial$ shows for $[x] \in K_*(A)$, $[\varphi] \in h(X^*_{DG}(A))$

$$d_{dR}(\langle ch(x), \varphi \rangle) = \langle ch(x), d_{dR} \circ \varphi \rangle = \langle ch(x), \varphi \circ \partial \rangle = \langle \partial(ch(x)), \varphi \rangle = 0$$

$\square$

Proof of Lemma 2.10:

Let ∂ be a graded derivation on A, $\partial \otimes 1 =: \partial'$ its extension to $M_\infty A =: A'$.

Even case:

Let $e = e^2 \in A'$. Then

$$\partial'(ch(e)) = \partial' e = \partial'(e^2) = (\partial' e)e + e(\partial' e) = [(\partial' e)e, e]_s + [e, e(\partial' e)]_s \in [A', A']_s$$

because

$$\partial' e = (\partial' e)e + e(\partial' e)$$

implies

$$e(\partial' e)e = e(\partial' e)e + e(\partial' e)e = 0$$

Odd case:

Let $uu^{-1} = u^{-1}u = 1 \in A'$. Then

$$\partial' ch(u) = \partial'(u^{-1}du) = \partial'(u^{-1})du + u^{-1}d(\partial' u) = -u^{-1}\partial' u u^{-1}du + u^{-1}d(\partial' u) =$$

$$= -[u^{-1}\partial' u, u^{-1}du]_s - (u^{-1}duu^{-1})\partial' u + u^{-1}d(\partial' u) =$$

$$= -[u^{-1}\partial' u, u^{-1}du]_s + d(u^{-1})\partial' u + u^{-1}d(\partial' u) =$$

$$-[u^{-1}\partial' u, u^{-1}du]_s + d(u^{-1}\partial' u) \in [A', \Omega^1(A')]_s + dA'$$

$\square$

The lemma above illustrates the role played by the higher homotopy data contained in the differential graded X-complex.

Chapter 3: Cyclic cohomology

To proceed in the construction of an asymptotic cyclic cohomology theory, i.e. a $\mathbb{Z}/2\mathbb{Z}$-graded chain complex for admissible Fréchet algebras that behaves functorially under asymptotic morphisms, it is necessary to make the complexes introduced in the last chapter functorial under (families) of linear maps.

This can be done in a universal way by replacing a unital algebra A by its tensor algebra RA. Working with tensor algebras also reduces the algebraic complexity of the involved chain complexes a lot: tensor algebras are of Hochschild cohomological dimension one and therefore the periodic de Rham complex of a tensor algebra becomes quasiisomorphic to the much smaller X-complex under the natural quotient map.

The cohomology of RA is uninteresting though because its X-complex depends only on the underlying based vector space of the given algebra. However, the tensor algebra of a given algebra comes equipped with a canonical adic filtration, defined by the kernel of the universal homomorphism $RA \to A$, which evidently allows to recover A from its filtered tensor algebra.

The filtration on RA induces a filtration on the associated X-complexes and a fundamental result of Cuntz and Quillen states that the X-complex of RA and the periodic de Rham complex of A with its Hodge filtration are quasiisomorphic as filtered complexes. So the complex $X_*(RA)$ is easy to manipulate algebraically on one hand but contains all information encoded in the whole periodic de Rham complex of A if its filtration is taken into account. Especially, the algebraic I-adic completion of this complex defines the well known periodic cyclic homology groups of Connes. While the X-complex of RA is evidently functorial with respect to linear maps of the underlying algebras, the completed X-complexes (and therefore the periodic cyclic groups) will only be functorial under homomorphisms of algebras, because a based, linear map induces a filtration preserving homomorphism of its tensor algebras iff it is multiplicative, i.e. an algebra homomorphism.

This chapter is taken from Cuntz-Quillen [CQ] (who studied the complexes $X_*(RA)$ from a somewhat different point of view.) The reason for repeating their proofs is that we work in a graded setting and we want to document the calculations in this case, too.

3-1 Extending functors

The problem of extending functors from a given to a larger category (in our case from the category of unital algebras to the category of unital algebras with based linear maps as morphisms) will now be formulated in full generality.

Let $j : \mathcal{C} \to \mathcal{C}'$ be a (covariant) functor between categories $\mathcal{C}, \mathcal{C}'$ and let $F : \mathcal{C} \to \mathcal{D}$ be any (contravariant) functor. By an **extension** of F from $\mathcal{C}$ to $\mathcal{C}'$ we mean a pair (F', ϕ), consisting of a functor $F' : \mathcal{C}' \to \mathcal{D}$ and a natural transformation $\phi : F \to F' \circ j$

$$
\begin{array}{ccc}
\mathcal{C} & \xrightarrow{j} & \mathcal{C}' \\
F \downarrow & & \downarrow F' \\
\mathcal{D} & \xrightarrow{\phi} & \mathcal{D}
\end{array}
$$

An extension (F', ϕ) is called **universal** if, given any extension (G, ψ) of F, there exists a unique transformation $\psi' : F' \to G$ making the following diagram commutative

$$
\begin{array}{ccccc}
\mathcal{C} & \xrightarrow{j} & \mathcal{C}' & & \\
F \downarrow & & \downarrow F' & \searrow G & \\
\mathcal{D} & \xrightarrow{\phi} & \mathcal{D} & \xrightarrow{\psi'} & \mathcal{D} \\
\| & & & & \| \\
\mathcal{D} & & \xrightarrow{\psi} & & \mathcal{D}
\end{array}
$$

The correspondence $\psi \leftrightarrow \psi'$ yields then a bijection

$$
Transf_{\mathcal{C}}(F, G \circ j) \simeq Transf_{\mathcal{C}'}(F', G)
$$

A sufficient criterion for the existence of a universal extension provides

Proposition 3.1:

Suppose that j admits a left adjoint (j^*, Φ): $j^* : \mathcal{C}' \to \mathcal{C}$

$$
\Phi : Hom_{\mathcal{C}'}(X, jY) \simeq Hom_{\mathcal{C}}(j^*X, Y)
$$

Then any functor $F : \mathcal{C} \to \mathcal{D}$ admits a universal extension to $\mathcal{C}'$. Moreover these extensions are natural with respect to F.

Instead of proving the proposition we will describe the construction of universal extensions in the following relevant case.

3-2 The algebra RA [CQ]

Let $\mathcal{C}, \mathcal{C}_\infty$ be the categories with augmented $\mathbb{C}$-algebras (Fréchet algebras) as objects and morphisms

$mor_{\mathcal{C}}(A, B) := $ based(=augmentation preserving) linear maps $A \to B$

$mor_{\mathcal{C}_\infty}(A, B) := $ based bounded linear maps $A \to B \otimes_\pi C^\infty(\mathbb{R}_+^\infty)$

Lemma 3.2:

The forgetful functors

$$j : Alg \to \mathcal{C} \ \ (Alg_\infty \to \mathcal{C}_\infty)$$

admit an adjoint

$$R : \mathcal{C} \to Alg \ \ (\mathcal{C}_\infty \to Alg_\infty)$$

$$Hom_{\mathcal{C}}(A, jB) \simeq Hom_{Alg}(RA, B) \quad Hom_{\mathcal{C}_\infty}(A, jB) \simeq Hom_{Alg_\infty}(RA, B)$$

Proof:

In the first case the quotient

$$RA := TA/(1_A - 1_{\mathbb{C}})$$

of the total tensor algebra $TA := \oplus_{n=0}^\infty A^{\otimes n}$ does the job. If $f : A \to B$ is linear and based, then the algebra homomorphism $Tf : TA \to B$ descends to an (augmentation preserving) homomorphism $RA \to B$.

In the second case, the goal is achieved by a completion of RA in the topology described in 3.6.

$\square$

Let

$$\varrho : A \to RA, \ \ \pi : RA \to A$$

be defined by

$$\varrho \leftrightarrow Id_{RA} \qquad\qquad\qquad Id_A \leftrightarrow \pi$$

$$Hom_{\mathcal{C}}(A, RA) \simeq Hom_{Alg}(RA, RA) \quad Hom_{\mathcal{C}}(A, A) \simeq Hom_{Alg}(RA, A)$$

One has

$$j(\pi) \circ \varrho = Id_A$$

and any based linear map $f \in mor_{\mathcal{C}}(A, B)$ admits a unique factorization

$$A \xrightarrow{\varrho} RA \to B$$

into the universal based linear map $\varrho : A \to RA$ and an algebra homomorphism $RA \to B$.

There is a canonical homomorphism of algebras

$$i : RA \to R(RA)$$

corresponding to the composition of universal based linear maps

$$A \xrightarrow{\varrho_A} RA \xrightarrow{\varrho_{RA}} R(RA)$$

under the bijection

$$Hom_{\mathcal{C}}(A, R(RA)) = Hom_{Alg}(RA, R(RA))$$

Proposition 3.1 yields in this case

Lemma 3.3:

The assignment

$$F \to (F \circ R, F(\pi))$$

yields a natural extension of any contravariant functor $F : Alg \to \mathcal{D}$ to $\mathcal{C}$.
If $G : \mathcal{C} \to \mathcal{D}$ is any contravariant functor the maps

$$Transf(F, Gj) \quad \leftrightarrow \quad Transf(F \circ R, G)$$

$$\varphi \qquad \to \qquad G(\varrho) \circ \varphi$$

$$\chi \circ F(\pi) \qquad \leftarrow \qquad \chi$$

are bijections inverse to each other. Therefore the above extensions are universal.

$$\square$$

The algebra RA depends only on the underlying vector space of A (and the augmentation map $A \to \mathbb{C}$). RA however admits a canonical filtration which enables one to recover the algebra structure of A.

3-3 I-adic filtrations [CQ]

Definition 3.4:

The **I-adic filtration** of RA is the filtration associated to the twosided ideal $IA \subset RA$ defined by the exact sequence

$$0 \to IA \to RA \xrightarrow{\pi} A \to 0$$

Lemma 3.5:

The commutative diagram of functors

$$
\begin{array}{ccc}
\text{Alg} & \xrightarrow{\text{(R,I-adic filt)}} & \text{Filtered Alg} \\
{\scriptstyle j}\downarrow & & \downarrow{\scriptstyle \text{forget}} \\
\mathcal{C} & \xrightarrow[R]{} & \text{Alg}
\end{array}
$$

induces on Hom-sets a pull-back(cartesian) square:

$$
\begin{array}{ccc}
\text{Hom}_{\text{alg}}(A,B) & \xrightarrow{\text{(R(-),I-adic filt.)}} & \text{Hom}_{\text{filt.pres.}}(RA,RB) \\
\downarrow & & \downarrow \\
\text{Hom}_{\mathcal{C}}(A,B) & \longrightarrow & \text{Hom}_{\text{alg}}(RA,RB)
\end{array}
$$

Proof:

Let (f,φ) be a pair consisting of $f \in Hom_{\mathcal{C}}(A,B)$ such that $Rf = \varphi$ preserves the I-adic filtrations of RA resp.RB. Then there is $g \in Hom_{alg}(A,B)$ given by

$$
\begin{array}{ccccccccc}
0 & \to & IA & \to & RA & \xrightarrow{\pi} & A & \to & 0 \\
& & \downarrow{\scriptstyle \varphi} & & \downarrow{\scriptstyle \varphi} & & \downarrow{\scriptstyle g} & & \\
0 & \to & IB & \to & RB & \xrightarrow{\pi} & B & \to & 0
\end{array}
$$

From the commutativity of

$$
\begin{array}{ccccc}
A & \xrightarrow{\varrho} & RA & \xrightarrow{\pi} & A \\
\downarrow{\scriptstyle f} & & Rf = \downarrow{\scriptstyle \varphi} & & \downarrow{\scriptstyle g} \\
B & \xrightarrow{\varrho} & RB & \xrightarrow{\pi} & B
\end{array}
$$

we obtain from the identity $j(\pi) \circ \varrho = Id$ the claimed equality $f = g \in Hom_{alg}(A,B)$.

$\square$

If the curvature (see 1.3) of the universal based linear map

$$
\varrho : A \to RA
$$

is denoted by ω

$$
\omega(a,b) := \varrho(ab) - \varrho(a) \otimes \varrho(b)
$$

the I-adic filtration on RA can be described explicitly as follows.

Proposition 3.6:[CQ]

There is an isomorphism of vector spaces

$$RA \quad \overset{\simeq}{\leftarrow} \quad \Omega^{ev} A$$

$$\varrho(a^0)\omega(a^1, a^2) \ldots \omega(a^{2n-1}, a^{2n}) \quad \leftarrow \quad a^0 da^1 \ldots da^{2n}$$

under which the I-adic filtration on RA corresponds to the degree (Hodge)-filtration on $\Omega^{ev} A$

$$(IA)^m \overset{\simeq}{\leftarrow} \bigoplus_{k=m}^{\infty} \Omega^{2k} A$$

The product on RA corresponds to the Fedosov product

$$\alpha * \alpha' := \alpha\alpha' - d\alpha d\alpha'$$

on $\Omega^{ev} A$.

If A is a Fréchet algebra then RA becomes a locally convex topological vector space under this isomorphism by giving $\Omega^{ev} A$ the topology of Chapter 2.

Proof:

Consider the subalgebra $\mathbb{C}[\varrho, \omega] \subset RA$ generated by the elements

$$\varrho(a), \omega(a', a''), \quad a. a'. a'' \in A$$

This subalgebra in fact equals RA because RA is generated by the elements $\varrho(a)$. By the Bianchi identity its elements can be written in the form

$$\sum_{finite} \varrho\omega^k$$

So the map under consideration is surjective. The injectivity follows from the fact that any sum

$$a^0 da^1 \ldots da^{2k} + \text{forms of lower degree}$$

is mapped (modulo tensors of degree $\leq$ 2k-1 in RA) to the element

$$\overline{a^0} \otimes \cdots \otimes \overline{a^{2k}} + (a^0 - \overline{a^0})\overline{a^1} \otimes \cdots \otimes \overline{a^{2k}}$$

in RA (see the explicit formula for $\omega(a, b)$). The formula for the powers of the ideal IA is clearly true for m=1; it follows in general by taking m-th powers and using the Bianchi identity to bring elements of RA into normal form.

The Fedosov product on differential forms is associative, so, by induction, it suffices to check that it corresponds to the product on RA for pairs of elements of the form $(a, a^0 da^1 \ldots da^{2n})$. For those it is clear from

$$a \rightarrow \varrho(a)$$

$$a^0 da^1 \ldots da^{2n} \rightarrow \varrho(a^0)\omega(a^1, a^2) \ldots \omega(a^{2n-1}, a^{2n})$$

$$a * a^0 da^1 \ldots da^{2n} = aa^0 da^1 \ldots da^{2n} - da da^0 da^1 \ldots da^{2n} \to$$
$$\to \varrho(aa^0)\omega^n - \omega(a, a^0)\omega^n = \varrho(a)\varrho(a^0)\omega^n$$

We still need the structure of the spaces $\Omega RA, \Omega^1 RA_\natural$.

Proposition 3.7:[CQ]

There is a canonical isomorphism of vector spaces

$$\Omega RA \leftarrow \bigoplus_{n,i_0,\ldots,i_n} (A \otimes \overline{A}^{\otimes 2i_0}) \otimes \overline{A} \otimes (A \otimes \overline{A}^{\otimes 2i_1}) \cdots \otimes \overline{A} \otimes (A \otimes \overline{A}^{\otimes 2i_n})$$

$$\varrho\omega^{i_0} \partial \varrho_1 \varrho\omega^{i_1} \ldots \partial \varrho_n \varrho\omega^{i_n} \leftarrow a^0 \otimes \overline{a^1} \otimes \cdots \otimes \overline{a^{n+2i_0+\cdots+2i_n}}$$

$\square$

The algebra ΩRA is canonically filtered by the powers of the ideal

$$0 \to I(\Omega RA) \to \Omega RA \xrightarrow{\Omega \pi} \Omega A \to 0$$

It is still called the I-adic filtration of ΩRA. Under the identification above the degree of an elementary tensor is $i_0 + i_1 + \cdots + i_n$.

The commutator quotient admits the following description:

Proposition 3.8:[CQ]

There is a canonical isomorphism of filtered vector spaces

$$\Omega^1 RA_\natural \qquad\qquad \leftarrow \qquad\qquad \Omega^{odd} A$$

$$(\varrho(a^0)\omega(a^1, a^2) \ldots \omega(a^{2k-1}, a^{2k}) d\varrho(a^{2k+1}))_\natural \leftarrow a^0 da^1 \ldots da^{2k+1}$$

Under this isomorphism the I-adic filtration of the RA bimodule $\Omega^1 RA_\natural$ on the left corresponds to the degree (Hodge)-filtration on the right side.

$\square$

3-4 Cyclic cohomology [CO],[CQ]

Let us reformulate what we obtained so far in this chapter

Proposition 3.9:

The universal extensions of the functors X_*, X_{DG}^* from the categories of algebras (Fréchet algebras with smooth families of morphisms) to the based linear categories $\mathcal{C}(\mathcal{C}_\infty)$ are given by

$$A \to X_*(RA) \quad A \to X_{DG}^*(RA)$$

Both of these complexes consist of filtered vector spaces (under the I-adic filtration). The morphisms of complexes

$$\varphi_* : X_*(RA) \to X_*(RB) \quad \psi^* : X_{DG}^*(RA) \leftarrow X_{DG}^*(RB)$$

induced by based linear maps $\varphi \in \mathcal{C}(A, B); \psi \in \mathcal{C}_\infty(A, B)$ preserve I-adic filtrations iff

φ, ψ are homomorphisms of algebras: $\varphi \in Hom(A, B), \psi \in Hom_\infty(A, B)$.

The behaviour of the differentials in the X-complex with respect to I-adic filtrations was determined by Cuntz and Quillen.

Theorem 3.10:[CQ]

Under the isomorphisms of vector spaces

$$RA \simeq \Omega^{ev} A \quad \Omega^1 RA_\natural \simeq \Omega^{odd} A$$

the complex $X_*(RA)$ is transformed into the complex

$$\mathcal{X}_*(RA) : \xrightarrow{\beta} \Omega^{ev} A \xrightarrow{\delta} \Omega^{odd} A \xrightarrow{\beta}$$

with

$$\delta = -\left(\sum_{i=0}^{n-1} \kappa^{2i}\right) b + \left(\sum_{j=0}^{2n} \kappa^j\right) d$$

on $\Omega^{2n} A$

$$\beta = b - (1 + \kappa) d$$

Both differentials lower the I-adic valuation degree by 1.

□

(For the definition of the Karoubi operator κ see the proof of theorem 3.11.)

The latter complex $\mathcal{X}_*(RA)$ is closely related to the periodic de Rham complex as well as to Connes normalized (b, B) bicomplex of A.

Theorem 3.11:[CQ]

a) The linear map

$$\Phi: \quad \Omega A \quad \rightarrow \quad \Omega A/(bNd + Ndb)\Omega A$$

$$\alpha^{2n} \quad \rightarrow \quad (-1)^n n! \, \overline{\alpha^{2n}}$$

$$\alpha^{2n+1} \quad \rightarrow \quad (-1)^n n! \, \overline{\alpha^{2n+1}}$$

induces a filtration preserving map of complexes

$$X_*(RA) \simeq \mathcal{X}_*(RA) \rightarrow \Omega_*^{PdR}(A)$$

b) Under this map the periodic de Rham complex of A becomes a deformation retract of $X_*(RA)$ in a canonical way, i.e. there exists a homomorphism of complexes

$$\Psi : \Omega_*^{PdR}(A) \rightarrow X_*(RA)$$

such that

$$\Phi \circ \Psi = Id_{\Omega_*^{PdR}(A)} \quad \Psi \circ \Phi = P_{CQ} \sim Id_{X_*(RA)}$$

where P_{CQ} is the Cuntz-Quillen projection [CQ].

c) The periodic de Rham complex of A is isomorphic to the normalized (b, B)-bicomplex of A ([CO]).

$\square$

Proof of theorem 3.10:

We repeat the calculation of [CQ] for the convenience of the reader and to verify them also in the graded case.

Odd degrees:

$$\beta(a^0 da^1 \ldots da^{2n+1}) = b_s(\varrho(a^0)\omega^n d\varrho(a^{2n+1})) =$$

$$= [\varrho\omega^n, \varrho]_s = \varrho(a^0)\omega^n \varrho(a^{2n+1}) - (-1)^{|\varrho\omega^n| \, ||\varrho|}\varrho(a^{2n+1})\varrho(a^0)\omega^n =$$

$$= \varrho\omega^{n-1}\omega(a^{2n-1}, a^{2n}a^{2n+1}) - \varrho\omega^{n-1}\omega(a^{2n-1}a^{2n}, a^{2n+1})$$

$$\vdots$$

$$+ \varrho(a^0)\omega(a^1, a^2a^3)\omega^{n-1} - \varrho(a^0)\omega(a^1a^2, a^3)\omega^{n-1}$$

$$+ \left(\varrho(a^0)\varrho(a^1)\omega^n - \varrho(a^0a^1)\omega^n \right) + \varrho(a^0a^1)\omega^n$$

$$\left(-(-1)^{|\varrho\omega^n| \, ||\varrho|}\varrho(a^{2n+1})\varrho(a^0)\omega^n + (-1)^{|\varrho\omega^n| \, ||\varrho|}\varrho(a^{2n+1}a^0)\omega^n \right)$$

$$-(-1)^{|\varrho\omega^n| \, ||\varrho|}\varrho(a^{2n+1}a^0)\omega^n$$

by the Bianchi identity. The two sums in brackets are equal to

$$-\omega(a^0, a^1)\omega^n \quad \text{and} \quad (-1)^{|\varrho\omega^n||\varrho|}\omega(a^{2n+1}, a^0)\omega^n \quad \text{respectively}$$

The whole sum corresponds therefore to

$$\sum_{i=0}^{2n}(-1)^i a^0 da^1 \ldots d(a^i a^{i+1}) \ldots da^{2n+1}$$

$$-(-1)^{|a^{2n+1}|(|a^0|+\cdots+|a^{2n}|)} a^{2n+1} a^0 da^1 \ldots da^{2n}$$

$$-da^0 da^1 \ldots da^{2n+1} + (-1)^{|a^{2n+1}|(|a^0|+\cdots+|a^{2n}|)} da^{2n+1} da^0 da^1 \ldots da^{2n}$$

$$= b_s(a^0 da^1 \ldots da^{2n+1}) - (1 - \kappa_s)(da^0 da^1 \ldots da^{2n+1})$$

Even degrees:

$$\delta(a^0 da^1 \ldots da^{2n}) = d(\varrho\omega^n) = d\varrho\omega^n + \sum_{i=0}^{n-1}\varrho\omega^i d\omega\omega^{n-i-1}$$

For a single summand of the latter sum one finds

$$\varrho\omega^i d\omega\omega^{n-i-1} = \varrho\omega^i d\varrho(a^{2i+1}a^{2i+2})\omega^{n-i-1}$$

$$-\varrho\omega^i d\varrho\varrho\omega^{n-i-1} - \varrho\omega^i \varrho d\varrho\omega^{n-i-1} =$$

$$= (-1)^{|\varrho\omega^i d\varrho(a^{2i+1}a^{2i+2})||\omega^{n-i-1}|}\omega^{n-i-1}\varrho\omega^i d\varrho(a^{2i+1}a^{2i+2})$$

$$-(-1)^{|\varrho\omega^i d\varrho||\omega^{n-i-1}|}\varrho\omega^{n-i-1}\varrho\omega^i d\varrho - (-1)^{|\varrho\omega^i \varrho d\varrho||\omega^{n-i-1}|}\omega^{n-i-1}\varrho\omega^i \varrho d\varrho$$

This corresponds (by denoting $(|a^0| + \cdots + |a^i|)(|a^{i+1}| + \cdots + |a^{2n}|)$ as $c_{i,i+1}$) to

$$(-1)^{c_{2i+2,2i+3}}\left((da^{2i+3}\ldots da^{2n}) * (a^0 da^1 \ldots da^{2i})\right) d(a^{2i+1}a^{2i+2})$$

$$-(-1)^{c_{2i+1,2i+2}}\left((a^{2i+2}da^{2i+3}\ldots da^{2n}) * (a^0 da^1 \ldots da^{2i})\right) d(a^{2i+1})$$

$$-(-1)^{c_{2i+2,2i+3}}(da^{2i+3}\ldots da^{2n} * a^0 * da^1 \ldots da^{2i} * a^{2i+1}) d(a^{2i+2})$$

$$= (-1)^{c_{2i+2,2i+3}}(da^{2i+3}\ldots a^0 \ldots da^{2i}d(a^{2i+1}a^{2i+2}) -$$

$$- da^{2i+3}\ldots a^0 \ldots da^{2i}a^{2i+1}da^{2i+2} + da^{2i+3}\ldots da^0 \ldots da^{2i}da^{2i+1}da^{2i+2})$$

$$- (-1)^{c_{2i+1,2i+2}}(a^{2i+2}da^{2i+3}\ldots a^0 \ldots da^{2i}da^{2i+1} -$$

$$- da^{2i+2}\ldots da^0 \ldots da^{2i}da^{2i+1})$$

$$= -b_s\left((-1)^{c_{2i+2,2i+3}}da^{2i+3}\ldots a^0 \ldots a^{2i+1}da^{2i+2}\right)$$

$$+ (-1)^{c_{2i+2,2i+3}}da^{2i+3}\ldots da^0 \ldots da^{2i+2}$$

$$+ (-1)^{c_{2i+1,2i+2}}da^{2i+2}\ldots da^0 \ldots da^{2i+1} =$$

$$= \left(-b_s(\kappa_s^{2(n-i-1)}) + \kappa_s^{2(n-i-1)} d + \kappa_s^{2(n-i-1)+1} d\right)(a^0 da^1 \ldots da^{2n})$$

Summing up yields

$$\delta = \kappa_s^{2n} d + \sum_{i=0}^{n-1} -b_s(\kappa_s^{2(n-i-1)}) + \kappa_s^{2(n-i-1)} d + \kappa_s^{2(n-i-1)+1} d =$$

$$= \kappa_s^{2n} d + \sum_{j=0}^{n-1} -b_s(\kappa_s^{2j}) + \kappa_s^{2j} d + \kappa_s^{2j+1} d = - \left(\sum_{j=0}^{n-1} \kappa_s^{2j}\right) b_s + \left(\sum_{i=0}^{2n} \kappa_s^i\right) d$$

on $\Omega^{2n} A$

□

Proof of theorem 3.11:

We recall from [CQ] that the Karoubi operator

$$\kappa_s := 1 - db_s - b_s d : \Omega A \to \Omega A$$

$$\kappa_s(a^0 da^1 \ldots da^n) = (-1)^{n-1}(-1)^{(|a^0|+\cdots+|a^{n-1}|)(|a^n|)} da^n a^0 da^1 \ldots da^{n-1}$$

satisfies the identity

$$(\kappa_s^n - 1)(\kappa_s^{n+1} - 1) = 0 \quad \text{on } \Omega^n A$$

The Karoubi operator commutes with the differentials β, δ of $\mathcal{X}_*(RA)$ so that $\mathcal{X}_*(RA)$ splits under the generalized eigenspace decomposition

$$\mathcal{X}_*(RA) \simeq \Omega A \simeq \ker(1 - \kappa_s)^2 \oplus \ker(1 + \kappa_s) \oplus \bigoplus_{\zeta \neq \pm 1} \ker(\kappa_s - \zeta)$$

into the direct sum of three complexes. The first is isomorphic to the periodic de Rham complex

$$\Omega_{PdR}^*(A) := (\Omega A / b_s(Nd) + (Nd)b_s, b + Nd)$$

In fact

$$b_s(Nd) + (Nd)b_s = (n+1)(b_s d + db_s) + (1 - db_s) - 1 = (n+1)(1 - \kappa_s) + \kappa_s^{n+1} - 1$$

on $\Omega^n A$ and the greatest common divisor of the polynomial

$$x^{n+1} - (n+1)x + n$$

and the minimal polynomial

$$(x^{n+1} - 1)(x^n - 1)$$

of κ_s equals

$$(x - 1)^2$$

So the canonical projection of ΩA onto the periodic de Rham complex Ω^*_{PdR} may be identified with the projection onto $ker(1 - \kappa_s)^2$ in the spectral decomposition of $\mathcal{X}_*(RA)$.

The differentials β, δ on this summand simplify to

$$\delta|_{ker(1-\kappa_s)^2} = -n\,b_s + (2n+1)\,d = -n\,b_s + Nd \text{ on } \Omega^{2n}A$$

$$\beta|_{ker(1-\kappa_s)^2} = b_s - 2d = b_s - \frac{1}{n+1}Nd \text{ on } \Omega^{2n+1}A$$

so that the map in a) defines in fact a map of complexes.

The inverse of the rescaling map in a) yields an inclusion of complexes

$$\psi := \Omega^{PdR}_* A \xrightarrow{\simeq} ker(1 - \kappa_s)^2 \subset \mathcal{X}_*(RA)$$

In order to show b) it suffices to prove that the complementary subcomplexes

$$ker(1 + \kappa_s) \text{ and } \bigoplus_{\zeta \neq \pm 1} ker(\kappa_s - \zeta)$$

in $\mathcal{X}_*(RA)$ are contractible. The differentials on the subcomplex $ker(1 + \kappa_s)$ are given by

$$\delta|_{ker(1+\kappa_s)} = -n\,b_s + d \text{ on } \Omega^{2n} \qquad \beta|_{ker(1+\kappa_s)} = b_s \text{ on } \Omega^{odd}$$

A contracting nullhomotopy is provided by

$$h = \begin{cases} Gd & \text{on } \Omega^{ev} \\ -\frac{1}{n+1}Gd & \text{on } \Omega^{2n+1} \end{cases}$$

(The notations are those of [CQ]).

The differentials on $ker(\kappa_s - \zeta); \zeta \neq \pm 1$ equal

$$\delta|_{ker(\kappa_s-\zeta)} = 0 \text{ on } \Omega^{ev} \qquad \beta|_{ker(\kappa_s-\zeta)} = b_s - (1+\zeta)d \text{ on } \Omega^{odd}$$

A contracting nullhomotopy is provided in this case by

$$h = G(d - (1+\zeta)^{-1}b_s)$$

The claim c) is clear because $Nd = B$ on $ker(1 - \kappa_s)^2$ by [CQ].

$\square$

Corollary 3.12:

a) The differentials on $X_*(RA)$ are continuous with respect to the I-adic topology.

Therefore one can define the following two complexes

b) The I-adic completion $\widehat{X}_*(RA)$ of the X-complex of RA. $\widehat{X}_*(RA)$ is canonically quasiisomorphic to the periodic cyclic Connes-bicomplex $CC_*^{per}(A)$ of A. Its homology equals the periodic cyclic homology of A:

$$h(\widehat{X}_*(RA)) \simeq PHC_*(A)$$

c) The complex $X_{fin}^*(RA)$ of linear functionals on $X_*(RA)$ vanishing on terms of high I-adic valuation. $X_{fin}^*(RA)$ is canonically quasiisomorphic to the cohomological Connes-bicomplex $CC^*(A)$ of cochains of finite support. Its cohomology groups equal the periodic cyclic cohomology groups of A:

$$h(X_{fin}^*(RA)) \simeq PHC^*(A)$$

□

Both complexes $\widehat{X}_*(RA), X_{fin}^*(RA)$ behave functorially under homomorphisms of algebras but not under arbitrary based, linear maps anymore.

Definition 3.13:[CO$_2$],[CQ]

a) Let

$$\varrho: \quad \widetilde{\mathbb{C}} = \mathbb{C}[e] \quad \to \quad R\widetilde{\mathbb{C}}$$
$$1_{\mathbb{C}} = e \quad \to \quad \varrho(e) = \varrho(1_{\mathbb{C}})$$

be the universal based linear map. According to 1.20 there exists an idempotent

$$F(\varrho(e)) \in \widehat{R\mathbb{C}}$$

in the I-adic completion of $R\mathbb{C}$ obtained from $\varrho(e)$ by functional calculus. Its image in $X_0(\widehat{R\widetilde{\mathbb{C}}})$ will be denoted by

$$ch(e) := \varrho(e) + \sum_{k=1}^{\infty} \binom{2k}{k}(\varrho(e) - \frac{1}{2})\,\omega(e,e)^k \in X_0(\widehat{R\widetilde{\mathbb{C}}})$$

b) Let $\varrho: \mathbb{C}[u, u^{-1}] \to R\mathbb{C}[u, u^{-1}]$ be the universal based linear map. Analogous to a) we put

$$ch(u) := \sum_{k=0}^{\infty} \varrho(u^{-1})\omega(u, u^{-1})^k d\varrho(u) \in X_1\widehat{R\mathbb{C}}[u, u^{-1}]$$

□

Chapter 4: Homotopy properties of X-complexes

In this chapter various Cartan homotopy operators expressing the triviality of the action of derivations on the cohomology of ordinary and differential graded X-complexes are constructed.

The homotopy formula is found by guessing it for the periodic de Rham complex of A in analogy with the classical homotopy formula for the Lie derivative along a vector field acting on the exterior differential forms on a manifold. This yields an operator h that works for algebraic differential forms of degree zero and one modulo error terms of higher degree:

$$\mathcal{L}_\delta - (h\partial + \partial h) = \psi : \Omega_*^{PdR} \to F^2\Omega_*^{PdR}$$

In the case of a tensor algebra however. the latter complex $F^2\Omega_*^{PdR}$ is contractible, so that ψ is nullhomotopic: $\psi = \partial h' + h'\partial$, which provides a true homotopy operator $H := h + h'$ on $\Omega_*^{PdR}(RA)$, respectively on the quasiisomorphic quotient complex $X_*(RA)$. Considering finally the I-adic filtration on $X_*(RA)$ allows via the identification

$$Gr_{I-adic}(X_*(RA)) \xrightarrow{qis} Gr_{Hodge}(\Omega_*^{pdR}(A))$$

to obtain a Cartan homotopy formula on the whole periodic de Rham complex of A which coincides in degrees zero and one with the formula guessed in the beginning.

We comment on this procedure in such detail because it is typical for the way one works with X-complexes and "lifts" constructions on algebraic differential forms of low degree to the full cyclic complexes. Another example for this technique will be the construction of exterior products on the chain level in chapter 8.

Beside this another homotopy formula for the action of a vector field on the asymptotic parameter space for the differential graded X-complex $X_{DG}^*(RA)$ is obtained which uses the higher homotopy information encoded in the differential graded X-complexes. Also we compare the differential graded to the ordinary X-complex which will be needed to construct natural transformations between the different cyclic theories encountered in later chapters.

Although all formulas and calculations are quite explicit it is not easy to develop a thorough understanding of the behaviour and the properties of the differential graded X-complex. This is provided in a final remark by calculating the cohomology of differential graded X-complexes using a universal coefficient spectral sequence in the abelian category of differential graded modules (DG-modules). The reason for not using the machinery of homological algebra from the beginning is that explicit formulas are needed as soon as topologies and growth properties are taken into account.

4-1 The Cartan homotopy formula

Let $\partial : A \to A$ be a graded derivation on A. It induces a graded derivation on RA which acts on $X_*(RA)$. This action is denoted by

$$\mathcal{L}_\partial : X_*(RA) \to X_*(RA)$$

The claim of this paragraph is to show that $\mathcal{L}_\partial$ acts trivially on the cohomology of $X_*(RA)$.

A naive attempt would be to generalize the homotopy formula

$$\mathcal{L}_X = i_X d + d i_X$$

for a vector field X acting on the de Rham complex of a smooth manifold. Here

$$i_X = i'_X \circ \text{antisymmetrization}$$

with

$$i'_X(f^0 df^1 \ldots df^n) = f^0 X(f^1) df^2 \ldots df^n$$

As antisymmetrization will yield reasonable results only for de Rham complexes of commutative algebras, it is better to start by generalizing the operator i'_X.

Definition 4.1:[CQ]

For ∂ a graded derivation on A put

$$i_\partial : \qquad \Omega^n A \qquad \to \qquad \Omega^{n-1} A$$

$$a^0 da^1 \ldots da^n \quad \to \quad (-1)^{n-1}(-1)^{|a^0||\partial|} a^0 \partial a^1 da^2 \ldots da^n$$

When we consider in how far this operator can be used as homotopy operator in the noncommutative de Rham complex respectively the homotopy equivalent (b, B) bicomplex the first observation is

Lemma 4.2:

$$[i_\partial, b_s] = 0$$

Proof:

$$[i_\partial, b_s](a^0 da^1 \nu da^n)$$

$$= b_s \left((-1)^{n-1}(-1)^{|a^0||\partial|} a^0 \partial a^1 da^2 \ldots da^n \right) - i_\partial \left((-1)^{n-1} [a^0 da^1 \nu, a^n]_s \right)$$

$$= +(-1)^{n-2}(-1)\, i_\partial(a^0 da^1 \nu)\, a^n$$

$$-(-1)^{n-1}(-1)^{|a^0||\partial|}(-1)^{n-2}(-1)^{(|a^0|+|\partial a^1|+\cdots+|a^{n-1}|)|a^n|} a^n a^0 \partial a^1 da^2 \ldots da^{n-1}$$

$$-(-1)^{n-1}\, i_\partial(a^0 da^1 \nu a^n)$$

$$+(-1)^{n-1}(-1)^{(|a^0|+|a^1|+\cdots+|a^{n-1}|)|a^n|}(-1)^{n-2}(-1)^{(|a^n|+|a^0|)|\partial|} a^n a^0 \partial a^1 da^2 \ldots da^{n-1}$$

$$= (-1)^{n-1} \left(i_\partial(a^0 da^1 \nu)a^n - i_\partial(a^0 da^1 \nu a^n) \right)$$

It has to be shown that the expression in brackets vanishes. One clearly may suppose n=2. In this case

$$i_\partial(a^0 da^1)a^2 - i_\partial(a^0 da^1 a^2) =$$

$$= (-1)^{|a^0||\partial|} a^0 \partial a^1 a^2 - (-1)^{|a^0||\partial|} a^0 \partial(a^1 a^2) + (-1)^{(|a^0|+|a^1|)|\partial|} a^0 a^1 \partial a^2 = 0$$

$\square$

With respect to the operator B however i_∂ behaves as a homotopy operator only in degrees less than two where the differential geometric formula is recovered. This would be not so bad, as in the X-complex of A only forms of degree less than two are considered, but unfortunately the operator i_∂ does not preserve the Hodge filtration on the noncommutative de Rham complex and does therefore not descend to an operator on the X-complex. For the algebras RA however this drawback can be overcome as the X-complex is not only a quotient of the de Rham complex but there is also a map of complexes

$$X_*(RA) \xrightarrow{X_*i} X_*(RRA) \to \Omega_*^{PdR}(RA)$$

defining a section of the quotient map and enabling one to construct a candidate for a homotopy operator on $X_*(RA)$.

Theorem 4.3:

Let δ be a graded derivation on A. Define

$$h_\delta : X_*(RA) \to X_{*+1}(RA)$$

to be the composition

$$X_*(RA) \xrightarrow{X_*(i)} X_*(RRA) \simeq \Omega RA \xrightarrow{i_\delta} \Omega RA \simeq X_{*+1}(RRA) \xrightarrow{X_*(\pi)} X_*(RA)$$

Then for the action $\mathcal{L}_\delta$ of δ on the complex $X_*(RA)$ the following homotopy formula is valid:

$$\mathcal{L}_\delta = h_\delta \partial_{X_*} + \partial_{X_*} h_\delta$$

The homotopy operator is given explicitly in terms of standard elements of $X_*(RA)$ (3.6, 3.8) as follows:

$$h_\delta : X_*(RA) \to X_{*+1}(RA)$$

$$\varrho\omega^n \to \delta(\varrho)d(\omega^n) + \sum_0^{n-2} \varrho\omega^j \delta(\omega)d(\omega^{n-1-j}) - \sum_0^{n-1} \varrho\omega^i \delta(\varrho)d\varrho\omega^{n-1-i}$$

$$\varrho\omega^n d\varrho \to \varrho\omega^n \delta(\varrho)$$

Proof:

The crucial step consists in showing the identity

$$X_*(\pi) \circ (i_\delta \partial + \partial i_\delta) = \mathcal{L}_\delta \circ X_*(\pi)$$

of operators on ΩRA. It is trivial in degree >3 because the operator $(i_\delta \partial + \partial i_\delta)$ shifts degrees by at most two and $X_*(\pi)$ vanishes in degrees >1.

We find

In degree 0:

$$X_*(\pi) \circ (i_\delta \partial + \partial i_\delta)(a) = X_*(\pi)(i_\delta da) =$$
$$= X_*(\pi)(\delta a) = \delta X_*(\pi)(a) = \mathcal{L}_\delta \circ X_*(\pi)(a)$$

In degree 1:

$$(i_\delta \partial + \partial i_\delta)(a^0 da^1) = (-i_\delta(1 + \kappa_s)d + di_\delta)(a^0 da^1) =$$
$$-i_\delta(da^0 da^1 - (-1)^{|a^0||a^1|}da^1 da^0) + (-1)^{|a^0||\delta|}d(a^0 \delta a^1)$$
$$= \delta a^0 da^1 - (-1)^{|a^0||a^1|}\delta a^1 da^0 + (-1)^{|a^0||\delta|}da^0 \delta a^1 + (-1)^{|a^0||\delta|}a^0 d\delta a^1$$
$$= \mathcal{L}_\delta(a^0 da^1) - (-1)^{|a^0||a^1|}[\delta a^1, da^0]_s$$

And so

$$X_*(\pi)(i_\delta \partial + \partial i_\delta)(a^0 da^1) = X_*(\pi)(\mathcal{L}_\delta(a^0 da^1)) = \mathcal{L}_\delta(X_*(\pi)(a^0 da^1))$$

In degree 2:

$$X_*(\pi) \circ (i_\delta \partial + \partial i_\delta) = X_*(\pi)\left((i_\delta(-b_s + (1 + \kappa_s + \kappa_s^2)d) + (b_s - (1 + \kappa_s)d)i_\delta\right)$$
$$= X_*(\pi)(-i_\delta b_s + b_s i_\delta) \text{ by degree considerations}$$
$$= 0 \text{ by the lemma.}$$

In degree 3 the reasoning is the same.

Finally one gets therefore

$$h_\delta \partial + \partial h_\delta = X_*(\pi) \circ i_\delta \circ X_*(i) \circ \partial + \partial \circ X_*(\pi) \circ i_\delta \circ X_*(i)$$
$$= X_*(\pi) \circ (i_\delta \circ \partial + \partial \circ i_\delta) \circ X_*(i) = \mathcal{L}_\delta \circ X_*(\pi) \circ X_*(i) = \mathcal{L}_\delta \circ X_*(\pi \circ i) = \mathcal{L}_\delta$$

The explicit form of h_δ will be calculated in 4.4.

$\square$

Corollary 4.4:

The homotopy operator constructed in the theorem above extends to an operator on the complexes

$$\widehat{X}_*(RA) \quad X^*_{fin}(RA)$$

of periodic chains (cochains of finite support). So the homotopy formula

$$\mathcal{L}_\delta = h_\delta \partial_{X_-} + \partial_{X_-} h_\delta$$

is valid on the complexes $\widehat{X}_*(RA)$, $X^*_{fin}(RA)$ calculating the periodic cyclic (co)homology of A.

Proof:

It has to be verified that the homotopy operator shifts I-adic valuations by a finite amount only, i.e.

$$h_\delta : F_I^N X_*(RA) \to F_I^{N-C} X_*(RA)$$

for some $C \in \mathbb{N}$

Denote the curvature of

$$\varrho_A : A \to RA$$

by ω, that of

$$\bar{\varrho} := \varrho_{RA} : RA \to RRA$$

by μ.

As $X_*(\pi) \circ i_\delta$ annihilates forms of degree >2 in $\Omega RA \simeq X_*(RRA)$ one obtains

$$h_\delta(\varrho\omega^n) = X_*(\pi) \circ i_\delta\left(\bar{\varrho}(\varrho)(\bar{\varrho}(\omega) + \mu(\varrho, \varrho))^n\right)$$

$$= X_*(\pi) \circ i_\delta\left(\bar{\varrho}(\varrho)\bar{\varrho}(\omega)^n + \sum_0^{n-1}\bar{\varrho}(\varrho)\bar{\varrho}(\omega)^i\mu(\varrho,\varrho)\bar{\varrho}(\omega)^{n-1-i}\right)$$

$$= X_*(\pi) \circ i_\delta\left(-\sum_0^{n-2}\bar{\varrho}(\varrho\omega^j)\mu(\omega,\omega^{n-1-j})\right) - X_*(\pi) \circ i_\delta\left(\mu(\varrho,\omega^n)\right) +$$

$$+ X_*(\pi) \circ i_\delta\left(\sum_0^{n-1}\bar{\varrho}(\varrho\omega^i)\mu(\varrho,\varrho)\bar{\varrho}(\omega^{n-1-i})\right)$$

$$= \sum_0^{n-2}\varrho\omega^j\delta(\omega)d(\omega^{n-1-j}) + \delta(\varrho)d(\omega^n) - \sum_0^{n-1}\varrho\omega^i\delta(\varrho)d\varrho\omega^{n-1-i}$$

which is of valuation $\geq n - 1$.

In odd dimensions the calculation is even simpler:

$$h_\delta(\varrho\omega^n d\varrho) = X_*(\pi) \circ i_\delta \left(\overline{\varrho}(\varrho)(\overline{\varrho}(\omega) + \mu(\varrho,\varrho))^n d\overline{\varrho}(\varrho)\right) = \varrho\omega^n \delta(\varrho)$$

This shows that the above claim holds for $C = 1$.

$\square$

As an application we show that the Cartan homotopy formula can be used to calculate the cyclic (co)homology of a direct sum of algebras.

Proposition 4.5:

Let A, B be unital. The canonical homomorphism

$$p : R(A \oplus B) \to RA \oplus RB$$

adjoint to the based, linear map

$$A \oplus B \xrightarrow{\varrho_A \oplus \varrho_B} RA \oplus RB$$

splits after I-adic completion, i.e. there exists a natural, continuous map

$$s : \widehat{R}A \oplus \widehat{R}B \to \widehat{R}(A \oplus B)$$

such that

$$\widehat{p} \circ s = Id_{\widehat{R}A \oplus \widehat{R}B}$$

and such that $s \circ \widehat{p}$ is canonically homotopic to the identity.

Corollary 4.6:

$$X_* \widehat{R}(A \oplus B) \xrightarrow{p_*} X_* \widehat{R}A \oplus X_* \widehat{R}B$$

is a quasiisomorphism.

$\square$

Corollary 4.7:

Let A be unital and let $\widetilde{A}$ be obtained from A by adjoining a unit. Then the canonical projection

$$\overline{X}_* \widehat{R}\widetilde{A} \to X_* \widehat{R}A$$

is a quasiisomorphism.

$\square$

Proof of Proposition 4.5:

The proof proceeds in several steps. First of all we define the splitting s.

1) Construction of a pair of orthogonal idempotents in $\widehat{R}(A \oplus B)$:
Put $\varrho(a) := \varrho((a,0)) \in R(A \oplus B)$, $\varrho(b) := \varrho((0,b)) \in R(A \oplus B)$. Let then

$$ch(1_A) := F(\varrho(1_A)) \in \widehat{R}(A \oplus B)$$
$$ch(1_B) := F(\varrho(1_B)) \in \widehat{R}(A \oplus B)$$

$$F(x) := x + \sum_{k=1}^{\infty} \binom{2k}{k}(x - \tfrac{1}{2})(x - x^2)^k$$

$ch(1_A)$, $ch(1_B)$ are idempotents in $\widehat{R}(A \oplus B)$ (1.20). In fact they are orthogonal.

Claim:
$$ch(1_A)\,ch(1_B) = 0 \quad ch(1_A) + ch(1_B) = 1$$

To verify the claim note first that $\varrho(1_A)$ and $\varrho(1_B)$ commute:

$$[\varrho(1_A), \varrho(1_B)] = [\varrho(1_A), 1 - \varrho(1_A)] = 0 \in R(A \oplus B)$$

Consequently $ch(1_A)$ and $ch(1_B)$ commute, too, so that $e' := ch(1_A)ch(1_B)$ is an idempotent in $\widehat{R}(A \oplus B)$ satisfying $\pi(e') = 1_A 1_B = 0 \in A \oplus B$. Thus $e' \in \widehat{I}(A \oplus B)$ and if $e' \in \widehat{I}^k(A \oplus B)$ for some $k > 0$ the equality $e' = e'^2 \in \widehat{I}^{2k}(A \oplus B)$ shows that

$$ch(1_A)ch(1_B) = e' \in \bigcap_{k=1}^{\infty} \widehat{I}^k(A \oplus B) = 0$$

This identity being established, it follows that $e'' := ch(1_A) + ch(1_B)$ is an idempotent, as well as $1 - e''$. The identity $\pi(e'') = 1_A + 1_B = 1$, $\pi(1 - e'') = 0$ leads as above to the conclusion $1 - e'' = 0$, i.e. $ch(1_A) + ch(1_B) = 1$

2) Construction of s:
We want to construct a homomorphism $s : RA \oplus RB \to \widehat{R}(A \oplus B)$ satisfying $s(1_A) = ch(1_A)$, $s(1_B) = ch(1_B)$. To do so, put first of all

$$\varphi_1 : \quad A \quad \to \qquad\qquad\qquad \widehat{R}(A \oplus B)$$
$$a \quad \to \quad \varrho(a) + \sum_{k=1}^{\infty} \binom{2k}{k}(\varrho(1_A) - \tfrac{1}{2})\omega(1_A, 1_A)^{k-1}\omega(1_A, a)$$

$$s_1 : \quad A \quad \to \qquad\qquad\qquad \widehat{R}(A \oplus B)$$
$$a \quad \to \quad \varphi_1(a)ch(1_A)$$

and define $\varphi_2 : B \to \widehat{R}(A \oplus B)$, $s_2 : B \to \widehat{R}(A \oplus B)$ by the analogous formulas.

Lemma 4.8:

$$s_1(a) = s_1(1_A)s_1(a) = s_1(a)s_1(1_A) \quad s_2(b) = s_2(1_B)s_2(b) = s_2(b)s_2(1_B)$$

Proof:

First of all note that

$$s_1(1_A) = \varphi_1(1_A)ch(1_A) = ch(1_A)ch(1_A) = ch(1_A)$$

This shows already

$$s_1(a)s_1(1_A) = \varphi_1(a)ch(1_A)ch(1_A) = \varphi_1(a)ch(1_A) = s_1(a)$$

Furthermore the Bianchi identity (1.3) implies that

$$ch(1_A)\varphi_1(a) =$$

$$= ch(1_A)\left(\varrho(a) + \sum_{k=1}^{\infty}\binom{2k}{k}(\varrho(1_A) - \frac{1}{2})\omega(1_A,1_A)^{k-1}\omega(1_A,a)\right)$$

$$= \varrho(a) + \sum_{k=1}^{\infty}(\lambda_k\varrho(1_A) + \mu_k)\omega(1_A,1_A)^{k-1}\omega(1_A,a)$$

with universal coefficients $\lambda_k, \mu_k \in \mathbb{R}$. Taking $a := 1_A$, the identity

$$ch(1_A)\varphi_1(1_A) = ch(1_A)ch(1_A) = ch(1_A)$$

shows

$$\lambda_k = -2\mu_k = \binom{2k}{k}$$

so that the whole sum equals in fact $\varphi_1(a)$. Consequently

$$s_1(1_A)s_1(a) = ch(1_A)\varphi_1(a)ch(1_A) = \varphi_1(a)ch(1_A) = s_1(a)$$

$\square$

Continuation of the proof of 4.5:

The lemma shows that $s_1(1_A)$, (resp. $s_2(1_B)$) acts as unit on the subalgebra of $\widehat{R}(A \oplus B)$ generated by $s_1(A)$, (resp. $s_2(B)$). Consequently there are homomorphisms of full tensor algebras

$$s_1 : TA \to \widehat{R}(A \oplus B) \qquad s_2 : TB \to \widehat{R}(A \oplus B)$$
$$a \to s_1(a) \qquad\qquad b \to s_2(b)$$
$$1_{\mathbb{C}} \to s_1(1_A) \qquad\qquad 1_{\mathbb{C}} \to s_2(1_B)$$

which annihilate $(1_A - 1_{\mathbb{C}})$, (resp. $(1_B - 1_{\mathbb{C}})$) and descend therefore to homomorphisms

$$s_1 : RA \to \widehat{R}(A \oplus B) \qquad s_2 : RB \to \widehat{R}(A \oplus B)$$

Furthermore, the images of these morphisms annihilate each other due to the identity

$$s_1(a)s_2(b) = s_1(a)s_1(1_A)s_2(1_B)s_2(b) = s_1(a)ch(1_A)ch(1_B)s_2(b) = 0$$

because $ch(1_A)$ and $ch(1_B)$ are orthogonal. Thus

$$s: \quad RA \oplus RB \quad \to \quad \widehat{R}(A \oplus B)$$
$$(x, y) \quad \to \quad s_1(x) + s_2(y)$$

is an algebra homomorphism. It is in fact unital because

$$s(1) = s((1_A, 1_B)) = s_1(1_A) + s_2(1_B) = ch(1_A) + ch(1_B) = 1$$

Finally s preserves I-adic filtrations as

$$s(\omega(a^0, a^1)) = s_1(a^0 a^1) - s_1(a^0) s_1(a^1) =$$
$$= \varphi_1(a^0 a^1) ch(1_A) - \varphi_1(a^0) ch(1_A) \varphi_1(a^1) ch(1_A)$$

projects to zero under $\pi : \widehat{R}(A \oplus B) \to A \oplus B$. Thus s extends to I-adic completions

$$s: \quad \widehat{R}(A) \oplus \widehat{R}(B) \quad \to \quad \widehat{R}(A \oplus B)$$

3)

$$\widehat{p} \circ s = Id_{\widehat{R}(A) \oplus \widehat{R}(B)}$$

As $\widehat{p} \circ s$ is a continuous homomorphism of algebras it suffices to check the identity on a set of generators of a dense subalgebra, i.e. on $\varrho(A) \subset \widehat{R}A$ (and similar for B). One finds

$$s(\varrho(a)) = s_1(a) = \varphi_1(a) ch(1_A) =$$
$$= \left(\varrho(a) + \sum_{k=1}^{\infty} x_k \omega(1_A, a) \right) \left(\varrho(1_A) + \sum_{l=1}^{\infty} y_l \omega(1_A, 1_A) \right)$$

for some $x_k, y_l \in R(A \oplus B)$. Now under the canonical map $p : R(A \oplus B) \to RA \oplus RB$
$\varrho(1_A) \to 1_{RA}$, $\varrho(1_B) \to 1_{RB}$ so that

$$p(\omega(1_A, a)) = \omega_{RA}(1, a) = 0, \, p(\omega(1_A, 1_A)) = \omega_{RA}(1, 1) = 0$$

and therefore

$$\widehat{p}(s(\varrho(a))) = p(\varrho(a)) p(\varrho(1_A)) = \varrho(a) 1_{RA} = \varrho(a)$$

which proves the claim.

4)

$$s \circ \widehat{p} \sim Id_{\widehat{R}(A \oplus B)}$$

As it suffices to describe a homotopy on the generators $\varrho(A \otimes B)$ of $\widehat{R}(A \oplus B)$ we consider $s \circ p(\varrho(a, b)) = s_1(a) + s_2(b)$. Put

$$F(-, t) := F_1(-, t) + F_2(-, t)$$
$$F_1(a, t) = \varrho(a) + t(s_1(a) - \varrho(a)) \quad F_1(b, t) = 0$$
$$F_2(b, t) = \varrho(b) + t(s_2(b) - \varrho(b)) \quad F_2(a, t) = 0$$

Then $F(-, t)$ defines a smooth family of endomorphisms of $\widehat{R}(A \oplus B)$ as is seen by the equality

$$F(1, t) = F_1(1_A, t) + F_2(1_B, t)$$

$$= \varrho(1_A) + t(ch(1_A) - \varrho(1_A)) + \varrho(1_B) + t(ch(1_B) - \varrho(1_B))$$

$$= 1 + t(ch(1_A) + ch(1_B) - 1) = 1$$

and the estimate

$$F(\omega(a, a'), t) = \varrho(aa') + t(s_1(aa') - \varrho(aa')) - (\varrho(a) + t\ldots)(\varrho(a') + t\ldots) =$$

$$= \varrho(aa') - \varrho(a)\varrho(a') = \omega(a, a') \mod \widehat{I}(A \oplus B)$$

$\square$

Proof of corollary 4.6:

If A and B are unital observe first that the map

$$X_*(A \oplus B) \xrightarrow{(X_*(\pi_0), X_*(\pi_1))} X_*A \oplus X_*B$$

is in fact an isomorphism of complexes. This is obvious for X_0 and follows for X_1 from the fact that the mixed terms adb, $b'da'$ of $\Omega^1(A \oplus B)$ vanish in the commutator quotient:

$$adb_\natural = ad(b1_B)_\natural = abd(1_B)_\natural + 1_Bad(b)_\natural = 0$$

From this we see that

$$X_*(\widehat{p}) : X_*\widehat{R}(A \oplus B) \to X_*\widehat{R}A \oplus X_*\widehat{R}B$$

$$X_*(s) : X_*\widehat{R}A \oplus X_*\widehat{R}B \to X_*\widehat{R}(A \oplus B)$$

satisfy $X_*(\widehat{p}) \circ X_*(s) = Id$ whereas $X_*(s) \circ X_*(\widehat{p}) = X_*(s \circ \widehat{p})$ is chain homotopic to the identity by the Cartan homotopy formula, applied to the smooth homotopy connecting $s \circ \widehat{p}$ and $Id_{\widehat{R}(A \oplus B)}$ and therefore a quasiisomorphism. It follows that $X_*(\widehat{p})$ is a quasiisomorphism itself.

$\square$

Proof of corollary 4.7:

For unital A one has $\widetilde{A} \simeq A \oplus \mathbb{C}$ as algebras. Thus

$$X_*\widehat{R}\widetilde{A} \simeq X_*\widehat{R}(A \oplus \mathbb{C}) \xrightarrow[qis]{X_*\widehat{p}} X_*\widehat{R}A \oplus X_*\widehat{R}\mathbb{C}$$

by corollary 4.6 and

$$X_*\widehat{R}A \oplus X_*\widehat{R}\mathbb{C} \xrightarrow[qis]{Id \oplus X_*\pi} X_*\widehat{R}A \oplus X_*\mathbb{C} = X_*\widehat{R}A \oplus \mathbb{C}$$

by Lemma 5.17 are natural quasiisomorphisms mapping the subcomplex

$X_*\mathbb{C} \simeq \mathbb{C}1_{RA} \subset X_*\widehat{\widetilde{RA}}$ to $\mathbb{C}(1_{RA}, 1_\mathbb{C}) \subset X_*\widehat{RA} \oplus \mathbb{C}$. Thus the composition of the maps above descends to a quasiisomorphism

$$\overline{X}_*\widehat{\widetilde{RA}} \xrightarrow{qis} X_*\widehat{RA} \oplus \mathbb{C}/\mathbb{C}(1_{RA}, 1_\mathbb{C}) = X_*\widehat{RA}$$

□

4-2 Homotopy formulas for differential graded X-complexes

Preliminary remarks:

For any unital algebra A one can form the differential graded algebras ΩRA and $R(\Omega A)$. They are related by canonical maps as follows:

Lemma 4.9:

a) There is a canonical homomorphism of DG-algebras

$$j : \Omega RA \to R(\Omega A)$$

defined by

$$Hom_{DGA}(\Omega RA, R(\Omega A)) \xrightarrow{\simeq} Hom_{Alg}(RA, R(\Omega A)_0)$$

$$j \qquad\qquad \leftrightarrow \quad Id : RA \to RA = R(\Omega A)_0$$

b) There is a canonical homomorphism of DG-algebras

$$k : R(\Omega A) \to \Omega RA$$

defined by

$$Hom_{DGA}(R(\Omega A), \Omega RA) \xrightarrow{\simeq} Hom_{DG-linear}(\Omega A, \Omega RA)$$

$$k \qquad\qquad \leftrightarrow \qquad\qquad k'$$

$$k'(a^0 da^1 \dots da^n) := \varrho(a^0)d(\varrho(a^1)) \dots d(\varrho(a^n))$$

c)

$$k \circ j = Id_{\Omega RA}$$

d) If A is a Fréchet algebra, then so are $\Omega RA, R(\Omega A)$ and the morphisms j, k are continuous.

e) The morphisms j, k preserve I-adic filtrations.

□

Now the first result of this section can be formulated:

Theorem 4.10:

Let $\delta : A \to A$ be a graded derivation on the Fréchet algebra A. Denote by δ' the corresponding derivation on ΩA commuting with the exterior differential on this algebra. Let

$$h'_\delta : X_*(R\Omega A) \to X_*(R\Omega A)$$

be the operator constructed in Theorem 4.3. Define

$$h_\delta : X^*_{DG}(RA) \to X^*_{DG}(RA)$$

to be induced by the composition

$$X_*(\Omega RA) \xrightarrow{X \cdot j} X_*(R\Omega A) \xrightarrow{h'_\delta} X_*(R\Omega A) \xrightarrow{X \cdot k} X_*(\Omega RA)$$

Then the following Cartan homotopy formula is valid in the differential graded X-complex $X^*_{DG}(RA)$:

$$\mathcal{L}_\delta = \partial_{X^*_{DG}} \circ h_\delta + h_\delta \circ \partial_{X^*_{DG}}$$

□

Proof:

It is clear that $h_{\delta'}$ commutes with the action of the exterior derivative ∂ and the number operator N on $X_*(\Omega RA)$ because j, k are homomorphisms of differential graded algebras and the differentials δ', ∂ on ΩA commute. Therefore h_∂ yields in fact a chain map on $X^*_{DG}(RA)$.

If we denote the differential of the periodic complex $X_*(\Omega RA)$ by d_{X_*}, we find

$$d_X.h_d + h_d d_{X_*} = d_{X_*} \circ X_*(k) \circ h_{\delta'} \circ X_*(j) + X_*(k) \circ h_{\delta'} \circ X_*(j) \circ d_{X_*} =$$

$$X_*(k)\circ(d_{X_*} \circ h_{\delta'} + h_{\delta'} \circ d_{X_*})\circ X_*(j) = X_*(k)\circ\mathcal{L}_{\delta'}\circ X_*(j) = X_*(k\circ j)\circ\mathcal{L}_{\delta'} = \mathcal{L}_{\delta'}$$

□

Now we prove a second homotopy formula for the differential graded X-complex. It concerns the homotopy properties in the "parameter space" $\mathbb{R}^\infty_+$ and corresponds to the condition, formulated by Connes-Moscovici [CM], that the "time derivative" of an asymptotic cocycle should be an asymptotic coboundary.

Theorem 4.11:

Let $V.$ be a DG-module,
$$i_Y : V. \to V._{-1}$$
a derivation of degree -1 and
$$\mathcal{L}_Y := \partial i_Y + i_Y \partial$$
the associated Lie-derivative. It is a derivation of degree 0.
Then $\mathcal{L}_Y$ acts on
$$Hom_{DG}(X_*(\Omega RA), V.) \quad \text{and} \quad Hom_{DG}^*(X_*(\Omega RA), X_*(\Omega RB) \otimes V.)$$
in an obvious way.

Put
$$h_Y : Hom_{DG}(X_*(\Omega RA), V.) \to Hom_{DG}(X_{*-1}(\Omega RA), V.)$$
$$(h_Y : Hom_{DG}^*(X_*(\Omega RA), X_*(\Omega RB) \otimes V.) \to Hom_{DG}^{*-1}(X_*(\Omega RA), X_*(\Omega RB) \otimes V.)$$

$$h_Y(\Phi) := (\partial i_Y) \circ \Phi \circ (\frac{1}{N} h_N) + i_Y \circ \Phi \circ (\frac{1}{N} h_N) \circ \partial$$

where $N : X_*(\Omega A) \to X_*(\Omega A)$ is the number operator and h_N is the Cartan homotopy operator associated to the action of N on $X_*(\Omega RA)$ via Theorem 4.3.

Then h_Y satisfies the Cartan homotopy formula
$$\mathcal{L}_Y = h_Y \circ \partial_{Hom_{DG}^*} + \partial_{Hom_{DG}^*} \circ h_Y$$

$\square$

Proof:

We treat the second case, the first being similar.
Let $\Phi \in Hom_{DG}^*(X_*(\Omega RA), X_*(\Omega RB) \otimes V.)$
Among the operators used to define h_Y, $\frac{1}{N}, h_N, \Phi$ preserve internal degrees, ∂ increases them by 1 and i_Y decreases them by 1 so that $h_Y(\Phi)$ is still degree preserving.

Furthermore
$$[\partial, h_Y(\Phi)] = (\partial i_Y) \circ \Phi \circ (\frac{1}{N} h_N) \circ \partial - (\partial i_Y) \circ \Phi \circ (\frac{1}{N} h_N) \circ \partial = 0$$
so that $h_Y(\Phi)$ is a DG-map.

Let us check the homotopy formula:

$$h_Y \circ \partial_{Hom_{DG}^*} + \partial_{Hom_{DG}^*} \circ h_Y =$$

$$= \partial_{Hom_{DG}^*} \left((\partial i_Y) \circ \Phi \circ (\frac{1}{N} h_N) + i_Y \circ \Phi \circ (\frac{1}{N} h_N) \circ \partial \right)$$

$$+ h_Y \left(\Phi \circ \partial_{X_*A} - (-1)^{deg\Phi} \partial_{X_*B} \circ \Phi \right) =$$

$$= (\partial i_Y) \circ \Phi \circ (\frac{1}{N} h_N) \circ \partial_{X_*A} + i_Y \circ \Phi \circ (\frac{1}{N} h_N) \circ \partial \circ \partial_{X_*A}$$

$$- (-1)^{deg\Phi+1} \partial_{X_*B} \circ (\partial i_Y) \circ \Phi \circ (\frac{1}{N} h_N) - (-1)^{deg\Phi+1} \partial_{X_*B} \circ i_Y \circ \Phi \circ (\frac{1}{N} h_N) \circ \partial$$

$$+ (\partial i_Y) \circ \Phi \circ \partial_{X_*A} \circ (\frac{1}{N} h_N) + i_Y \circ \Phi \circ \partial_{X_*A} \circ (\frac{1}{N} h_N) \circ \partial$$

$$- (-1)^{deg\Phi} (\partial i_Y) \circ \partial_{X_*B} \circ \Phi \circ (\frac{1}{N} h_N) - (-1)^{deg\Phi} i_Y \circ \partial_{X_*B} \circ \Phi \circ (\frac{1}{N} h_N) \circ \partial$$

$$= (\partial i_Y) \circ \Phi \circ \frac{1}{N}(h_N \circ \partial_{X_*A} + \partial_{X_*A} \circ h_N)$$

$$+ i_Y \circ \Phi \circ \frac{1}{N}(h_N \circ \partial_{X_*A} + \partial_{X_*A} \circ h_N) \circ \partial$$

as

$$[\partial_{X_*}, \partial] = [\partial_{X_*}, \frac{1}{N}] = [\partial_{X_*B}, i_Y] = 0$$

$$= (\partial i_Y) \circ \Phi + i_Y \circ \Phi \circ \partial = (\partial i_Y + i_Y \partial) \circ \Phi = \mathcal{L}_Y \Phi$$

as

$$[\Phi, \partial] = 0$$

□

Corollary 4.12:

Let $U \subset \mathcal{R}^n$ be an open submanifold and let $Y \in \Gamma(TU)$ be a smooth vector field on U. Denote by

$$i_Y : \Gamma(\Lambda^k T^*U) \to \Gamma(\Lambda^{k-1} T^*U)$$

the contraction by Y and by

$$\mathcal{L}_Y : \mathcal{E}(U) \to \mathcal{E}(U)$$

the Lie derivative along Y. (See 2.7.)
Then $\mathcal{L}_Y$ acts trivially on the cohomology groups

$$h(Hom_{DG}^*(X_*(\Omega RA), X_*(\Omega RB) \otimes_\pi \mathcal{E}(U))) \text{ and } h(Hom_{DG}(X_*(\Omega RA), \mathcal{E}(U)))$$

□

Corollary 4.13:

Let

$$X^*_{DG,fin}(RA) \subset X^*_{DG}(RA)$$

be the subcomplex of functionals vanishing on elements of high I-adic filtration (i.e. on $F^N_I(\Omega RA)$ for some $N >> 0$).

Let as before Y be a smooth vector field on $I\!R^\infty_+$.

. Then $\mathcal{L}_Y$ preserves the subcomplex of cochains with finite support

$$\mathcal{L}_Y : X^*_{DG,fin}(RA) \to X^*_{DG,fin}(RA)$$

and acts trivially on its cohomology.

$\square$

We will now study the natural projection of the differential graded onto the ordinary X-complex and will exhibit a natural contracting homotopy on the kernel of this projection.

Theorem 4.14:

Let A, B be algebras and let V. be a DG-module.

a) There exist natural maps of complexes

$$Hom^*_{DG}(X_*(\Omega RA), X_*(\Omega RB)) \to Hom^*(X_*(RA), X_*(RB))$$

$$Hom^*_{DG}(X_*(\Omega RA), X_*(\Omega RB) \otimes V.) \to Hom^*_{DG}(X_*(\Omega RA), X_*(RB) \otimes V.)$$

obtained by restricting functionals to the degree zero subspace $X_*(RA)$ of $X_*(\Omega RA)$, and by projecting $X_*(\Omega RB)$ onto $X_*(RB)$, respectively

b) The underlying maps of vector spaces split naturally.

c) Denote the kernels of the maps in a) by

$$Hom^*_{DG,+}(X_*(\Omega RA), X_*(\Omega RB)) \subset Hom^*_{DG}(X_*(\Omega RA), X_*(\Omega RB))$$

and

$$Hom^*_{DG,+}(X_*(\Omega RA), X_*(\Omega RB) \otimes V.) \subset Hom^*_{DG}(X_*(\Omega RA), X_*(\Omega RB) \otimes V.)$$

respectively. Then there exist contracting homotopies

$$H : Hom^*_{DG,+}(X_*(\Omega RA), X_*(\Omega RB)) \to Hom^{*-1}_{DG,+}(X_*(\Omega RA), X_*(\Omega RB))$$

$$H' : Hom^*_{DG,+}(X_*(\Omega RA), X_*(\Omega RB) \otimes V.) \to Hom^{*-1}_{DG,+}(X_*(\Omega RA), X_*(\Omega RB) \otimes V.)$$

$$Id = \partial_{Hom^\cdot} \circ H + H \circ \partial_{Hom^\cdot}$$

$$Id = \partial_{Hom^\cdot} \circ H' + H' \circ \partial_{Hom^\cdot}$$

which are natural in A, B (resp. $A, B, V.$).

$\square$

Proof:

. We treat the second case, the first being similar.

The complex $Hom^*_{DG,+}(X_*(\Omega RA), X_*(\Omega RB) \otimes V.)$ is naturally filtered by the subcomplexes

$$X^{*,j}_{DG,+}(RA, RB)_{V.} := \{ \Phi \in X^*_{DG,+}(RA, RB)_{V.} | \Phi(X_*(\Omega RA)) \subset F^j(X_*(\Omega RB) \otimes V.) \}$$

where

$$F^j(W.) := \bigoplus_{n=j}^{\infty} W_n = \{ x \in W., \, deg \, x \geq j \}$$

for any DG-module W.

$$X^{*,1}_{DG,+}(RA, RB)_{V.} = X^*_{DG,+}(RA, RB)_{V.}$$

Define a natural map

$$\begin{array}{ccc} h_j : & X^{*,j}_{DG}(RA, RB)_{V.} & \rightarrow & X^{*-1,j}_{DG}(RA, RB)_{V.} \\ & \Phi & \rightarrow & h_j\Phi \end{array}$$

by the commutative diagram

$$h_j\Phi : \quad \begin{array}{ccc} X_*(\Omega RA)_{j+2} & \overset{0}{\rightarrow} & (X_*(\Omega RB) \otimes V)_{j+2} \\ \partial \uparrow & & \uparrow \partial \\ X_*(\Omega RA)_{j+1} & \xrightarrow{\partial \circ \frac{1}{N}(h_N \circ \Phi) \circ h_\partial} & (X_*(\Omega RB) \otimes V)_{j+1} \\ \partial \uparrow & & \uparrow \partial \\ X_*(\Omega RA)_j & \xrightarrow{\frac{1}{N}(h_N \circ \Phi)} & (X_*(\Omega RB) \otimes V)_j \\ \partial \uparrow & & \uparrow \partial \\ X_*(\Omega RA)_{j-1} & \overset{0}{\rightarrow} & (X_*(\Omega RB) \otimes V)_{j-1} \end{array}$$

Here

$$h_\partial : X_*(\Omega RA). \rightarrow X_*(\Omega RA)._{-1}$$

is the natural contracting homotopy on the DG-module $(\bigoplus_i X_i(\Omega RA), \partial)$ and

$$h_N : X_*(\Omega RB) \rightarrow X_{*-1}(\Omega RB)$$

is the homotopy operator associated to the number operator acting on ΩRB via Theorem 4.3.

Put now

(*)
$$f_j := Id - (\partial_{Hom^\bullet} \circ h_j + h_j \circ \partial_{Hom^\bullet}) : X^{*,j}_{DG,+}(RA, RB)_V. \to X^{*,j+1}_{DG,+}(RA, RB)_V.$$

f_j is a natural map of chain complexes homotopic to the identity. These maps can be used to obtain successively the desired nullhomotopy. Define

$$H : X^*_{DG,+}(RA, RB)_V. \to X^{*-1}_{DG,+}(RA, RB)_V.$$

$$H := \sum_{j=1}^{\infty} h_j \circ f_{j-1} \circ f_{j-2} \circ \cdots \circ f_1$$

Each term of the sum is well defined by (*) and no problems of convergence arise because the maps f_k preserve degrees and h_j vanishes except in degrees $j, j+1$.

Finally

$$\partial_{Hom^\bullet} \circ H + H \circ \partial_{Hom^\bullet} =$$

$$= \sum_{j=1}^{\infty} \partial_{Hom^\bullet} \circ h_j \circ f_{j-1} \circ f_{j-2} \circ \cdots \circ f_1 + h_j \circ f_{j-1} \circ f_{j-2} \circ \cdots \circ f_1 \circ \partial_{Hom^\bullet}$$

$$= \sum_{j=1}^{\infty} (\partial_{Hom^\bullet} \circ h_j + h_j \circ \partial_{Hom^\bullet}) \circ f_{j-1} \circ f_{j-2} \circ \cdots \circ f_1$$

$$= \sum_{j=1}^{\infty} (Id - f_j) \circ f_{j-1} \circ f_{j-2} \circ \cdots \circ f_1 = Id - \cdots \circ f_j \circ \cdots \circ f_1 = Id$$

because $\cdots \circ f_j \circ \cdots \circ f_1$ maps $X^*_{DG,+}(RA, RB)_V.$ into

$$\bigcap_{j=1}^{\infty} X^{*,j}_{DG,+}(RA, RB)_V. = 0$$

□

Theorem 4.15:

For any Fréchet algebra A the natural maps of complexes

$$X^*(RA) \to X^*_{DG}(RA)$$

$$X^*(RA, RB) \leftarrow Hom^*_{DG}(X_*(\Omega RA), X_*(\Omega RB)) \to X^*_{DG}(RA, RB)$$

are quasiisomorphisms, i.e. induce isomorphisms on cohomology groups.

Proof:

This is a consequence of the contractibility of the "parameter space" $I\!\!R_+^\infty$.

If one chooses a contraction, for example

$$F_t(x) := (1-t)x + tx_0$$

for some interior point $x_0 \in I\!\!R_+^\infty$, the second Cartan homotopy formula yields for any $\Phi \in X_{DG}^*(RA)$

$$F_1^*\Phi - \Phi = F_1^*\Phi - F_0^*\Phi = \int_0^1 \mathcal{L}_{\frac{\partial}{\partial t}}(F_t^*\Phi)dt$$

$$= \partial \left(\int_0^1 h_{\frac{\partial}{\partial t}} F_t^*\Phi dt \right) + \int_0^1 h_{\frac{\partial}{\partial t}} F_t^*(\partial\Phi)dt = \partial H\Phi + H\partial\Phi$$

with

$$H\Phi := \int_0^1 h_{\frac{\partial}{\partial t}} F_t^*\Phi dt$$

As F_1^* retracts $X_{DG}^*(RA)$ onto the subcomplex $X^*(RA)$, this shows that $X^*(RA)$ is a deformation retract of the differential graded X-complex and the assertion follows. In the bivariant case Theorem 4.14. has also to be used.

$\square$

Corollary 4.16:

The inclusion

$$X_{fin}^*(RA) \to X_{DG,fin}^*(RA)$$

is a quasiisomorphism and therefore

$$h_*(X_{DG,fin}^*(RA)) \simeq PHC^*(A)$$

$\square$

Remark:

The comparison theorems between the ordinary and differential graded X-complexes were formulated so explicitly in order to be able to translate them into a topologized setting in the next chapters. If we stay in a purely algebraic context however, the results above can be explained in a more conceptual way.

It is shown easily that the category of DG-modules is an abelian category with enough injectives and projectives: a DG-module (V, ∂) is injective (projective) iff (V, ∂) is contractible as chain complex and the subobjects $V_j, j \in \mathbb{N}$ are injective (projective) in the underlying category of $\mathbb{C}$-vector spaces. Especially $X_*(\Omega RA)$ is a $\mathbb{Z}/2$-graded chain complex of projective DG-modules. It is therefore possible to calculate the cohomology of the bivariant differential graded X-complexes via the universal coefficient spectral sequence [GO]

$$(E_r^{p,q}): \quad E_2^{p,q} := Ext^q(H_{*+p}(C_*), H_*(D_*)) \Rightarrow Gr^p H^{p+q}(Hom^*(C_*, D_*))$$

In our case this yields the E_2-terms

$$Ext_{DG}^q(H_{*+p}(X_*(\Omega RA)), H_*(X_*(\Omega RB)))$$

$$\Rightarrow Gr^p H^{p+q}(Hom_{DG}^*(X_*(\Omega RA), X_*(\Omega RB)))$$

$$Ext_{DG}^q(H_{*+p}(X_*(\Omega RA)), H_*(X_*(\Omega RB) \otimes V.))$$

$$\Rightarrow Gr^p H^{p+q}(Hom_{DG}^*(X_*(\Omega RA), X_*(\Omega RB) \otimes V.))$$

As DG-module $H_*(X_*(\Omega RA))$ is concentrated in degree zero because the number operator N acts trivially on it by the Cartan homotopy formula 4.3.

To calculate the desired Ext-groups, we note that any DG-module M_0 concentrated in degree zero has a canonical projective resolution by acyclic DG-modules of length two (i.e. concentrated in two consecutive degrees). Using this resolution provides isomorphisms

$$Ext_{DG}^*(M_0, (V, \partial)) \simeq H^*(Hom_{\mathbb{C}}(M_0, V), \partial) \simeq Hom_{\mathbb{C}}(M_0, H^*(V, \partial))$$

Applying this to our examples yields

$$Ext_{DG}^q(H_{*+p}(X_*(\Omega RA)), H_*(X_*(\Omega RB))) \simeq$$

$$\begin{cases} Hom_{\mathbb{C}}^*(H_{*+p}(X_*(RA)), H_*(X_*(RB))) & q = 0 \\ 0 & q > 0 \end{cases}$$

$$Ext_{DG}^q(H_{*+p}(X_*(\Omega RA)), H_*(X_*(\Omega RB) \otimes V.)) \simeq$$

$$Hom_{\mathbb{C}}^*(H_{*+p}(X_*(RA)), H_*(X_*(RB)) \otimes H^q((V, \partial)))$$

So we see that the projections

$$Hom_{DG}^*(X_*(\Omega RA), X_*(\Omega RB)) \to Hom^*(X_*(RA), X_*(RB))$$

$$Hom_{DG}^*(X_*(\Omega RA), X_*(\Omega RB) \otimes V.) \to Hom_{DG}^*(X_*(\Omega RA), X_*(RB) \otimes V.)$$

induce isomorphisms on the E_2-terms of the associated universal coefficient spectral sequences. Therefore their kernels have to be acyclic.

$\square$

Chapter 5: The analytic X-complex

The complexes $X_*(RA)$ studied up to now have a rich algebraic structure but are uninteresting from a cohomological point of view: they behave functorially under linear maps of algebras and the Cartan homotopy formulas imply then the vanishing of their cohomology groups because any linear map is linearly homotopic to zero. Already in the algebraic setting it was necessary not to consider the X-complex of the tensor algebra RA, but that of its algebraic completion $\hat{R}A$ with respect to the I-adic topology. In this chapter we suppose that A itself comes equipped with a (Fréchet)-topology and will construct a (formal) topological I-adic completion $\mathcal{R}A$ of the tensor algebra RA in the case that A is admissible.

The choice of topology on RA is dictated by the demand that

1) The completed tensor algebra $\mathcal{R}A$ should still be of cohomological dimension one and the cohomology of the completed X-complexes $X_*(\mathcal{R}A)$, $X_{DG}^*(\mathcal{R}A)$ should be nontrivial.

2) A linear map $f : A \to B$ which is almost multiplicative should still induce a continuous homomorphism $f : \mathcal{R}A \to \mathcal{R}B$ (at least on a subalgebra that depends on the deviation of f from being multiplicative, i.e. its curvature).

The difficulty with 2) is that the induced homomorphism $Rf : RA \to RB$ of a linear map does not preserve I-adic filtrations unless f is multiplicative. In fact it may move the I-adic valuation of tensors by an arbitrary large amount. On the other hand the norm of these "correction terms" of different degrees decays exponentially fast with the I-adic valuation if the curvature of f becomes small:

If $a \in I^n A / I^{n+1} A$, then one obtains

$$Rf(a) = \sum_{k=0}^{n} b_k \quad b_k \in I^k B / I^{k+1} B$$

$$\text{with } \| b_k \| \leq C^{n-k} \| a \|$$

where C depends only on the maximum of the curvature of f on the entries of a and becomes arbitrarily small if the curvature does so.

Thus an almost multiplicative map will "almost" preserve I-adic filtrations if norms are taken into account.

This suggests the following construction: Fix a multiplicatively closed subset K of A and consider tensors over A with entries in K. Expand a given element of this subalgebra of RA in a standard basis (Chapter 3) with respect to the I-adic filtration. A weighted L^1-norm for the coefficients of such an expansion is then introduced allowing the coefficients to grow exponentially to the basis N with respect to the I-adic valuation. Denote the corresponding completion by $RA_{(K,N)}$. It is a Fréchet algebra and possesses the following crucial property: If $f : A \to B$ is linear with curvature uniformly bounded on $K \subset A$ by a sufficiently small constant, then Rf induces a continuous homomorphism $Rf : RA_{(K,N)} \to RB_{(K',N')}$ for a suitable multiplicatively closed subset $K' \subset B$ and $N' \geq 1$. In practice f will be an asymptotic morphism as studied in chapter one. As the curvature of an asymptotic morphism is uniformly bounded only over compact sets, the multiplicatively closed subsets $K \subset A$ used for the construction above will throughout taken to be relatively compact. To guarantee the existence of sufficiently many multiplicatively closed compact sets the underlying Fréchet algebra will be supposed to be admissible.

In this case, the completed tensor algebras $RA_{(K,N)}$ will be admissible Fréchet algebras, too. Moreover, as the algebraic I-adic completion $\widehat{R}A$, the algebras $RA_{(K,N)}$ are of cohomological dimension one, i.e.quasifree.

The study of these completed tensor algebras will make up most of this chapter. The topological I-adic completion will finally be the formal inductive limit

$$\mathcal{R}A := \text{ " } \varinjlim_{\to (K,N)} \text{ "} RA_{(K,N)}$$

of the algebras constructed above. The kernel of the projection $\pi : \mathcal{R}A \to A$ is formally topologically nilpotent, so that $\mathcal{R}A$ defines in fact a formally topologically nilpotent extension of A.

The chapter ends with the introduction of analytic cyclic (co)homology of A as the (co)homology of the complexes $X_*(\mathcal{R}A)$, $X^*(\mathcal{R}A)$. This is justified by the cohomological dimension of $\mathcal{R}A$ being equal to one. The resulting complex turns out to be closely related to the entire cyclic bicomplex of Connes $[CO_2]$.

5-1 Behaviour of I-adic filtrations under based linear maps

The isomorphism $RA \simeq \Omega^{ev}A$ of (3.6) allows one to expand tensors over A in a sum of standard elements corresponding to homogeneous differential forms. The algebra structure of RA and the behaviour of the I-adic filtration under homomorphisms induced by linear maps will now be analyzed with respect to this standard presentation. The first and most basic result is

Lemma 5.1:

Let $f : A \to B$ be a based, linear map. Denote the curvature of f by

$$\kappa(a, a') = f(aa') - f(a)f(a')$$

Then

$$Rf\left(\varrho(a^0)\omega(a^1, a^2) \dots \omega(a^{2n-1}, a^{2n})\right) =$$

$$= \varrho(f(a^0)) \prod_1^n (\varrho(\kappa(a^i, a^{i+1})) + \omega(f(a^i), f(a^{i+1}))) = \sum_{\alpha=1}^M \varrho(c_\alpha^0) \prod_1^{n_\alpha} \omega(c_\alpha^{2i-1}, c_\alpha^{2i})$$

where

1) the entries c_α^i are products of terms

$$\kappa(a^{2j-1}, a^{2j}) \text{ and } f(a^k)$$

and each entry contains at most 2 factors of the form $f(a^k)$.

2) In each summand $\varrho(c_\alpha^0)\prod_1^{n_\alpha} \omega(c_\alpha^{2i-1}, c_\alpha^{2i})$ of the right hand side, if we put $\sharp\omega := n_\alpha$ and denote by $\sharp\kappa$ the total number of factors $\kappa(a^{2j-1}, a^{2j})$ in the entries c_α, one has

$$\sharp\kappa + \sharp\omega \geq n \quad \sharp\omega \leq n$$

3) The total number of summands M is bounded by

$$M = \sharp\alpha \leq 8^n$$

Proof:

One has

$$\varrho(b^0)\prod_1^n (\varrho(\kappa_i) + \omega(b^{2i-1}, b^{2i})) =$$

(*)

$$\sum_{2^n \text{ terms}} \varrho(b^0)(\prod_1^{n_1} \varrho(\kappa_{\alpha_i^1}))(\prod_1^{n_2} \omega(b^{\alpha_i^2}, b^{\alpha_{i+1}^2}))\ldots(\prod_1^{n_{2k-1}} \varrho(\kappa_{\alpha_i^{2k-1}}))(\prod_1^{n_{2k}} \omega(b^{\alpha_i^{2k}}, b^{\alpha_{i+1}^{2k}}))$$

If such a product is reduced via the Bianchi identity

$$\varrho(a)\varrho(b) = -\omega(a,b) + \varrho(ab)$$

to standard form (r.h.s. of 5.1.) the number of κ-factors remains constant while the number of ω-factors may increase, but certainly cannot decrease. As in the initial term we have

$$\sharp\kappa + \sharp\omega = n$$

the first part of 2) is proved. The second follows from induction over n.

By having a closer look at the Bianchi identity, it becomes also clear, that the entries c_α of the r.h.s.of 5.1. are of the form claimed in 1).

It remains to estimate the number of terms of (*) in standard form. We proceed by induction over k. First we apply the Bianchi identity to

$$(\prod_1^{n_{2k-1}} \varrho(\kappa_{\alpha_i^{2k-1}}))$$

and obtain a sum

$$\sum_{\leq 2^{n_{2k-1}} \text{ terms}} \varrho(c_{2k-1})\prod_j \omega(c_j, c_j')$$

Next, the Bianchi identity applied to

$$(\prod_1^{n_{2k-2}} \omega(b^{\alpha_i^{2k-2}}, b^{\alpha_{i+1}^{2k-2}}))\varrho(c_{2k-1})$$

yields a sum

$$\sum_{\leq 2n_{2k-2}+1 \text{ terms}} \varrho(c_{2k-2})\prod_\beta \omega_\beta$$

The initial expression of a summand of the r.h.s. of $(*)$ is thus reduced to a sum of

$$2^{n_{2k-1}}(2n_{2k-2}+1)$$

terms of the form

$$\varrho(b^0)(\prod_1^{n_1}\varrho(\kappa_{\alpha_i^1}))(\prod_1^{n_2}\omega(b^{\alpha_i^2},b^{\alpha_{i+1}^2}))\dots(\prod_1^{n_{2k-3}}\varrho(\kappa_{\alpha_i^{2k-3}}))\varrho(c_{2k-2})(\prod\omega_{\beta'})$$

which establishes the induction step.

We end up with at most

$$2.2^{n_1+1}(2n_2+1)2^{n_3+1}\dots(2n_{2k-2}+1)2^{n_{2k-1}}$$

terms. Using

$$2k+1\leq 2^{k+1}\leq 4^k,\ k\geq 1$$

we obtain

$$2.2^{n_1+1}(2n_2+1)2^{n_3+1}\dots(2n_{2k-2}+1)2^{n_{2k-1}}\leq 4^{n_1+n_2+\dots+n_{2k-1}}\leq 4^n$$

As the initial expression $(*)$ consists of 2^n summands of the form above we find 3).

$\square$

It will turn out to be necessary to do functional calculus on the ideal IA. First of all, the standard presentation of high powers of a given tensor is needed.

Lemma 5.2:

Let A be an augmented algebra and let

$$\{a_j = \sum_\beta \lambda_\beta\varrho_\beta\omega^{i_\beta}\mid j\in J\}$$

be a set of elements of IA with entries in $K\cup\{1\}\subset A$ and $\lambda_\beta\in\mathbb{C}$ such that

$$\sum_{i_\beta=n}|\lambda_\beta|\leq CN^n$$

for some $C>0, N>1$. Then $\prod_1^k a_j$ can be represented as a sum

$$\prod_1^k a_j = \sum_\gamma \lambda_\gamma\varrho_\gamma\omega^{i_\gamma}$$

with entries in the multiplicative closure of $K\cup\{1\}$ and

$$C(k,n) := \sum_{i_\gamma=n}|\lambda_\gamma|$$

satisfies

$$C(k, n) \leq (2C)^k (2N)^n \frac{n^k}{k!}$$

□

Proof:

Modulo $(IA)^{n+1}$ one finds

$$\prod_1^k a_j = \sum_{\substack{(\beta_1, \ldots, \beta_k) \\ i_{\beta_1} + \cdots + i_{\beta_k} \leq n \\ i_{\beta_j} \geq 1}} \lambda^1_{\beta_1} \varrho_{\beta_1} \omega^{i_{\beta_1}} \ldots \lambda^k_{\beta_k} \varrho_{\beta_k} \omega^{i_{\beta_k}}$$

which, by the proof of Lemma 5.1. equals a sum

$$\sum_{i_{\beta_1} + \cdots + i_{\beta_k} \leq n; \, i_{\beta_j} \geq 1} (\lambda^1_{\beta_1} \ldots \lambda^k_{\beta_k}) \left(\sum_{c(i_{\beta_1}, \ldots, i_{\beta_k}) \, terms} \varrho_\gamma \omega^{i_\gamma} \right)$$

where

$$c(i_{\beta_1}, \ldots, i_{\beta_k}) \leq (2i_{\beta_1} + 2) \ldots (2i_{\beta_k} + 2) \quad i_\gamma \geq i_{\beta_1} + \cdots + i_{\beta_k}$$

and the entries of $\varrho_\gamma \omega^{i_\gamma}$ belong to the multiplicative closure of $K \cup \{1\}$. It remains to estimate the number of terms and the coefficients λ_γ. One finds

$$\sum_{i_\gamma = n} |\lambda_\gamma| \leq \sum_{\substack{(\beta_1, \ldots, \beta_k) \\ i_{\beta_1} + \cdots + i_{\beta_k} \leq n \\ i_{\beta_j} \geq 1}} |\lambda^1_{\beta_1} \ldots \lambda^k_{\beta_k}| c(i_{\beta_1}, \ldots, i_{\beta_k})$$

$$\leq \sum_{i_{\beta_1} + \cdots + i_{\beta_k} \leq n; \, i_{\beta_j} \geq 1} C^k N^{i_{\beta_1} + \cdots + i_{\beta_k}} (2i_{\beta_1} + 2) \ldots (2i_{\beta_k} + 2)$$

$$\leq C^k N^n 2^k \sum_{i_{\beta_1} + \cdots + i_{\beta_k} \leq n; \, i_{\beta_j} \geq 1} (i_{\beta_1} + 1) \ldots (i_{\beta_k} + 1)$$

$$\leq (2C)^k N^n \sum_{i_{\beta_1} + \cdots + i_{\beta_k} \leq n; \, i_{\beta_j} \geq 1} 2^{i_{\beta_1} + \cdots + i_{\beta_k}} \leq (2C)^k (2N)^n F(k, n)$$

where

$$F(k, n) = \#\{(i_1, \ldots, i_k) \in \mathbb{N}^k | i_j \geq 1, \, i_1 + \cdots + i_k \leq n\}$$

We claim

$$F(k, n) \leq \frac{n^k}{k!}$$

the case $k = 1$ being obvious. For $k > 1$

$$F(k, n) \leq F(k-1, 1) + \cdots + F(k-1, n-1) \leq \frac{1^{k-1}}{(k-1)!} + \cdots + \frac{(n-1)^{k-1}}{(k-1)!}$$

by the induction hypothesis

$$\leq \int_0^n \frac{x^{k-1}}{(k-1)!} dx = \frac{n^k}{k!}$$

$\square$

Then power series can be treated

Lemma 5.3:

Let

$$f(z) = \sum_{n=1}^{\infty} c_n z^n$$

be a complex power series with radius of convergence $R > 0$. Let

$$a = \sum_{\beta} \lambda_\beta \varrho_\beta \omega^{i_\beta} \in \widehat{I}A \ (i_\beta \geq 1)$$

be an element of the I-adic completion of IA with entries in the multiplicatively closed set $K \cup \{1\} \subset A$ and $\lambda_\beta \in \mathbb{C}$. Assume that

$$\sum_{i_\beta = n} |\lambda_\beta| \leq CN^n$$

for some $C > 0, N > 1$. Then

$$f(a) := \sum_{n=1}^{\infty} c_n a^n \in \widehat{I}A$$

is well defined and can be represented as

$$f(a) = \sum_{\gamma} \lambda_\gamma \varrho_\gamma \omega^{i_\gamma}$$

with entries in $K \cup \{1\}$ and where for any $R' < R$

$$\sum_{i_\gamma = n} |\lambda_\gamma| \leq C'(2N)^n exp(\frac{2C}{R'}n)$$

for some $C' > 0$ depending only on f and R'.

Similarly, if $a \in \hat{I}A$, $b \in \hat{I}B$ are of the same form as above with entries in $K \subset A$ (resp. $K' \subset B$), then

$$f(a \otimes b) \in \hat{I}(A \otimes B)$$

is well defined and can be written as

$$f(a \otimes b) = \sum_{\gamma} \lambda_{\gamma} (\varrho_{\gamma_1} \omega^{i_{\gamma_1}} \otimes \varrho_{\gamma_2} \omega^{i_{\gamma_2}})$$

where the entries of $\varrho_{\gamma_1} \omega^{i_{\gamma_1}}$ (resp. $\varrho_{\gamma_2} \omega^{i_{\gamma_2}}$) belong to $K \cup \{1\} \subset A$ (resp. $K' \cup \{1\} \subset B$) and

$$\sum_{i_{\gamma} : i_{\gamma_1} + i_{\gamma_2} = n} |\lambda_{\gamma}| \leq C'(2N)^n exp(\frac{2C}{R'}n)$$

$\square$

Proof:

The sum $f(a) = \sum_{n=1}^{\infty} c_n a^n$ converges in $\hat{I}A$ because $a \in \hat{I}A$. If one brings $f(a)$ in standard form using the Bianchi identity as in Lemma 5.1 and Lemma 5.2 one finds

$$f(a) = \sum_{k=1}^{\infty} c_k (\sum_{\gamma_k} \lambda_{\gamma_k} \varrho_{\gamma_k} \omega^{i_{\gamma_k}})$$

with entries in the multiplicative closure of $K \cup \{1\}$ and such that

$$\sum_{i_{\gamma_k} = n} |\lambda_{\gamma_k}| \leq (2C)^k (2N)^n \frac{n^k}{k!}$$

It follows that

$$f(a) = \sum_{k, \gamma_k} c_k \lambda_{\gamma_k} \varrho_{\gamma_k} \omega^{i_{\gamma_k}} = \sum_{\gamma} \lambda_{\gamma} \varrho_{\gamma} \omega^{i_{\gamma}}$$

with

$$\sum_{i_{\gamma} = n} |\lambda_{\gamma}| \leq \sum_{k=1}^{\infty} |c_k| (2C)^k (2N)^n \frac{n^k}{k!}$$

The radius of convergence of f being equal to R, one has

$$\overline{\lim_{k}} |c_k|^{\frac{1}{k}} = \frac{1}{R}$$

which yields

$$|c_k| \leq C'(\frac{1}{R'})^k$$

for any $R' < R$ and some $C'(f, R') > 0$ so that

$$\sum_{i_{\gamma} = n} |\lambda_{\gamma}| \leq C'(2N)^n \sum_{k=1}^{\infty} \frac{(\frac{2Cn}{R'})^k}{k!}$$

$$\leq C'(2N)^n exp(\frac{2Cn}{R'})$$

which yields the claim. The proof in the case of a tensor product is similar. ·

□

5-2 Locally convex topologies on subalgebras of RA

Formal inductive limits

Definition 5.4:

Let C be a category. The category $Ind\,C$ of **Ind-objects** or **formal inductive limits** over C is the category with functors from ordered sets to C as objects, i.e.

$$ob(Ind\,C) = \{\mathcal{X} = " \lim_I " X_i\}$$

$$\mathcal{X} = \{X_i,\, i \in I,\, f_{i,i'} : X_i \to X_{i'},\, i \leq i'\}$$

where I is an ordered set, $X_i \in ob(C)$, $f_{i,i'} \in mor(C)$ and $f_{i',i''} \circ f_{i,i'} = f_{i,i''}$ for $i \leq i' \leq i''$. If

$$\mathcal{Y} = " \lim_J " Y_j : (Y_j, g_{jj'};\, j \leq j',\, j, j' \in J)$$

is another object in $Ind\,C$, the morphisms in $Ind\,C$ from $\mathcal{X}$ to $\mathcal{Y}$ are defined as

$$Hom_{Ind\,C}(\mathcal{X}, \mathcal{Y}) := \lim_{\leftarrow I} \lim_{\to J} Hom_C(X_i, Y_j)$$

where the projective (resp.inductive) limit above is taken in the category of sets.

□

It is easily verified that with this definition $Ind\,C$ is in fact a category, i.e. that morphisms compose well.

The topological I-adic completion of RA

Definition 5.5:

Let A be an admissible Fréchet algebra and fix a "small" open, convex neighbourhood U of zero in A (see 1.13.). Denote by $\mathcal{K} = \mathcal{K}(A, U)$ the set of pairs (K, N), where $K \subset A$ is a compact subset of U and $N \geq 1$ is a real number. $\mathcal{K}$ becomes a partially ordered set by putting

$$(K, N) \leq (K', N') \quad \Leftrightarrow \quad K \subset K' \text{ and } N \leq N'$$

To any pair $(K, N), (K', N') \in \mathcal{K}$ there exists $(K'', N'') \in \mathcal{K}$ such that $(K, N) \leq (K'', N'')$ and $(K', N') \leq (K'', N'')$

Definition and Proposition 5.6: (Topological I-adic completion)

Let A be an admissible Fréchet algebra and let $(K, N) \in \mathcal{K}(A, U)$.

1) Put

$$RA_K := \{ a \in RA \,|\, a = \sum_\beta \lambda_\beta \varrho_\beta \omega^{k_\beta} \}$$

where $\lambda_\beta \in \mathbb{C}$ and the entries of $\varrho_\beta \omega^{k_\beta}$ belong to $K^\infty \cup \{1\}$ where K^∞ denotes the multiplicative closure of K. (It is relatively compact by definition.) RA_K is a unital subalgebra of RA.

2) For any $m \in I\!N$ the functional

$$\| - \|_{N,m}^K : \quad RA_K \quad \to \qquad\qquad\qquad I\!R_+$$

$$a \quad \to \quad \inf_{a = \sum \lambda_\beta \varrho_\beta \omega^{k_\beta}} \sum_\beta |\lambda_\beta| \, (1 + k_\beta)^m \, N^{-k_\beta}$$

defines a seminorm on RA_K where the infimum is taken over all presentations of a with entries in $K^\infty \cup \{1\}$ as in 1). Denote the completion of RA_K by these seminorms as

$$RA_{(K,N)} := Compl(RA_K, \| - \|_{N,m}^K , m \in I\!N)$$

It is a Fréchet space.

3) The multiplication on RA_K satisfies

$$\| xy \|_{N,m}^K \leq 2^{m+1} \| x \|_{N,m+1}^K \| y \|_{N,m}^K$$

so that $RA_{(K,N)}$ becomes a Fréchet algebra in fact.

4) The Fréchet algebra $RA_{(K,N)}$ is admissible. A "small" open set is given by

$$U := \{ x \in RA_{(K,N)} |\, \| x \|_{N,1} < \frac{1}{2} \}$$

5) For $(K, N) \leq (K', N')$ the natural inclusion

$$RA_K \to RA_{K'}$$

of subalgebras of RA extends to a continuous homomorphism of Fréchet algebras

$$RA_{(K,N)} \to RA_{(K',N')}$$

6) The Ind-Fréchet algebra

$$\mathcal{R}A := \text{"} \lim_{\to \mathcal{K}(A,U)} \text{"} RA_{(K,N)}$$

is called the **topological I-adic completion** of RA. It is independent of the choice of the small open set $U \subset A$.

$\square$

Proof:

1): follows from the Bianchi-identity (5.1).

2): Evident.

3): We show that for $x, y \in RA_K$ the inequality

$$\| \, xy \, \|_{m-1} \leq 2^m \, \| \, x \, \|_m \, \| \, y \, \|_{m-1}$$

is valid where we write $\| - \|_m$ instead of $\| - \|_{N,m}^K$.

Choose presentations

$$x = \sum \lambda_\beta \varrho_\beta \omega^{k_\beta} \qquad y = \sum \lambda_{\beta'} \varrho_{\beta'} \omega^{k_{\beta'}}$$

such that

$$\sum_\beta |\lambda_\beta| \, (1 + k_\beta)^m \, N^{-k_\beta} \leq \| \, x \, \|_m + \epsilon$$

$$\sum_{\beta'} |\lambda_{\beta'}| \, (1 + k_{\beta'})^{m-1} \, N^{-k_{\beta'}} \leq \| \, y \, \|_{m-1} + \epsilon$$

Then

$$xy = \sum \lambda_\beta \lambda_{\beta'} \varrho_\beta \omega^{k_\beta} \varrho_{\beta'} \omega^{k_{\beta'}}$$

and we find

$$\varrho_\beta \omega^{k_\beta} \varrho_{\beta'} \omega^{k_{\beta'}} = \sum \varrho_\gamma \omega^{k_\gamma}$$

where $\varrho_\gamma \omega^{k_\gamma}$ has entries in $K^\infty \cup \{1\}$, the number of summands equals $2k_\beta + 2$ and the valuation k_γ satisfies $k_\beta + k_{\beta'} \leq k_\gamma \leq k_\beta + k_{\beta'} + 1$. Therefore

$$\| \, xy \, \|_{m-1} \leq \sum_{\beta, \beta'} |\lambda_\beta \lambda_{\beta'}| (2k_\beta + 2)(1 + k_\gamma)^{m-1} N^{-k_\gamma}$$

$$\leq \sum_{\beta, \beta'} |\lambda_\beta| |\lambda_{\beta'}| \, 2(1 + k_\beta)((1 + k_\beta) + (1 + k_{\beta'}))^{m-1} N^{-k_\beta - k_{\beta'}}$$

$$\leq \sum_{\beta, \beta'} |\lambda_\beta| |\lambda_{\beta'}| \, 2(1 + k_\beta) 2^{m-1}(1 + k_\beta)^{m-1}(1 + k_{\beta'})^{m-1} N^{-k_\beta} N^{-k_{\beta'}}$$

$$\leq 2^m (\sum_\beta |\lambda_\beta| \, (1 + k_\beta)^m N^{-k_\beta})(\sum_{\beta'} |\lambda_{\beta'}| \, (1 + k_{\beta'})^{m-1} N^{-k_{\beta'}})$$

$$\leq 2^m (\| \, x \, \|_m + \epsilon)(\| \, y \, \|_{m-1} + \epsilon)$$

where we have used $a + b \leq 2ab$ for $a, b \geq 1$.

4): By Lemma 1.14. it suffices to check that the multiplicative closure of any compact subset $V \subset RA_K \cap U$ is relatively compact. Moreover it is easily seen that one may suppose that every element of V is homogeneous, i.e. corresponds to a differential form of a single degree under the isomorphism $RA \simeq \Omega^{ev} A$. Let $m \in \mathbb{Z}_+$ and put $C(m) := \sup_{x \in V} \| \, x \, \|_{N,m}$. Clearly $C(1) < \epsilon < \frac{1}{2}$ for some ϵ. Let $x_1, \ldots, x_n \in V$. Then $x_i = \sum \lambda_{ij} \varrho_{ij} \omega_{ij}^{k_i}$ with entries in K^∞. Fix i_0 such that $k_{i_0} = \max_i k_i$.

Choose presentations $x_i = \sum \lambda_{\beta_i} \varrho_{\beta_i} \omega_{\beta_i}^{k_i}$ s.t.

$$\| x_i \|_{N,1} \leq \sum_{\beta_i} |\lambda_{\beta_i}|(1+k_i)N^{-k_i} \leq \epsilon \ i \neq i_0$$

$$\| x_{i_0} \|_{N,m+1} \leq \sum_{\beta_{i_0}} |\lambda_{\beta_{i_0}}|(1+k_{i_0})^{m+1}N^{-k_{i_0}} \leq C(m+1)+1$$

then

$$\| x_1 \ldots x_n \|_{N,m} \leq \sum_{\beta_1,\ldots,\beta_n} |\lambda_{\beta_1}|\ldots|\lambda_{\beta_n}| \, \| \varrho_{\beta_1}\omega_{\beta_1}^{k_1}\cdots\varrho_{\beta_n}\omega_{\beta_n}^{k_n} \|_{N,m}$$

$$\leq \sum_{\beta_1,\ldots,\beta_n} |\lambda_{\beta_1}|\ldots|\lambda_{\beta_n}|(2k_1+2)\ldots(2k_n+2)\, \| \varrho_\gamma\omega^{k_\gamma} \|_{N,m}$$

where $\varrho_\gamma\omega^{k_\gamma}$ has entries in K^∞ and $k_1+\cdots+k_n \leq k_\gamma \leq k_1+\cdots+k_n+(n-1)$. Therefore

$$\| x_1 \ldots x_n \|_{N,m} \leq$$

$$\leq 2^n \sum_{\beta_1,\ldots,\beta_n} |\lambda_{\beta_1}|\ldots|\lambda_{\beta_n}|(1+k_1)\ldots(1+k_n)(n+k_1+\cdots+k_n)^m N^{-(k_1+\cdots+k_n)}$$

$$\leq 2^n n^m \sum_{\beta_1,\ldots,\beta_n} |\lambda_{\beta_1}|\ldots|\lambda_{\beta_n}|(1+k_1)\ldots(1+k_n)(1+k_{i_0})^m N^{-(k_1+\cdots+k_n)}$$

$$\leq 2^n n^m \epsilon^{n-1}(C(m+1)+1) = \epsilon^{-1}(C(m+1)+1)(2\epsilon)^n n^m$$

As $\epsilon < \frac{1}{2}$ this shows

$$\lim_{n\to\infty}(\sup_{x_i \in V} \| x_1 \ldots x_n \|_{N,m}) = 0$$

for all $m \in \mathbb{Z}_+$.

5): Evident.

6): First of all note that every algebra of the form $RA_{(K,N)}$ is isomorphic to an algebra $RA_{(K',N')}$ for which K' is a multiplicatively closed, compact subset of any given neighbourhood of 0. It suffices to choose $M \geq 1$ large enough that $[0,\frac{1}{M}]\overline{K^\infty} =: K' \subset W$. K' is then a multiplicatively closed, compact set. Moreover, the subalgebras $RA_K \subset RA$, $RA_{K'} \subset RA$ coincide and the identity $RA_K \xrightarrow{\Rightarrow} RA_{K'}$ extends to a topological isomorphism $RA_{(K,N)} \xrightarrow{\cong} RA_{(K',NM^2)}$ as is readily seen by applying the obvious homothety to the entries of tensors in RA_K.

Choosing $W := U$ the argument above shows that, as long as the order structure on $\mathcal{K}(A,U)$ is ignored, one may suppose that for an algebra of type $RA_{(K,N)}$ that in fact $K \subset U$ is multiplicatively closed.

Putting $W := U \cap U'$ where U' is another "small" convex neighbourhood of 0 the argument proves that

$$\text{"}\lim_{\mathcal{K}(A,U)}\text{"} RA_{(K,N)} \xleftarrow{\cong} \text{"} \lim_{\mathcal{K}(A,U\cap U')} \text{"} RA_{(K'',N'')} \xrightarrow{\cong} \text{"} \lim_{\mathcal{K}(A,U')} \text{"} RA_{(K',N')}$$

as claimed.

$\square$

The next two lemmas concern technical results that will be needed for the study of the cohomology of a direct sum and for the construction of exterior products.

Lemma 5.7:

The notations are those of 5.6.

Let $e = e^2 \in A$ be an idempotent acting as a unit on the compact set $K \subset A$ ($e \in \mathbb{C}K$) and let $x \in RA_{(K,N)}$

$$x = \sum \lambda_\beta \varrho_\beta \omega^{k_\beta}, \quad \sum |\lambda_\beta| (1 + k_\beta)^m N^{-k_\beta} \leq \| x \|_{N,m}^K + \epsilon$$

satisfy the following condition: For every β all except at most N_0 entries of $\varrho_\beta \omega^{k_\beta}$ are equal to e. Then for any $y \in RA_{(K,N)}$

$$\| xy \|_{N,m}^K \leq (2N_0 + 2) \| x \|_{N,m}^K \| y \|_{N,m}^K$$

Proof:

The same as for 5.6.3). The better estimates are due to the fact that almost all ω-terms in x equal $\omega(e,e)$ and if the product xy is brought into standard form via the Bianchi identity the majority of the arising terms cancels due to the identity

$$\omega(e,e)\varrho(a) = \omega(e, ea) - \omega(e^2, a) + \varrho(e)\omega(e, a)$$

$$= \omega(e, a) - \omega(e, a) + \varrho(e)\omega(e, a) = \varrho(e)\omega(e, a)$$

$\square$

Lemma 5.8:

Let A be an admissible Fréchet algebra and $\mathcal{A} \subset A$ a dense subalgebra. Let $U \subset A$ be a "small" neighbourhood of zero. Define

$$\mathcal{K}'(A) := \{(K', N')\}$$

where K' is a nullsequence in $\mathcal{A} \cap U$ and $N' > 1$. Then

$$\text{"}\lim_{\mathcal{K}'}\text{"} RA_{(K',N')} \xrightarrow{\cong} \text{"}\lim_{\mathcal{K}}\text{"} RA_{(K,N)}$$

Proof:

We are going to construct an inverse of the obvious morphism

$$\text{"}\lim_{\to \mathcal{K}'}\text{"}RA_{\mathcal{K}'} \to \text{"}\lim_{\to \mathcal{K}}\text{"}RA_{\mathcal{K}}$$

So let $(K, N) \in \mathcal{K}(A)$. The notations of Lemma 1.7. are used throughout. For a subset $S \subset A$ we denote by $\langle S \rangle$ its linear span and by $Con(S)$ its convex closure.

Construct a nullsequence $K''(= \mathcal{B})$ for K^∞ as in the proof of Lemma 1.7. Choose

$$Y_N := \{y_1^N, \dots, y_m^N\} \subset \langle x_1^N, \dots, x_{n_N}^N \rangle$$

for all $N >> 0$ such that

$$Y_N \subset B(0, 2\lambda_N) \quad \langle x_1^N, \dots, x_{n_N}^N \rangle \cap B(0, \lambda_N) \subset Conv(Y_N)$$

where $B(0, r)$ denotes the r-ball around 0 in a suitable translation-invariant metric on A.

Let

$$Z_N := \{\frac{x_i^N x_j^N - x_k^N}{\lambda_N} | d(x_i^N x_j^N, x_k^N) < \lambda_N^{\frac{3}{2}}\}$$

and put finally

$$K' := K'' \cup \bigcup_{N>>0} (Y_N \cup Z_N)$$

K' is a nullsequence. Replacing K' by $\frac{1}{C}K'$ for some $C > 1$ we may assume $K' \subset \mathcal{A} \cap U$. The desired map

$$l : RA_{(K,N)} \to RA_{(K',N')}$$

(for a suitable N') will be defined on generators $\varrho(y), y \in K^\infty$ by

$$l(\varrho(y)) := \sum_{n=0}^\infty \lambda_n \varrho(b_n) \in RA_{(K',N')}$$

where

$$y = \sum_{n=0}^\infty \lambda_n b_n \quad b_n = \frac{x_{k_{n+1}(y)}^{n+1} - x_{k_n(y)}^n}{\lambda_n}$$

is a presentation of y as in 1.7.

We note first that $l(\varrho(y))$ does not depend on the choice of such a presentation: Let

$$i(y) = \lambda_0 \frac{x_{k_0(y)}^0}{\lambda_0} + \sum_{j=1}^\infty \lambda_{j-1} \left(\frac{x_{k_j(y)}^j - x_{k_{j-1}(y)}^{j-1}}{\lambda_{j-1}} \right)$$

$$i'(y) = \lambda_0 \frac{x_{k_0'(y)}^0}{\lambda_0} + \sum_{j=1}^\infty \lambda_{j-1} \left(\frac{x_{k_j'(y)}^j - x_{k_{j-1}'(y)}^{j-1}}{\lambda_{j-1}} \right)$$

be two presentations of y constructed as in the proof of Lemma 1.7. Let N' be such that

$$\lambda_{N'-1} < \epsilon, \quad \sum_{n=N'}^{\infty} \lambda_n < \epsilon$$

Then

$$\| i(y) - i'(y) \|_{N,m}^{K''} = \| x_{k_{N'}(y)}^{N'} - x_{k'_{N'}(y)}^{N'} + \sum_{j=N'+1}^{\infty} \cdots - \sum_{j=N'+1}^{\infty} \cdots \|$$

$$\leq \| x_{k_{N'}(y)}^{N'} - x_{k'_{N'}(y)}^{N'} \| + 2C\epsilon$$

Now

$$d_A(x_{k_{N'}(y)}^{N'}, x_{k'_{N'-1}(y)}^{N'-1}) \leq d_A(x_{k_{N'}(y)}^{N'}, y) + d_A(y, x_{k'_{N'-1}(y)}^{N'-1})$$

$$\leq \frac{\lambda_{N'}^2}{2} + \frac{\lambda_{N'-1}^2}{2} \leq \lambda_{N'-1}^2$$

so that

$$\frac{x_{k_{N'}(y)}^{N'} - x_{k'_{N'-1}(y)}^{N'-1}}{\lambda_{N'-1}} \in K'$$

and

$$\| i(y) - i'(y) \|$$

$$\leq \| \lambda_{N'-1} \frac{x_{k_{N'}(y)}^{N'} - x_{k'_{N'-1}(y)}^{N'-1}}{\lambda_{N'-1}} \| + \| (-\lambda_{N'-1}) \frac{x_{k'_{N'}(y)}^{N'} - x_{k'_{N'-1}(y)}^{N'-1}}{\lambda_{N'-1}} \| + 2C\epsilon$$

$$\leq 2C\lambda_{N'-1} + 2C\epsilon \leq 4C\epsilon$$

Let $\mathbb{C}\langle K^{\infty} \rangle$ be the free vector space generated by the set K^{∞} and let $l : \mathbb{C}\langle K^{\infty} \rangle \to RA_{(K',N')}$ be the linear extension of l. We claim that it factors

$$l : \mathbb{C}\langle K^{\infty} \rangle \to A_{K^{\infty}} \to RA_{(K',N')}$$

where $A_{K^{\infty}} \subset A$ is the linear span of K^{∞} inside A.

So let $a_1, \ldots, a_m \in K^{\infty}$ be such that $\sum \mu_i a_i = 0$ in A. We have to show that $\sum \mu_i l(\varrho(a_i)) = 0$ in $RA_{(K',N')}$. Now

$$\sum \mu_i l(\varrho(a_i)) = \sum_i \mu_i \sum_{j=0}^{\infty} \lambda_j b^j(a_i) =$$

$$= \sum_i \mu_i x_{k_N(a_i)}^N + \sum_i \mu_i \sum_{j=N+1}^{\infty} \lambda_j b^j(a_i)$$

and thus

$$\| \sum \mu_i l(\varrho(a_i)) \|_{(K',N')} \leq \| \sum_i \mu_i x_{k_N(a_i)}^N \|_{(K',N')} + (\sum_i |\mu_i|) \left(\sum_{j=N+1}^{\infty} \lambda_j \right)$$

for all N. Moreover, for any seminorm on A

$$\| \sum_i \mu_i x^N_{k_N(a_i)} \|_A = \| \sum_i \mu_i (x^N_{k_N(a_i)} - a_i) \|_A$$

$$\leq (\sum_i |\mu_i|) max_i \| x^N_{k_N(a_i)} - a_i \|_A$$

which is small. Thus

$$\sum_i \mu_i (x_{k_N(a_i)} = \sum_k \nu^N_k y^N_k$$

with $y^N_k \in Y_N$ and $\lim_{N \to \infty} \sum_k |\nu^N_k| = 0$ so that, by the choice of K'

$$\| \sum \mu_i l(\varrho(a_i)) \|_{(K',N')} \leq \left(\sum_k |\nu^N_k| \right) + (\sum_i |\mu_i|) \left(\sum_{j=N+1}^{\infty} \lambda_j \right) < \epsilon$$

for $N >> 0$. A similar estimate, using $Z_N \subset K'$ shows

$$l(kk') = l(k)l(k') \quad \forall k, k' \in K^\infty$$

Moreover, the map l is bounded: Let $a \in RA_K$ be represented as

$$a = \sum_\beta \mu_\beta \varrho_\beta \omega^{k_\beta}$$

with entries in $K^\infty \cup \{1\}$ and such that

$$\sum_\beta |\mu_\beta| (1 + k_\beta)^m N^{-k_\beta} \leq \| a \|^K_{N,m} + \epsilon$$

Then

$$\| i(a) \|^{K''}_{C^2 N,m} = \| \sum_\beta \mu_\beta \varrho (\sum \lambda_{i_0} y_{i_0}) \omega (\sum \ldots, \sum \ldots)^{k_\beta} \|^{K''}_{C^2 N,m}$$

$$= \sum_{\beta, i_0, \ldots, i_{2k_\beta}} \| \mu_\beta \lambda_{i_0} \ldots \lambda_{i_{2k_\beta}} \varrho (y_{i_0}) \ldots \omega (y_{i_{2k_\beta}-1}, y_{i_{2k_\beta}}) \|^{K''}_{C^2 N,m}$$

$$\leq \sum_\beta |\mu_\beta| C^{2k_\beta + 1} (1 + k_\beta)^m (C^2 N)^{-k_\beta}$$

$$\leq C(\sum_\beta |\mu_\beta| (1 + k_\beta)^m N^{-k_\beta} \leq C(\| a \|^K_{N,m} + \epsilon)$$

This implies finally, that $l : A_{K^\infty} \to RA_{(K',N')}$ extends to a continuous homomorphism of algebras

$$l : RA_{(K,N)} \to RA_{(K',N')}$$

provided that $N' \geq N$ is choosen large enough. It is easily checked that this construction provides a morphism of Ind-algebras

$$\text{"}\lim_{\to K}\text{"} RA_K \to \text{"} \lim_{\to K'}\text{"} RA_{K'}$$

inverse to the obvious inclusion.

□

Lemma 5.9:

a) The canonical quotient map

$$\pi : RA \to A$$

extends to a continuous map

$$\pi : RA_{(K,N)} \to A$$

for all $(K, N) \in \mathcal{K}$.

b) The curvature of the map

$$\varrho : l^1(K^\infty \cup \{1\}) \to RA_{(K,N)}$$

is bounded by

$$\| \omega_\varrho(x, y) \|_{(N,m)}^K \leq 2^m N^{-1} \ \forall x, y \in K^\infty$$

□

It was mentioned in the introduction that the topological I-adic completion is a (formally) topologically nilpotent extension of A. We want to make this precise. For an ideal in a Banach algebra topological nilpotence means that the spectra of all its elements reduce to zero. Consequently topological nilpotence is equivalent to the condition that holomorphic functional calculus can be applied with any function holomorphic near zero. This means that if f is a complex power series of strictly positive radius of convergence and z belongs to the topologically nilpotent ideal $\mathcal{I}$, the series $f(z)$ will converge in A. It is this condition for topological nilpotence that can be verified for the kernel of the projection $RA \to A$. This result will be used when we investigate the compatibility of the Chern character with the multiplicative structures on K-theory and cyclic cohomology.

Lemma 5.10:

Let

$$f(z) = \sum_{n=1}^{\infty} a_n z^n$$

be a complex power series of radius of convergence $R > 0$. Let A, B be admissible and

$$B(0, C)_{N,0} := \{x \in RA_{(K,N)}, \| x \|_{N,0}^K < C\}$$

$$B'(0, C)_{N,0} := \{y \in RB_{(K',N)}, \| y \|_{N,0}^{K'} < C\}$$

Then f defines continuous maps

$$f: \ IA_{(K,N)} \cap B(0,C)_{N,0} \ \rightarrow \ IA_{(K,M)}$$

$$x \ \rightarrow \ \sum_{n=1}^{\infty} a_n x^n$$

$$f: \ IA_{(K,N)} \cap B(0,C)_{N,0} \times IB_{(K',N)} \cap B'(0,C)_{N,0} \ \rightarrow \ I(A \otimes_\pi B)_{(K \otimes K',M)}$$

$$(x,y) \ \rightarrow \ \sum_{n=1}^{\infty} a_n (x \otimes y)^n$$

for any

$$M > 2N \, exp(\frac{2C}{R})$$

where $IA_{(K,N)}$ denotes the closure of $IA_K \subset RA_K$ in $RA_{(K,N)}$. $\qquad \square$

Proof:

Choose $\epsilon > 0$ and let $a \in IA_{(K,N)}$ be presented as a sum

$$a = \sum_{\beta} \lambda_\beta \varrho_\beta \omega^{k_\beta}$$

with entries in $K^\infty \cup \{1\}$ and such that

$$\sum_{\beta} |\lambda_\beta| N^{-k_\beta} \leq \| a \|_{N,0} + \epsilon$$

and therefore

$$\sum_{k_\beta = n} |\lambda_\beta| \leq (\| a \|_{N,0} + \epsilon) N^n$$

Lemma 5.3 shows then that

$$f(a) = \sum_{\gamma} \lambda_\gamma \varrho_\gamma \omega^{k_\gamma}$$

with entries in $K^\infty \cup \{1\}$ and such that for $R' < R$

$$\sum_{k_\gamma = n} |\lambda_\gamma| \leq C'(2N)^n exp(\frac{2(\| a \|_{N,0} + \epsilon)}{R'})^n$$

for some C' depending only on f, R' (and not on a).

Let M be any real number satisfying

$$M > 2n \, exp(\frac{2 \| a \|_{N,0}}{R})$$

If $\epsilon > 0$, $R' < R$ are choosen such that

$$M' = 2N exp(\frac{2(\| a \|_{N,0} + \epsilon)}{R'}) < M$$

one finds

$$\| f(a) \|_{M,m}^K \leq \sum_\gamma |\lambda_\gamma| (1 + k_\gamma)^m M^{-k_\gamma}$$

$$\leq \sum_{n=0}^\infty C' M'^n (1 + n)^m M^{-n} = C' \sum_{n=0}^\infty (1 + n)^m (\frac{M'}{M})^n < \infty$$

which proves that $f(a) \in RA_{(K,M)}$.

To show continuity choose $M > 2N exp(\frac{2\|a\|_{N,0}}{R})$ and let $b \in RA_{(K,N)}$ be close to a. Then $f(b) \in RA_{(K,M)}$ is well defined and

$$f(a) - f(b) = \sum_{k=1}^\infty c_k \sum_{j=0}^{k-1} a^j (a - b) b^{k-1-j}$$

so that

$$\| f(a) - f(b) \|_{M,m}^K \leq$$

$$\sum_{k=1}^\infty |c_k| \sum_{j=0}^{k-1} 2^{m+1} \| a^j \|_{M,m+1} 2^{m+1} \| (a - b) \|_{M,m+1} \| b^{k-1-j} \|_{M,m}$$

by Lemma 5.3

$$\leq 2^{2(m+1)} \| (a - b) \|_{M,m+1} C' \sum_{k=1}^\infty \sum_{j=0}^{k-1} (\frac{1}{R'})^k \| a^j \|_{M,m+1} \| b^{k-1-j} \|_{M,m}$$

(for the choice of C', R' see 5.3.)

$$\leq 4^{(m+1)} \| (a-b) \|_{M,m+1} \frac{C'}{R'} \left(\sum_{j'=0}^\infty (\frac{1}{R'})^{j'} \| a^{j'} \|_{M,m+1} \right) \left(\sum_{k'=0}^\infty (\frac{1}{R'})^{k'} \| b^{k'} \|_{M,m} \right)$$

The estimates of Lemma 5.2 allow now to conclude for

$$a = \sum_\beta \lambda_\beta \varrho_\beta \omega^{k_\beta}$$

and a similar presentation

$$b = \sum_{\beta'} \lambda_{\beta'} \varrho_{\beta'} \omega^{k_{\beta'}}$$

that

$$\sum_{j'=0}^\infty (\frac{1}{R'})^j \| a^j \|_{M,m+1} \leq \sum_{j'=0}^\infty (\frac{1}{R'})^j \sum_\gamma |\lambda_\gamma| (1 + k_\gamma)^{m+1} M^{-k_\gamma}$$

$$\leq \sum_{j=0,n=1}^\infty (\frac{1}{R'})^j C(j,n)(1 + n)^{m+1} (2N)^{-n} exp(-\frac{2(\| a \|_{N,0} + \epsilon)}{R'} n) C''^n$$

where

$$C'' = 2N exp(\frac{2(\| a \|_{N,0} + \epsilon)}{R'}) M^{-1} < 1$$

$$\leq \sum_{j=0,n=1}^{\infty} (\frac{1}{R'})^j (2 \parallel a \parallel_{N,0} +\epsilon)^j \frac{n^j}{j!} (1+n)^{m+1} exp(-\frac{2(\parallel a \parallel_{N,0} +\epsilon)}{R'} n) C''^n$$

by Lemma 5.2

$$\leq \sum_{n=1}^{\infty} exp(\frac{2(\parallel a \parallel_{N,0} +\epsilon)}{R'} n) exp(-\frac{2(\parallel a \parallel_{N,0} +\epsilon)}{R'} n)(1+n)^{m+1} C''^n$$

$$\leq \sum_{n=1}^{\infty} (1+n)^{m+1} C''^n = C'''(f,R',\epsilon, \parallel a \parallel_{N,0}, M, m) < \infty$$

If b is very close to a a similar calculation holds for the second sum

$$\sum_{k=0}^{\infty} (\frac{1}{R'})^k \parallel b^k \parallel_{M,m}$$

and the same choice of

$$M > 2N exp(\frac{2 \parallel a \parallel_{N,0}}{R})$$

so that finally

$$\parallel f(a) - f(b) \parallel_{M,m}^{K} \leq C''''(f,R',\epsilon,M,m) C' \parallel a - b \parallel_{N,m+1}$$

which is what we wanted. The proof in the tensor product case is similar. $\square$

It will now be verified that the algebras $RA_{(K,N)}$ are quasifree, resp. of cohomological dimension one in the category of Fréchet algebras. This means that they possess the lifting property with respect to nilpotent extensions of Fréchet algebras with continuous linear splitting. The result guarantees that the cyclic (co)homology of the algebras $RA_{(K,N)}$ can be calculated by their X-complex.

This gives the theoretical justification for the definition of the analytic resp. asymptotic cyclic cohomology of admissible Fréchet algebras presented later in this and the next chapter. The proposition is also needed for establishing the topological versions of the Cartan homotopy formulas and the exterior products.

Proposition 5.11:

The natural homomorphism

$$i : RA \to R(RA)$$

extends to a continuous homomorphism

$$i : RA_{(K,N)} \to \widehat{R}(RA_{(K,N)})$$

i.e. for each $k \geq 0$ the natural map

$$RA \xrightarrow{i} R(RA) \simeq \Omega^{ev}(RA) \xrightarrow{p_{2k}} \Omega^{2k}(RA)$$

extends to a continuous linear map

$$RA_{(K,N)} \to \Omega^{2k}(RA_{(K,N)}) \simeq RA_{(K,N)} \otimes_\pi \overline{RA_{(K,N)}}^{\otimes_\pi^{2k}}$$

Proof:

The homomorphism $i : RA \to R(RA)$ is given on standard elements by

$$i : \varrho\omega^n \to \overline{\varrho} \prod_1^n (\overline{\varrho}(\omega) + \overline{\omega}(\varrho, \varrho))$$

where $\varrho : A \to RA, \overline{\varrho} : RA \to R(RA)$ are the natural inclusions. we are interested only in the part of $I(RA)$-adic valuation k if the right hand side is written in standard form. Modulo $I(RA)^{k+1}$ we find

$$i(\varrho\omega^n) = \sum \overline{\varrho}\overline{\varrho}(\omega)^{i_0}\overline{\omega}(\varrho, \varrho)^{j_0} \ldots \overline{\varrho}(\omega)^{i_l}\overline{\omega}(\varrho, \varrho)^{j_l}$$

where $j_0 + \ldots j_l \leq k$ and $i_0 + j_0 + \ldots i_l + j_l \leq n$ so that the number of terms is bounded by

$$\sum_{j=0}^k \binom{n}{k} \leq (k+1)n^k$$

It is not difficult to verify that a power $\overline{\varrho}^m$ of a simple tensor can be written modulo $I(RA)^{k+1}$ as

$$\overline{\varrho}^m = \sum_{j=0}^k \sum_{\leq c(j)terms} \overline{\varrho}^{k+1-j}\overline{\omega}^j$$

with $c(j) \leq \binom{m-k-1}{j}$. Thus

$$\overline{\varrho}(\omega)^m = \sum_{j=0}^k \sum_{\leq c(j)terms} \overline{\varrho}(\omega)^{k+1-j}\overline{\omega}(\omega^l, \omega^{l'})^j$$

$$= \sum \overline{\varrho}(\omega^l)\overline{\omega}(\omega^{l'}, \omega^{l''})^j, \quad j \leq k$$

where the number of terms grows for fixed k only polynomially in m. Using the calculations and estimates of 5.1. one sees further that the original expression for $i(\varrho\omega^n)$ can be written as

$$i(\varrho\omega^n) = \sum \overline{\varrho}(\varrho\omega^{l_0})\overline{\omega}(\varrho\omega^{l_1}, \varrho\omega^{l_2}) \ldots \overline{\omega}(\varrho\omega^{l_{2j-1}}, \varrho\omega^{l_{2j}})$$

where $j \leq k$ and the number of terms grows for fixed k polynomially in n. This however implies

$$\| i(\alpha) \|_{(N,m)\otimes_\pi^{2k+1}} \leq C(m, m') \| \alpha \|_{N,m'}$$

for $\alpha \in RA_K$ and $m' >> 0$ which proves the claim.

□

The crucial property of the completed tensor algebras $RA_{(K,N)}$, which motivated their construction is derived in

Proposition 5.12:

Let

$$f : A \to B$$

be a bounded, based, linear map of admissible Fréchet algebras. Let $K \subset A$, $K' \subset B$ be compact sets contained in "small" convex balls $U \subset A$, $U' \subset B$ and $0 < \lambda < 1$, $\mu \geq 1$
such that

$$f(aa') - f(a)f(a') \in \lambda K' \ \forall a, a' \in K^\infty$$

$$f(a) \in \mu K' \ \forall a \in K^\infty$$

Suppose that λ satisfies

$$\lambda < \frac{1}{8N}$$

Then for $M > 8N\mu^2$ the linear map f induces a continuous morphism of Fréchet algebras

$$Rf : RA_{(K,N)} \to RB_{(K',M)}$$

Proof:

Suppose first that $\mu = 1$. Let $a \in RA_K$ be presented as a sum

$$a = \sum_\beta \lambda_\beta \varrho_\beta \omega^{k_\beta}$$

with entries in $K^\infty \cup \{1\}$ such that

$$\sum_\beta |\lambda_\beta| (1 + k_\beta)^m N^{-k_\beta} \leq \| a \|_{N,m} + \epsilon$$

According to Lemma 5.1 $Rf(\varrho_\beta \omega^{k_\beta})$ can be represented as

$$Rf(\varrho_\beta \omega^{k_\beta}) = \sum_{\gamma=0}^{C_\beta} \lambda_\gamma \varrho_\gamma \omega^{k_\gamma}$$

with entries in $K'^\infty \cup \{1\}$ and $C_\beta \leq 8^{k_\beta}$ and

$$|\lambda_\gamma| \leq \lambda^{j_\gamma} \text{ with } j_\gamma + k_\gamma = \sharp\kappa_\gamma + \sharp\omega_\gamma \geq k_\beta, \ k_\gamma \leq k_\beta$$

We therefore find the estimate

$$\| Rf(a) \|_{M,m}^{K'} \leq \sum_\beta |\lambda_\beta| \sum_\gamma |\lambda_\gamma| (1 + k_\gamma)^m M^{-k_\gamma}$$

$$\leq \sum_\beta |\lambda_\beta| (1 + k_\beta)^m \sum_\gamma \lambda^{j_\gamma} M^{-k_\gamma}$$

$$\leq \sum_\beta |\lambda_\beta|(1 + k_\beta)^m \sum_\gamma (8N)^{-j_\gamma}(8N)^{-k_\gamma}$$

$$\leq \sum_\beta |\lambda_\beta|(1 + k_\beta)^m (\#\gamma)(8N)^{-k_\beta}$$

$$\leq \sum_\beta |\lambda_\beta|(1 + k_\beta)^m (N)^{-k_\beta} \leq \| a \|_{N,m}^K + \epsilon$$

For the general case let $\overline{K}(\overline{K}')$ be the cone over $\frac{1}{\mu} K(K')$ with vertex 0. Then $\overline{K} \subset U, \overline{K}' \subset U'$ are compact sets satisfying

$$\kappa_f(\overline{K}^\infty \times \overline{K}^\infty) \subset \frac{\lambda}{\mu^2} \overline{K}', \varrho_f(\overline{K}^\infty) \subset \overline{K}'$$

Thus Rf induces for $M > 8N\mu^2$ by the first case considered above a continuous morphism

$$Rf : RA_{(K,N)} \simeq RA_{(\overline{K},N\mu^2)} \to RB_{(\overline{K}',M)} \simeq RB_{(K',M)}$$

which proves the claim.

$\square$

Remark:

The proposition shows how the seminorms $\| - \|_{N,m}^K$ come up by starting from the L^1 norm $\| - \|_{1,0}^K$ and looking for a natural, locally convex topology on RA such that for any almost multiplicative, linear map $f : A \to B$ the induced morphism $Rf : RA \to RB$ becomes continuous.

$\square$

Our constructions suggest two possible natural topologies on the commutator quotient $(\Omega^1 RA)_\natural$, one induced from the topology of $RA_{(K,N)}$, the other obtained by expanding elements of $(\Omega^1 RA)_\natural$ in a series of standard elements (3.7) and taking weighted l^1-norms of the coefficients. We show that both topologies coincide.

Lemma 5.13:

Consider the linear space

$$(\Omega^1 RA_K)_\natural := \langle \varrho\omega^n d\varrho \rangle$$

with entries in K. Define seminorms $\| - \|_{N,m}^K$ on $(\Omega^1 RA_K)_\natural$ by

$$\| a \|_{N,m}^K := \inf \sum_\beta |\lambda_\beta|(1 + k_\beta)^m N^{-k_\beta}$$

where the infimum is taken over all presentations

$$a = \sum_{\beta} \lambda_{\beta} \varrho_{\beta} \omega^{k_{\beta}} d\varrho_{\beta}$$

with entries in $K^{\infty} \cup \{1\}$. Denote by $\overline{\Omega}^1(RA_K)_{\natural,N}$ the completion of $(\Omega^1 RA_K)_{\natural}$ with respect to the topology defined by these seminorms. Then

$$\overline{\Omega}^1(RA_K)_{\natural,N} \simeq \Omega^1(RA_{(K,N)})_{\natural}$$

as Fréchet spaces.

Proof:

It is clear that the natural map

$$\overline{\Omega}^1(RA_K)_{\natural,N} \to \Omega^1(RA_{(K,N)})_{\natural}$$

is continuous. We construct an inverse map on the dense subspace

$$\Omega^1(RA_K)_{\natural} \subset \Omega^1(RA_{(K,N)})_{\natural}$$

and show that it is continuous providing thus a global inverse.

So let

$$\omega = \sum_{finite} a_i db_i \in \Omega^1(RA_K)_{\natural}$$

with

$$a_i = \sum_{\beta_i} \lambda_{\beta_i} \varrho_{\beta_i} \omega^{k_{\beta_i}} \quad b_i = \sum_{\gamma_i} \mu_{\gamma_i} \varrho_{\gamma_i} \omega^{k_{\gamma_i}}$$

Assume that this presentation of ω satisfies

$$\sum_i \| a_i \|_{N,m+2}^K \| b_i \|_{N,m+2}^K \leq \| \omega \|_{N,m+2,\Omega^1(RA_{K,N})_{\natural}}^K + \epsilon$$

$$\sum_{\beta_i} |\lambda_{\beta_i}|(1+k_{\beta_i})^{m+1} N^{-k_{\beta_i}} \leq \| a_i \|_{N,m+1} + \epsilon'$$

$$\sum_{\gamma_i} |\mu_{\gamma_i}|(1+k_{\gamma_i})^{m+2} N^{-k_{\gamma_i}} \leq \| b_i \|_{N,m+2} + \epsilon'$$

Then

$$a_i db_i = \sum_{\beta_i,\gamma_i} \lambda_{\beta_i} \mu_{\gamma_i} \varrho_{\beta_i} \omega^{k_{\beta_i}} d(\varrho_{\gamma_i} \omega^{k_{\gamma_i}})$$

reduces modulo commutators to a sum

$$\sum_{\beta_i,\gamma_i} \lambda_{\beta_i} \mu_{\gamma_i} \sum_{\delta} \varrho_{\delta} \omega^{k_{\delta}} d\varrho_{\delta}$$

with entries in $K^{\infty} \cup \{1\}$ where for β, γ given the number of interior summands can be estimated by

$$\#\delta \leq C(1+k_{\beta})(1+k_{\gamma})^2$$

for some universal constant $C > 0$. The latter sum defines an element of $(\Omega^1 RA_K)_\natural$ in canonical form and for the norms we find

$$\| \omega \|^K_{N,m,\overline{\Omega}^1(RA_K)_\natural . N} = \| \sum_i a_i db_i \|^K_{N,m,\overline{\Omega}^1(RA_K)_\natural . N}$$

$$\leq C \sum_{i,\beta_i,\gamma_i} |\lambda_{\beta_i}| |\mu_{\gamma_i}|(1+k_\beta)(1+k_\gamma)^2 (1+k_\beta+k_\gamma)^m N^{-(k_\beta+k_\gamma-1)}$$

$$\leq CN \sum_i \left(\sum_{\beta_i} |\lambda_{\beta_i}|(1+k_{\beta_i})^{m+1} N^{-k_{\beta_i}} \right) \left(\sum_{\gamma_i} |\mu_{\gamma_i}|(1+k_{\gamma_i})^{m+2} N^{-k_{\gamma_i}} \right)$$

as $1 + k_\beta + k_\gamma \leq (1+k_\beta)(1+k_\gamma)$

$$\leq CN \sum_i (\| a_i \|_{N,m+1} + \epsilon')(\| b_i \|_{N,m+2} + \epsilon')$$

$$\leq \left(CN \sum_i \| a_i \|_{N,m+2} \| b_i \|_{N,m+2} \right) + \epsilon$$

if ϵ' is choosen correctly

$$\leq CN \| \omega \|^K_{N,m+2,\Omega^1(RA_{K.N})_\natural} + (1+CN)\epsilon$$

$\square$

The differentials of the X-complex extend to continuous operators on $X_* RA_{(K,N)}$. Their norm satisfies the following estimates

Lemma 5.14:

For $a \in RA_{(K,N)}$ and $\omega \in \overline{\Omega}^1(RA_K)_{\natural,N}$

$$\| da \|_{N,m} \leq 2 \| a \|_{N,m+1} + \frac{2}{3}N \| a \|_{N,m+3}$$

$$\| b\omega \|_{N,m} \leq 2 \| \omega \|_{N,m+1} + 2^{m+1} N^{-1} \| \omega \|_{N,m}$$

$\square$

5-3 The analytic X-complex

Let A, B be admissible Fréchet algebras, and let $\tilde{A}, \tilde{B}$ be obtained from A, B by adjoining a unit.

Denote by

$$\mathcal{R}A := \text{"}\lim_K\text{"} RA_{(K,N)} \quad \mathcal{R}B := \text{"}\lim_{K'}\text{"} RB_{(K',N')}$$

the topological I-adic completions of RA, RB. They are formal inductive limits of families of admissible Fréchet algebras of Hochschild-cohomological dimension one.

Let V, W be locally convex, topological vector spaces.

Definition 5.15:

a) The **homological analytic X-complex** of A with coefficients in (V, W) is defined as

$$X_*^{\epsilon, V, W}(A) := Hom(V, \overline{X}_*(\mathcal{R}\tilde{A}) \otimes_\pi W) = \varinjlim_{\mathcal{K}} Hom(V, \overline{X}_*(R\tilde{A}_{(K,N)}) \otimes_\pi W)$$

We put

$$X_*^\epsilon(A) := X_*^{\epsilon, \mathbb{C}, \mathbb{C}}(A) = \varinjlim_{\mathcal{K}} \overline{X}_*(R\tilde{A}_{(K,N)})$$

b) The **cohomological analytic X-complex** of A with coefficients in (V, W) is defined to be

$$X_{\epsilon, V, W}^*(A) := Hom(\overline{X}_*(\mathcal{R}\tilde{A}) \otimes_\pi V, W) = \varprojlim_{\mathcal{K}} Hom(\overline{X}_*(R\tilde{A}_{(K,N)}) \otimes_\pi V, W)$$

We put

$$X_\epsilon^*(A) := X_{\epsilon, \mathbb{C}, \mathbb{C}}^*(A) = Hom(\overline{X}_*(\mathcal{R}\tilde{A}), \mathbb{C}) = \varprojlim_{\mathcal{K}} Hom(\overline{X}_*(R\tilde{A}_{(K,N)}), \mathbb{C})$$

c) The **bivariant analytic X-complex** of the pair (A, B) with coefficients in (V, W) is defined as

$$X_{\epsilon, V, W}^*(A, B) := Hom_*(\overline{X}_*(\mathcal{R}\tilde{A}) \otimes_\pi V, \overline{X}_*(\mathcal{R}\tilde{B}) \otimes_\pi W)$$

$$= \varprojlim_{\mathcal{K}} \varinjlim_{\mathcal{K}'} Hom_*(\overline{X}_*(R\tilde{A}_{(K,N)}) \otimes_\pi V, \overline{X}_*(R\tilde{B}_{(K',N')}) \otimes_\pi W)$$

The homological, (cohomological, bivariant) X-complexes define covariant,
(contravariant, bivariant) functors from the category of admissible Fréchet algebras to the category of $\mathbb{Z}/2$-graded chain complexes of vector spaces.

□

The **composition product** defines a bilinear map of chain complexes

$$X_{\epsilon, V, W}^*(A, B) \otimes X_{\epsilon, W, F}^*(B, C) \to X_{\epsilon, V, F}^*(A, C)$$

As a special case, there exist **pairings** of complexes

$$X_*^{\epsilon, V, W}(A) \otimes X_{\epsilon, W, F}^*(A) \to Hom(V, F)$$

$$X_*^\epsilon(A) \otimes X_\epsilon^*(A) \to \mathbb{C}$$

Definition 5.16:

The homology groups

$$HC_*^{\epsilon, V, W}(A) := h(X_*^{\epsilon, V, W}(A)) \quad HC_{\epsilon, V, W}^*(A) := h(X_{\epsilon, V, W}^*(A))$$

$$HC_{\epsilon, V, W}^*(A, B) := h(X_{\epsilon, V, W}^*(A, B))$$

are called the **analytic cyclic homology, (cohomology, bivariant cohomology) groups** of A, resp. (A, B).

□

The coefficient groups of these theories are nontrivial:

Lemma 5.17:[CO$_2$]

a) The universal Chern character cycles of an idempotent (resp. an invertible) element of definition 3.13. satisfy

$$ch(e) \in R\mathbb{C}[e]_{(\{1,e\}, 4+\delta)} = R\widetilde{\mathbb{C}}_{(\{1,e\}, 4+\delta)} \quad ch(u) \in R\mathbb{C}[u, u^{-1}]_{(\{u,u^{-1}\}, 1+\delta)}$$

for all $\delta > 0$

b) The homomorphisms of Ind-Fréchet algebras

$$s : \widetilde{\mathbb{C}} \to R\widetilde{\mathbb{C}} \quad \pi : R\widetilde{\mathbb{C}} \to \widetilde{\mathbb{C}}$$
$$e \to ch(e)$$

satisfy

$$\pi \circ s = Id_{\widetilde{\mathbb{C}}} \quad s \circ \pi =: p$$

where p is an idempotent endomorphism of $R\widetilde{\mathbb{C}}$ canonically homotopic to the identity whose image is the one-dimensional subalgebra generated by $ch(e) \in R\widetilde{\mathbb{C}}$.

c) The complex

$$\overline{X}_*(\widetilde{\mathbb{C}}) = X_*(\mathbb{C}) = \begin{cases} \mathbb{C} \simeq \langle ch(e) \rangle & * = 0 \\ 0 & * = 1 \end{cases}$$

is a deformation retract of $X_*(R\widetilde{\mathbb{C}})$, i.e.

$$X_*\pi \circ X_*s = Id, \ X_*s \circ X_*\pi = X_*p$$

and X_*p is homotopic to the identity:

$$X_*id - X_*p = h\partial + \partial h$$

for some operator

$$h : X_*(R\widetilde{\mathbb{C}}) \to X_{*-1}(R\widetilde{\mathbb{C}})$$

Proof:

b) The homotopy between the identity and p is defined by

$$R\widetilde{\mathbb{C}} \ni \varrho(e) \to (1-s)\,\varrho(e) + s\,ch(e) \in R\widetilde{\mathbb{C}}, \quad s \in [0,1]$$

which extends to a continuous homomorphism $F : R\widetilde{\mathbb{C}} \to C^\infty([0,1], R\widetilde{\mathbb{C}})$.

c) The homotopy operator h is provided by the composition

$$X_*(R\widetilde{\mathbb{C}}) \xrightarrow{X_*i} X_*R(R\widetilde{\mathbb{C}}) \xrightarrow{X_*RF} X_*R(C^\infty([0,1], R\widetilde{\mathbb{C}}))$$

$$\xrightarrow{h'} X_{*-1}R(R\widetilde{\mathbb{C}}) \xrightarrow{X_*\pi} X_{*-1}(R\widetilde{\mathbb{C}})$$

where X_*i, X_*RF and $X_*\pi$ are continuous by 5.11., 5.17.b), and 5.9. The homotopy operator h' is constructed in Theorem 6.12. where also its continuity is shown.

$\square$

Lemma 5.18:

There exist canonical homotopy equivalences of chain complexes

$$X_*^{\epsilon,V,W}(A) \xrightarrow{\sim} X_{\epsilon,V,W}^*(\mathbb{C}, A)$$

$$X_{\epsilon,V,W}^*(A) \xrightarrow{\sim} X_{\epsilon,V,W}^*(A, \mathbb{C})$$

which show that the analytic homological, resp. cohomological X-complexes are deformation retracts of the corresponding bivariant complexes.

$\square$

Proof:

By Lemma 5.17 the maps

$$X_*^\epsilon(A) = Hom^*(\overline{X}.\widetilde{\mathbb{C}}, \overline{X}.R\widetilde{A}) \xrightarrow{-\circ X_*\pi} Hom^*(\overline{X}.R\widetilde{\mathbb{C}}, \overline{X}.R\widetilde{A}) = X_\epsilon^*(\mathbb{C}, A)$$

$$X_\epsilon^*(A, \mathbb{C}) = Hom^*(\overline{X}.R\widetilde{A}, \overline{X}.R\widetilde{\mathbb{C}}) \xrightarrow{X_*\pi\circ-} Hom^*(\overline{X}.R\widetilde{A}, \overline{X}.\widetilde{\mathbb{C}}) = X_\epsilon^*(A)$$

are quasiisomorphisms and $X_*^\epsilon(A)$, $X_\epsilon^*(A)$ become deformation retracts of the bivariant complexes $X_\epsilon^*(\mathbb{C}, A)$, $X_\epsilon^*(A, \mathbb{C})$.

$\square$

Corollary 5.19:

$$HC_\epsilon^*(\mathbb{C}, \mathbb{C}) \simeq HC_\epsilon^*(\mathbb{C}) \simeq HC_*^\epsilon(\mathbb{C}) \simeq \begin{cases} \mathbb{C} & *=0 \\ 0 & *=1 \end{cases}$$

$\square$

Theorem 5.20: (Homotopy invariance)

Bivariant analytic cyclic cohomology is a smooth homotopy bifunctor on the category of admissible Fréchet algebras, i.e. if

$$f, g : A \to B$$

are smoothly homotopic morphisms of admissible Fréchet algebras, then

$$f_* = g_* \in HC^0_{\epsilon,V,V}(A,B)$$

and consequently

$$f^* = g^* : HC^*_{\epsilon,V,W}(B,C) \to HC^*_{\epsilon,V,W}(A,C)$$

$$f_* = g_* : HC^*_{\epsilon,V,W}(D,A) \to HC^*_{\epsilon,V,W}(D,B)$$

$\square$

This is an immediate consequence (see also 6.15.) of

Theorem 5.21:(Cartan homotopy formula)

Let A be an admissible Fréchet algebra and

$$\delta : A \to A$$

a bounded derivation.

a) Let $K \subset U \subset A$ be a multiplicatively closed, compact subset of A contained in a "small" ball U around 0. Choose $M \geq 1$ large enough that $K \cup \frac{1}{M}\delta(K) \subset K' \subset U$ for some compact set K'.

Then the Lie derivative

$$\mathcal{L}_\delta : X_*(RA) \to X_*(RA)$$

and the associated Cartan homotopy operator

$$h_\delta : X_*(RA) \to X_{*-1}(RA)$$

of Chapter 4 provide continuous operators

$$\mathcal{L}_\delta : X_*(RA_{(K,N)}) \to X_*(RA_{(K',N)})$$

$$h_\delta : X_*(RA_{(K,N)}) \to X_{*-1}(RA_{(K',N)})$$

b) Therefore they define elements

$$\mathcal{L}_\delta \in X_\epsilon^0(A, A) \quad \text{and} \quad h_\delta \in X_\epsilon^1(A, A)$$

satisfying

$$\partial_{Hom} \cdot h_\delta = \mathcal{L}_\delta$$

and consequently, the Cartan homotopy formula

$$\mathcal{L}_\delta = h_\delta \circ \partial_{X_*} + \partial_{X_*} \circ h_\delta$$

is valid.

$\square$

Proof:

b): Follows from a) and the definition of the bivariant analytic X-complex.

a): From the definition of the Lie derivative

$$\mathcal{L}_\delta(\varrho\omega^n(a^0, \ldots, a^{2n})) = \sum_0^{2n} \varrho\omega^n(a^0, \ldots, \delta a^i, \ldots, a^{2n})$$

$$\mathcal{L}_\delta(\varrho\omega^n d\varrho(a^0, \ldots, a^{2n+1})) = \sum_0^{2n+1} \varrho\omega^n d\varrho(a^0, \ldots, \delta a^i, \ldots, a^{2n+1})$$

the estimate

$$\| \mathcal{L}_\delta(a) \|_{N,m}^{K'} \le 2M \| a \|_{N,m+1}^K$$

follows readily for $a \in X_*(RA_K)$. Especially δ acts as bounded derivation on $RA_{(K,N)}$.

The Cartan homotopy operator is given by the composition (see 4.3)

$$X_*(RA) \xrightarrow{X_* i} X_*(RRA) \simeq \Omega RA \xrightarrow{i_\delta} \Omega RA \simeq X_{*-1}(RRA) \xrightarrow{X_* \pi} X_{*-1}RA$$

If we take topologies into account, we have

$$X_* i : X_*(RA_{(K,N)}) \to \Omega(RA_{(K,8N)})$$

by Proposition 5.11.

$$i_\delta : \Omega(RA_{(K,8N)}) \to \Omega(RA_{(K',8N)})$$

by what we just showed about δ and

$$X_* \pi : \Omega(RA_{(K',8N)}) \to X_*(RA_{(K',8N)})$$

by Lemma 5.9.

Together this yields

$$h_\delta : X_*(\mathcal{R}A) \to X_{*-1}(\mathcal{R}A)$$

which is what we wanted.

Proposition 5.22:

Let A, B be unital, admissible Fréchet algebras. Let

$$p : \mathcal{R}(A \oplus B) \xrightarrow{(\mathcal{R}p_1, \mathcal{R}p_2)} \mathcal{R}A \oplus \mathcal{R}B$$

be the canonical homomorphism of topological I-adic completions induced by the projection of $A \oplus B$ onto its factors.

There exists a homomorphism $s : \mathcal{R}A \oplus \mathcal{R}B \to \mathcal{R}(A \oplus B)$ which splits p: $p \circ s = Id$ and such that $s \circ p$ is smoothly homotopic to the identity.

□

Corollary 5.23:

The natural maps

$$X_* \mathcal{R}(A \oplus B) \to X_* \mathcal{R}A \oplus X_* \mathcal{R}B$$

$$\overline{X}_* \mathcal{R}(\widetilde{A}) \to X_* \mathcal{R}A$$

are quasiisomorphisms.

Corollary 5.24:

Let A, B, C, D be (not necessarily unital) admissible Fréchet algebras. Then

$$HC_\epsilon^*(C, A \oplus B) \xrightarrow{\cong} HC_\epsilon^*(C, A) \oplus HC_\epsilon^*(C, B)$$

$$HC_\epsilon^*(A, D) \oplus HC_\epsilon^*(B, D) \xrightarrow{\cong} HC_\epsilon^*(A \oplus B, D)$$

□

Proof of 5.22-5.24:

A splitting s with the desired properties has been constructed in a purely algebraic setting in Proposition 4.5. We show that this splitting extends to topological I-adic completions as well as the homotopy connecting $s \circ p$ and the identity. The two corollaries follow from this as in 4.6, 4.7. From now on the notations are those of Proposition 4.5 and its demonstration. We consider

$$s : RA \to \widehat{R}(A \oplus B)$$

$$s(\varrho(a)) = s_1(a) = \varphi_1(a)ch(1_A) \quad (s_1(1_A) = ch(1_A))$$

$$\varphi_1(a) := \varrho(a) + \sum_{k=1}^{\infty} \binom{2k}{k}(\varrho(1_A) - \frac{1}{2})\omega(1_A, 1_A)^{k-1}\omega(1_A, a)$$

and

$$s(\omega(a, a')) = s_1(aa') - s_1(a)s_1(a') = s_1(aa') - \varphi_1(a)s_1(1_A)s_1(a') =$$

$$= s_1(aa') - \varphi_1(a)s_1(a') = (\varphi_1(aa') - \varphi_1(a)\varphi_1(a'))ch(1_A)$$

Let $K \subset U \subset A$ be compact and contained in a "small" ball $U \subset A$ and denote by K' its image in $A \oplus B$. Suppose $1_A \in CK$ for some $C \geq 1$. The formula for $\varphi_1(a)$ above and the estimate $\binom{2k}{k} \leq (1+1)^{2k} = 4^k$ show that

$$ch(1_A), \; \varphi_1(a), \; \varphi_1(aa') - \varphi_1(a)\varphi_1(a') \; \in R(A \oplus B)_{N,m}^{K'}$$

for $a, a' \in K^{\infty}$ and $N > 4C$. Moreover all except at most two entries of the elements above equal 1_A which is an idempotent in $A \oplus B$ that acts as a unit on $A \subset A \oplus B$. Therefore Lemma 5.7. applies for $a_i \in K^{\infty}$ and provides the estimates

$$\| s(\varrho\omega^n(a^0, \ldots, a^{2n})) \|_{N,m}^{K'} =$$

$$= \| \varphi_1(a^0)ch(1_A) \prod_1^n (\varphi_1(a^{2i-1}a^{2i}) - \varphi_1(a^{2i-1})\varphi_1(a^{2i}))ch(1_A) \|_{N,m}^{K'}$$

$$\leq 6^{2n+1} \| \varphi_1(a^0) \|_{N,m}^{K'} \left(\| ch(1_A) \|_{N,m}^{K'} \right)^{n+1} \prod_1^n \| \omega_{\varphi_1}(a^{2i-1}, a^{2i}) \|_{N,m}^{K'}$$

with

$$\| \varphi_1(a^0) \|_{N,m}^{K'}, \; \| ch(1_A) \|_{N,m}^{K'}, \; \| \omega_{\varphi_1}(a^{2i-1}, a^{2i}) \|_{N,m}^{K'} \leq C(K, N, m)$$

for some constant C. Then the estimate

$$\| x \|_{NM,m}^{K'} \leq M^{-k} \| x \|_{N,m}^{K'} \quad \forall x \in I^k(A \oplus B)_{(K', N)}$$

shows that

$$\| s(\varrho\omega^n) \|_{NM,m}^{K'} \leq 6C^2 \left(\frac{36C^2}{M} \right)^n$$

which proves that given $(K, N') \in \mathcal{K}(A)$ $s : RA_{(K,N')} \to R(A \oplus B)_{(K',M')}$ is continuous provided that M' is large enough. The map s being defined naturally, it extends over the formal inductive limit

$$s : \mathcal{R}A \oplus \mathcal{R}B \to \mathcal{R}(A \oplus B)$$

A similar reasoning shows that the algebraic homotopy F connecting $s \circ p$ to the identity extends to a continuous homomorphism

$$F : \mathcal{R}(A \oplus B) \to C([0,1], \mathcal{R}(A \oplus B))$$

If we denote by t the coordinate function on $[0,1]$, then in fact the image of $x \in R(A \oplus B)_{(K,N)}$ can be expanded in a formal power series in t:

$$F(x,t) = x + \sum_{n=1}^{\infty} x_n t^n$$

where $x_n \in I^n(A \oplus B)_{(K,N)}$ by definition of F and the assignments $x \to x_n$ form a bounded family of continuous selfmaps of $R(A \oplus B)_{(K,N)}$. One finds therefore for the time derivatives of F:

$$\| \frac{\partial^k}{\partial t^k} F(x,t) \|_{N,m}^{K} \leq \sum_{n=1}^{\infty} n^k \| t^{n-k} x_n \|_{N,m}^{K}$$

$$\leq \sum_{n=1}^{\infty} (1+n)^{-2} \| (1+n)^{k+2} x_n \|_{N,m}^{K} \leq \sum_{n=1}^{\infty} (1+n)^{-2} \| x_n \|_{N,m+k+2}^{K}$$

$$\leq \sum_{n=1}^{\infty} (1+n)^{-2} C(k) \| x \|_{N,m+k+2}^{K}$$

which shows that the homotopy F is in fact smooth.

$\square$

5-4 The analytic X-complex and entire cyclic cohomology

Before discussing the relation between analytic and entire cyclic cohomology, the normalization procedure of Cuntz-Quillen has to be extended to the analytic situation

Lemma 5.25:

The Cuntz-Quillen projection defines a natural continuous map of complexes

$$P_{CQ} : X_*(RA_{(K,N)}) \to X_*(RA_{(K,N)})$$

The subcomplex

$$(1 - P_{CQ}) X_*(RA_{(K,N)}) \subset X_*(RA_{(K,N)})$$

is naturally contractible.

$\square$

Proof:

We use throughout the notations of Theorem 3.11. Especially, the complex $X_*(RA)$ is identified with the differential graded envelope ΩA of A as in (3.6), (3.7) and all operators (d, b, κ) are supposed to act on ΩA. The norm of a differential form is meant to be the norm of the corresponding element in $X_* RA_{(K,N)}$.

The Cuntz-Quillen projection P_{CQ} equals the spectral projection onto the generalized 1-eigenspace of the Karoubi-operator κ:

Each $a_n \in \Omega^n$ decomposes as

$$a_n = \sum_{(\lambda^n - 1)(\lambda^{n+1} - 1) = 0} a_n^\lambda$$

with

$$\kappa(a_n^\lambda) = \lambda a_n^\lambda \quad \lambda \neq 1 \quad (\kappa - 1)^2(a_n^1) = 0 \quad \lambda = 1$$

As

$$\kappa : RA_K \to RA_K$$

satisfies for $a \in RA_K$

$$\| \kappa^j(a) \|_{N,m}^K \leq (j + 1) \| a \|_{N,m}^K$$

we find

$$\| a_n^\lambda \|_{N,m}^K = \| \frac{(\kappa^n - 1)(\kappa^{n+1} - 1)}{\kappa - \lambda}(a_n) \|_{N,m}^K \leq \begin{cases} O(n^2) \| a_n \|_{N,m}^K & \lambda \neq 1 \\ O(n^3) \| a_n \|_{N,m}^K & \lambda = 1 \end{cases}$$

so that

$$\| P_{CQ}(a) \|_{N,m}^K \leq C \| a_n \|_{N,m+3}^K$$

which proves the continuity of the Cuntz-Quillen projection.

The contracting homotopies for the complex

$$(1 - P_{CQ})X_*(RA)$$

on the eigenspaces of the Karoubi-operator for the eigenvalues $\lambda \neq 1$ are

$$h_\lambda := \begin{cases} Gd & \text{for } \lambda = -1 \text{ on } X_0 \\ -\frac{1}{n+1}Gd & \text{for } \lambda = -1 \text{ on } X_1 \\ G(d - (1 + \lambda)^{-1}b) & \text{for } \lambda \neq -1 \end{cases}$$

where Greens operator equals

$$G|_{ker(\kappa - \lambda)} = \frac{1}{\lambda} \quad \lambda \neq 1$$

so that

$$\| h_{-1}(a_n^{-1}) \|_{N,m}^K \leq 2^m N \| a_n^{-1} \|_{N,m}^K$$

and

$$\| h_\lambda(a_n^\lambda) \|_{N,m}^K \leq O(n^2)2^m N \| a_n^\lambda \|_{N,m}^K$$

for all $\lambda \neq -1$.

Altogether

$$\| h(a_n) \|_{N,m}^K = \| \sum_{\lambda \neq 1} h_\lambda(a_n^\lambda) \|_{N,m}^K \leq O(n^2) 2^m N \sum_{\lambda \neq 1} \| a_n^\lambda \|_{N,m}^K$$

$$\leq O(n^2) O(n^2) 2^m N \sum_{\lambda \neq 1} \| a_n \|_{N,m}^K \leq O(n^5) 2^m N \| a_n \|_{N,m}^K$$

so that we may conclude

$$\| h(a) \|_{N,m}^K \leq C(N,m) \| a \|_{N,m+5}^K$$

for $a \in RA_K$

□

Definition 5.26:

(The notations are those of $[CO_2]$.)
Let A be a Banach algebra. For $\varphi \in C^n(A, A^*)$ and $K \subset A$ put

$$\| \varphi \|_K := \sup_{a^i \in K} |\varphi(a^0, \ldots, a^n)|$$

A cochain

$$(\phi_{2n}) \in C^{ev}(A) \quad ((\psi_{2n+1}) \in C^{odd}(A))$$

is called **locally entire** if

$$\sum \| \varphi_{2n} \|_K \frac{z^n}{n!} \quad \left(\sum \| \psi_{2n+1} \|_K \frac{z^n}{n!} \right)$$

is an entire function of z for every compact subset $K \subset A$.

The locally entire cochains form a $\mathbb{Z}/2$-graded chain complex under Connes's differential ∂ which is denoted by $CC^*_{\epsilon,loc}(A)$.

□

Theorem 5.27:

For any Banach algebra A the normalized, locally entire
Connes-bicomplex is a natural deformation retract of the cohomological analytic
X-complex.

$$X^*_\epsilon(A) \xrightarrow[quis]{\sim} CC^*_{\epsilon,loc}(A)$$

A natural retraction is provided by the maps Φ, Ψ of Theorem 3.11.

□

Proof:

Because the algebras $RA_{(K,N)}$ are completions of the subalgebras $RA_K \subset RA$ of RA and

$$RA = \langle \bigcup_K RA_K \rangle, \quad \langle \rangle = \text{linear span}$$

an element

$$\chi \in X_\epsilon^*(A) = \varprojlim_{\leftarrow K} Hom_*(\overline{X}_*(R\tilde{A}_{(K,N)}), \mathbb{C})$$

is uniquely determined by the functionals

$$\chi_{2n}(a^0, \ldots, a^{2n}) := \chi(\varrho\omega^n(a^0, \ldots, a^{2n}))$$

$$\chi_{2n+1}(a^0, \ldots, a^{2n+1}) := \chi(\varrho\omega^n d\varrho(a^0, \ldots, a^{2n+1}))$$

On the other hand, a cochain

$$\eta \in CC_{\epsilon,loc}^*(A)$$

is by definition given by a family

$$\eta = (\eta_n) \quad \eta_n \in C^n(A, A^*)$$

The maps Φ, Ψ between these two complexes can be described as follows (see Theorem 3.11. and $[CO_2]$)

$$\Phi: \quad CC_{\epsilon,loc}^*(A) \quad \rightarrow \quad X_\epsilon^*(A)$$

$$\eta_{2n} \quad \rightarrow \quad (-1)^n \tfrac{n!}{(2n)!} \eta_{2n}$$

$$\eta_{2n+1} \quad \rightarrow \quad (-1)^n \tfrac{n!}{(2n+1)!} \eta_{2n+1}$$

$$\Psi: \quad X_\epsilon^*(A) \quad \rightarrow \quad CC_{\epsilon,loc}^*(A)$$

$$\chi_{2n} \quad \rightarrow \quad (-1)^n \tfrac{(2n)!}{n!} P_{CQ}(\chi_{2n})$$

$$\chi_{2n+1} \quad \rightarrow \quad (-1)^n \tfrac{(2n+1)!}{n!} P_{CQ}(\chi_{2n+1})$$

so that

$$\Psi \circ \Phi = Id \quad \Phi \circ \Psi = P_{CQ}$$

Let now

$$\eta = (\eta_n) \in CC_{\epsilon,loc}^*(A)$$

Choose as basic "small" open set the open unit ball $B(0,1)$ in A and let K be a compact subset of $B(0,1)$. Then there exists for any $N > 1$ some $C(N) > 0$ such that

$$\| \eta_{2n} \|_{K^\infty} \leq C(N, \eta) N^{-2n} n! \quad \| \eta_{2n+1} \|_{K^\infty} \leq C(N, \eta) N^{-(2n+1)} n!$$

Let $a \in X_0(RA_K)$ be presented as a sum

$$a = \sum_{\beta} \lambda_\beta \varrho_\beta \omega^{k_\beta}$$

with entries in $K^\infty \cup \{1\}$ such that

$$\sum_{\beta} |\lambda_\beta| N^{-2k_\beta} \leq \| a \|_{N^2,0} + \epsilon$$

Then

$$|(\Phi\eta)(a)| = | \sum_{n=0}^{\infty} \sum_{k_\beta=n} \lambda_\beta \, (\Phi\eta)_{2n}(\varrho_\beta \omega^{k_\beta})|$$

$$\leq \sum_{n=0}^{\infty} \sum_{k_\beta=n} |\lambda_\beta| \frac{n!}{(2n)!} \, \| \eta_{2n} \|_{K^\infty} \leq C(N,\eta) \sum_{n=0}^{\infty} \sum_{k_\beta=n} |\lambda_\beta| N^{-2n} \binom{2n}{n}^{-1}$$

$$\leq C(N,\eta)(\| a \|_{N^2,0} + \epsilon)$$

so that $\Phi(\eta) \in X_\epsilon^*(A)$. The odd case is similar.

To show that Ψ maps the cohomological analytic X-complex to the locally entire Connes-bicomplex it suffices in light of the preceeding proposition to prove that for

$$\mu = P_{CQ}\chi \in X_\epsilon^*(A)$$

the functionals

$$\mu'_{2n}(a^0,\ldots,a^{2n}) := (-1)^n \frac{(2n)!}{n!} \mu(\varrho\omega^n(a^0,\ldots,a^{2n}))$$

$$\mu'_{2n+1}(a^0,\ldots,a^{2n+1}) := (-1)^n \frac{(2n+1)!}{n!} \mu(\varrho\omega^n d\varrho(a^0,\ldots,a^{2n+1}))$$

form the components of a locally entire cochain.

First of all

$$\sup_{a^i \in K} |\mu'_n(a^0,\ldots,a^n)| < \infty$$

for every compact set $K \subset A$ and every $n \in I\!N$ which already implies that the maps μ'_n extend to bounded, linear functionals on $A_\pi^{\otimes(n+1)}$.

For fixed K there exists $C > 0$ such that

$$\frac{1}{C} K \subset K' \subset B(0,1)$$

for some multiplicatively closed, compact set K'.
Then

$$\frac{1}{n!}|\mu'_{2n}(a^0,\ldots,a^{2n})| \leq \frac{1}{n!} C^{2n+1} \frac{(2n)!}{n!} |\mu(\varrho\omega^n(\frac{a^0}{C},\ldots,\frac{a^{2n}}{C}))|$$

$$\leq C^{2n+1} \binom{2n}{n} \| \mu \|_{N,m}^{K'} \| \varrho\omega^n(\frac{a^0}{C},\ldots,\frac{a^{2n}}{C}) \|_{N,m}^{K'}$$

for some $m \in \mathbb{N}$ because μ is continuous on $RA_{(K,N)}$ for all $N \geq 1$

$$\leq C^{2n+1} 4^n \parallel \mu \parallel_{N,m}^{K'} (1+n)^m N^{-n} \leq \parallel \mu \parallel_{N,m}^{K'} C (1+n)^m \left(\frac{N}{4C^2}\right)^{-n}$$

so that

$$\overline{\lim}_n \left(\frac{1}{n!} \parallel \mu'_{2n} \parallel_K\right)^{\frac{1}{n}} \leq \frac{4C^2}{N}$$

for every $N \geq 1$. The conclusion follows.

$\square$

Corollary 5.28:

There exists a natural map of complexes

$$CC_\epsilon^*(A) \to X_\epsilon^*(A)$$

from Connes's entire cyclic-bicomplex to the cohomological analytic X-complex given by the composition

$$CC_\epsilon^* \to CC_{\epsilon,loc}^* \xrightarrow{\Phi} X_\epsilon^*$$

$\square$

5-5 The analytic Chern character

Definition and Proposition 5.29:(Chern Character) [CO$_2$]

Let A be a unital admissible Fréchet algebra.

The **Chern Character**

$$ch : K_*(-) \to HC_\epsilon^*(\mathbb{C}, -) = HC_*^\epsilon(-)$$

is defined by

$$ch_0 : \quad K_0(A) = [\mathbb{C}[e], M_\infty(A)] \quad \to \quad HC_0^\epsilon(M_\infty(A)) \xrightarrow{\times Tr} HC_0^\epsilon(A)$$

$$[f] \quad\quad \to \quad [f_*(ch(e))]$$

$$ch_1 : \quad K_1(A) = [\mathbb{C}[u, u^{-1}], M_\infty(A)] \quad \to \quad HC_1^\epsilon(M_\infty(A)) \xrightarrow{\times Tr} HC_1^\epsilon(A)$$

$$[f] \quad\quad \to \quad [f_*(ch(u))]$$

The Chern character is a natural transformation of smooth homotopy functors. The exterior product with the trace is defined in Chapter 8 and coincides with that of [CO].

Proof:

In the definition of the Chern character care has to be taken of the fact that $\mathbb{C}[e], \mathbb{C}[u, u^{-1}]$ are not equipped with topologies. The analytic X-complexes are still defined for arbitrary algebras however by declaring

$$\mathcal{K}(A) := \{(K, N) | K \subset A, \ K \text{ a finite set}\}$$

Therefore $[ch(e)], [ch(u)]$ may be identified with classes

$$[ch(e)] \in HC_\epsilon^0(\mathbb{C}, \mathbb{C}) \quad [ch(u)] \in HC_\epsilon^1(\mathbb{C}, \mathbb{C}[u, u^{-1}])$$

The only thing to be shown is that any continuous homotopies $e_0 \sim e_1 : \mathbb{C}[e] \to A$ (resp. $u_0 \sim u_1 : \mathbb{C}[u, u^{-1}] \to A$) may be replaced by piecewise smooth ones. The rest follows then from the fact that bivariant analytic cohomology is a smooth homotopy functor. The assertion is clear for morphisms $\mathbb{C}[u, u^{-1}] \to A$ because the invertible elements of A form an open set and the case $\mathbb{C}[e] \to A$ follows by applying functional calculus to a smooth path close to a continuous path of idempotents connecting e_0 and e_1.

$\square$

Corollary 5.30:

There exists a natural **pairing** for unital admissible Fréchet algebras

$$K_* \otimes HC_\epsilon^* \xrightarrow{ch \otimes id} HC_*^\epsilon \otimes HC_\epsilon^* \to \mathbb{C}$$

$\square$

For an explicit description of the pairing see 6-2.

Chapter 6: The asymptotic X-complex

Our main goal, the construction of covariant (contravariant, bivariant) chain complexes on the category of admissible Fréchet algebras that behave functorially under linear asymptotic morphisms will be achieved in this chapter. The cohomological asymptotic X-complex of an admissible Fréchet algebra A is the differential graded X-complex of the topological I-adic completion $\mathcal{R}A$ of the algebra RA of tensors over A with coefficients in the DG-module of germs around infinity of smooth differential forms on the asymptotic parameter space $\mathcal{R}_+^\infty$.

Asymptotic cocycles in the sense of Connes-Moscovici [CM] do not yield cocycles in our sense and vice versa, but many concrete examples turn out to be asymptotic cocycles with respect to both definitions. Asymptotic cocycles in the sense of Connes-Moscovici are globally defined (on $X_*(RA)$) functionals with uniform growth conditions imposed only on tensors with entries belonging to a given finite set. Our cochains "live" only on virtual neighbourhoods of infinity in the parameter space and in fact their components in the corresponding (b, B)-bicomplex will not be families of globally defined, but only of densely defined multilinear functionals in general. Uniform growth conditions are imposed however on tensors with entries in a given multiplicatively closed compact set.

The bivariant asymptotic X-complex is then introduced in analogy with the differential graded bivariant X-complex. The composition product of bivariant X-complexes carries over to the asymptotic setting. This is no formal matter anymore, because it corresponds to the Kasparov product in bivariant K-theory as we will see later. The fundamental observation is that every linear asymptotic morphism defines an even cocycle in the bivariant asymptotic X-complex whose cohomology class depends only on the continuous (!) homotopy class of the morphism. This correspondence is compatible with compositions, i.e. functorial. Consequently the (bivariant) asymptotic X-complex defines a (bi)functor on the linear asymptotic category and asymptotic cohomology is an asymptotic homotopy (bi)functor. It turns out to be a nontrivial theory as the calculation of the asymptotic cohomology of the algebra of complex numbers shows. Finally the Chern character and the natural pairing between K-theory and asymptotic cohomology is treated and explicit formulas for the values of the pairing are derived.

6-1 The asymptotic X-complex

Before coming to the definition of the asymptotic X-complexes some notations and definitions have to be fixed. First of all, the topological I-adic completion of chapter 5 has to be extended to the differential graded setting.

Definition 6.1:

Let A be an admissible Fréchet algebra and let $(K, N) \in \mathcal{K}(A, U)$.

a) Define

$$\Omega \mathcal{R} A := \text{"}\lim_{\mathcal{K}}\text{"}\Omega R A_{(K,N)}$$

where the differential graded algebras $\Omega R A_{(K.N)}$ are topologized as in 2.4, i.e. by declaring $\Omega B \simeq \bigoplus B \otimes \overline{B}^{\otimes k}$ to be a topological isomorphism.

b) Put

$$R(\Omega A)_K := \{\sum \lambda_\beta \varrho_\beta \omega_\beta^{k_\beta}, \lambda_\beta \in \mathbb{C}\}$$

where the entries of ϱ_β, ω_β are of the form $a^0 \partial a^1 \ldots \partial a^j$ with $a^i \in K^\infty \cup \{1\}$. Denote the subspace of elements of degree k by $R(\Omega A)_K^{(k)}$. Analogous to 5.6 there are weighted l^1-seminorms $\| - \|_{N,m}^K$ on $R(\Omega A)_K^{(k)}$. The algebra $R(\Omega A)_{(K,N)}$ is then defined by declaring

$$R(\Omega A)_{(K,N)} \xrightarrow{\simeq} \bigoplus_k R(\Omega A)_{(K,N)}^{(k)}$$

to be a topological isomorphism where the direct sum on the right hand side is given the product topology. Finally we put

$$\mathcal{R}(\Omega A) := \text{"}\lim_{\mathcal{K}}\text{"}R(\Omega A)_{(K,N)}$$

$\square$

We still have to treat a technical point, (necessarily coming up when the bivariant asymptotic X-complex is defined), which concerns the notion of a function with values in a formal inductive limit.

If X is a topological space (a smooth manifold) and if A denotes a topological vector space, then the assignment

$$U \to C(U, A) \quad (U \to \mathcal{C}^\infty(U, A))$$

defines a sheaf of topological vector spaces on X and $C(X, A)$ (resp. $\mathcal{C}^\infty(X, A)$) coincides with its module of global sections. On the contrary, for a formal inductive limit of topological vector spaces "$\lim_I A_i$" the presheaf

$$(*) \quad U \to \text{"}\lim_I\text{"}C(U, A_i) \quad (\text{"}\lim_I\text{"}\mathcal{C}^\infty(U, A_i))$$

will not be a sheaf in general. Therefore we define a continuous (smooth) function on X with values in "$\lim_I A_i$" to be a global section of the sheaf associated to the presheaf above.

Definition 6.2:

Let X be a topological space (a smooth manifold) and let $"\lim_I A_i"$ be a formal inductive limit of topological vector spaces.

a) Define

$$C(X," \lim_I "A_i) \quad := \quad \Gamma(sh(U \to " \lim_I "C(U, A_i)))$$

$$\mathcal{C}^\infty(X," \lim_I "A_i) \quad := \quad \Gamma(sh(U \to " \lim_I "\mathcal{C}^\infty(U, A_i)))$$

$$\cong\downarrow \qquad\qquad\qquad\qquad \downarrow\cong$$

$$\mathcal{C}^\infty(X) \otimes_\pi " \lim_I "A_i \quad := \quad \Gamma(sh(U \to " \lim_I "\mathcal{C}^\infty(U) \otimes_\pi A_i)))$$

where $\Gamma(sh(*))$ denotes the space of global sections of the sheaf associated to the presheaf $*$.

b) If X is a locally compact space (manifold) then the module of global sections of the sheaf above can be described explicitly as

$$C(X," \lim_I "A_i) = \lim_I \lim_{\leftarrow V \to I} C(V, A_i)$$

where V runs over all relatively compact open subsets of X ordered by inclusion.

c) Similarly, if $\mathcal{E}(X)$ denotes the algebra of smooth differential forms on the manifold X equipped with the topology of 2.6., we define

$$" \lim_I "A_i \otimes_\pi \mathcal{E}(X) := \Gamma(sh(U \to " \lim_I "A_i \otimes_\pi \mathcal{E}(U)))$$

$\square$

Sometimes we will write $" \lim_I "A_i \otimes_\pi^{sh} \mathcal{E}(X)$ instead of $" \lim_I "A_i \otimes_\pi \mathcal{E}(X)$ to remind the reader that a sheafification has been carried out.

Finally let us recall the asymptotic parameter space $\mathcal{R}_+^\infty = \bigcup_n \mathcal{R}_+^n$ (see (1.1).) The punctured neighbourhoods of $\infty \in \mathcal{R}_+^n$ form a directed set by inclusion, denoted by $\mathcal{U}$.

Definition 6.3:

$$\mathcal{C}^\infty(\mathcal{U}) := " \lim_\mathcal{U} "\mathcal{C}^\infty(U) \quad \mathcal{E}(\mathcal{U}) := " \lim_\mathcal{U} "\mathcal{E}(U)$$

Now we can construct the central objects of our study,
the asymptotic X-complexes of an admissible Fréchet algebra.

Definition 6.4:

Let A, B be admissible Fréchet algebras.

a) The **cohomological asymptotic X-complex** of A is defined as

$$X_\alpha^*(A) := Hom_{DG}(\overline{X}_*(\Omega\mathcal{R}\widetilde{A}), \mathcal{E}(\mathcal{U})) = \varprojlim_{\leftarrow\mathcal{K}} \varinjlim_{\rightarrow\mathcal{U}} Hom_{DG}(\overline{X}_*(\Omega\mathcal{R}\widetilde{A})_{(K,N)}, \mathcal{E}(U))$$

b) The cohomology of the asymptotic X-complex of A

$$HC_\alpha^*(A) := h(X_\alpha^*(A))$$

is called the **asymptotic cyclic cohomology** of A.

c) The **bivariant asymptotic X-complex** of the pair (A, B) is given by

$$X_\alpha^*(A, B) := Hom_{DG}(\overline{X}_*(\Omega\mathcal{R}\widetilde{A}), \overline{X}_*(\Omega\mathcal{R}\widetilde{B}) \otimes_\pi^{sh} \mathcal{E}(\mathcal{U}))$$

$$:= \varprojlim_{\leftarrow\mathcal{K}} \varinjlim_{\rightarrow\mathcal{U}} Hom_{DG}(\overline{X}_*(\Omega\mathcal{R}\widetilde{A})_{(K,N)}, \overline{X}_*(\Omega\mathcal{R}\widetilde{B}) \otimes_\pi^{sh} \mathcal{E}(U))$$

d) The cohomology of the bivariant asymptotic X-complex of (A, B)

$$HC_\alpha^*(A, B) := h(X_\alpha^*(A, B))$$

is called the **bivariant asymptotic cohomology** of (A, B).

$\square$

The cohomological (bivariant) asymptotic X-complex defines a contravariant functor (bifunctor) from the category of admissible Fréchet algebras to the category of $\mathbb{Z}/2$-graded chain complexes of vector spaces.

Every morphism $f : A \to B$ of admissible Fréchet algebras provides a cocycle

$$f_* := X_*(\Omega\mathcal{R}f) \in \varprojlim_{\leftarrow\mathcal{K}} \varinjlim_{\rightarrow\mathcal{K}'} Hom_{DG}^0(X_*(\Omega RA_{(K,N)}), X_*(\Omega RB_{(K',N')}))$$

Theorem 6.5:

The **composition product** for the bivariant, differential graded X-complex (2.6) yields a natural and associative composition product

$$X_\alpha^*(A, B) \otimes X_\alpha^*(B, C) \to X_\alpha^*(A, C)$$

and consequently a natural and associative product

$$HC_\alpha^*(A, B) \otimes HC_\alpha^*(B, C) \to HC_\alpha^*(A, C)$$

on cohomology.

$\square$

This pairing makes $HC^*_\alpha(A, A)$ into an associative algebra with unit $Id^A_* \in HC^0_\alpha(A, A)$.

Under this product $f : A \to B$, $g : B \to C$ and $\varphi \in HC^*_\alpha(B, D)$, $\psi \in HC^*_\alpha(E, B)$ satisfy

$$f_* \otimes g_* = g_* \circ f_* = (g \circ f)_* \quad f_* \otimes \varphi = \varphi \circ f_* =: f^*\varphi \quad \psi \otimes g_* = g_* \circ \psi =: g_*\psi$$

Theorem 6.6:

There is a similar natural composition product

$$X^*_\alpha(A, B) \otimes X^*_\alpha(B) \to X^*_\alpha(A)$$

inducing a natural product

$$HC^*_\alpha(A, B) \otimes HC^*_\alpha(B) \to HC^*_\alpha(A)$$

on cohomology.

□

This product makes $X^*_\alpha(A)$ into a module under $X^*_\alpha(A, A)$.

Under this product again we put

$$f_* \otimes \varphi = \varphi \circ f_* =: f^*\varphi$$

for $f : A \to B$ and $\varphi \in HC^*_\alpha(B, D)$.

Definition 6.7:

A cohomology class $x \in HC^*_{\alpha,\epsilon}(A, B)$ is called **asymptotically (analytically) HC-invertible** or an **asymptotic (analytic) HC-equivalence** if there exists $y \in HC^*_{\alpha,\epsilon}(B, A)$ such that

$$y \circ x = Id^A_* \in HC^0_{\alpha,\epsilon}(A, A) \quad x \circ y = Id^B_* \in HC^0_{\alpha,\epsilon}(B, B)$$

□

Proof of Theorem 6.5:

Fix $\Phi \in X_\alpha^*(A,B), \Psi \in X_\alpha^*(B,C)$. Let Φ be represented by functionals

$$\Phi_{(K,N)} \in Hom_{DG.loc}^*(\overline{X}_*(\Omega R\widetilde{A}_{(K,N)}), \overline{X}_*(\Omega R\widehat{B}) \otimes_\pi \mathcal{E}(U))$$

for some punctured neighbourhoods $U(K,N)$ of $\infty \in \mathcal{R}_+^\infty$ Let $U_{f_1,\ldots,f_m} := U' \subset \overline{U'} \subset U$ be a fundamental punctured neighbourhood of $\infty \in \mathcal{R}_+^m$ as described in (1.1). By Lemma 1.2 the sets

$$U_n := U' \cap \pi_m^{-1}([0,n])$$

are relatively compact in U, so that the restrictions of $\Phi_{(K,N)}$ to U_n are of the form

$$\Phi_{(K,N)}^n := (Id \otimes i_{U_n,U'}^*) \circ \Phi_{(K,N)}$$

$$\in Hom_{DG}^*(\overline{X}_*(\Omega R\widetilde{A}_{(K,N)}), \overline{X}_*(\Omega R\widetilde{B}_{(K'_n,N'_n)}) \otimes_\pi \mathcal{E}(U_n))$$

for some $(K'_n, N'_n) \in \mathcal{K}'$ as follows from 6.2.b).

Choose now representatives

$$\Psi_{(K'_n,N'_n)} \in Hom_{DG}^*(\overline{X}_*(\Omega R\widetilde{B}_{(K'_n,N'_n)}), \overline{X}_*(\Omega R\widetilde{C}) \otimes_\pi^{sh} \mathcal{E}(V_n))$$

for suitable fundamental neighbourhoods

$$V_n = V_{g_1^n,\ldots,g_{m'}^n} \subset \mathcal{R}_+^{m'}$$

of ∞.

Construct an open set $W_{(K,N)} \subset \mathcal{R}_+^{m+m'}$ as follows:
Define inductively piecewise continuous,
strictly monotone increasing functions $k_1,\ldots,k_{m'}$ on $\mathbb{R}_+$ by

$$k_1|_{[n-1,n[} = \sup_{n' \leq n} g_1^{n'}$$

$$k_2|_{[k_1(n-1),k_1(n)[} = \sup_{n' \leq n} g_2^{n'}$$

$$\vdots$$

$$k_{m'}|_{[k_{m'-1}\circ\cdots\circ k_1(n-1),k_{m'-1}\circ\cdots\circ k_1(n)[} = \sup_{n' \leq n} g_{m'}^{n'}$$

and choose $h_1,\ldots,h_{m'}$ continuous, strictly monotone increasing, convex and satisfying $h_j \geq k_j$; $1 \leq j \leq m'$. Put

$$W_{(K,N)} := U_{f_1,\ldots,f_m,h_1,\ldots,h_{m'}} \subset \mathcal{R}_+^{m+m'}$$

We claim that

$$\Psi \circ \Phi_{(K,N)} \in Hom_{DG}^*(\overline{X}_*(\Omega R\widetilde{A}_{(K,N)}), \overline{X}_*(\Omega R\widetilde{C}) \otimes_\pi^{sh} \mathcal{E}(W_{(K,N)}))$$

Choose $x = (x_1, \ldots, x_m, x_1', \ldots, x_{m'}') \in W$ and a relatively compact neighbourhood W' of x in W satisfying

$$\pi_m(W') \subset [0, [x_m + 1]] =: [0, M]$$

Then

$$\pi(W') \subset U \cap \pi_m^{-1}([0, M]) = U_M \quad \pi'(W') \subset V_M$$

are relatively compact. So

$$W' \subset U_M \times V_M$$

and on this set

$$(Id \otimes i_{W', U_M \times V_M}^*)(\Psi \circ \Phi_{(K,N)}) = \Psi_{(K_M', N_M')} \circ \Phi_{(K,N)}^M$$

$$\in Hom_{DG}^*(\overline{X}_*(\Omega R \widetilde{A}_{(K,N)}), \overline{X}_*(\Omega R \widetilde{C}) \otimes_\pi^{sh} \mathcal{E}(U_M \times V_M))$$

as

$$\Phi_{(K,N)}^M \in Hom_{DG}^*(\overline{X}_*(\Omega R \widetilde{A}_{(K,N)}), \overline{X}_*(\Omega R \widetilde{B}_{(K_M', N_M')}) \otimes_\pi \mathcal{E}(U_M))$$

and

$$\Psi_{(K_M', N_M')} \in Hom_{DG}^*(\overline{X}_*(\Omega R \widetilde{B}_{(K_M', N_M')}), \overline{X}_*(\Omega R \widetilde{C}) \otimes_\pi^{sh} \mathcal{E}(V_M))$$

Restriction to possibly smaller open sets shows that the construction is independent of all choices. Passing to the limit over (K, N) the conclusion follows.

The demonstration of Theorem 6.6 is similar.

□

6-2 Comparison with other cyclic theories

Proposition 6.8:

a) There exist natural transformations of (bi)functors

$$HC_\epsilon^*(-) \to HC_\alpha^*(-) \quad HC_\epsilon^*(-,-) \to HC_\alpha^*(-,-)$$

compatible with composition products.

b) Moreover, there is a natural equivalence of functors

$$HC_\alpha^*(-, \mathbb{C}) \xrightarrow{\simeq} HC_\alpha^*(-)$$

Proof:

a)

The complex $X_\epsilon^*(A) = \lim_{\leftarrow K} Hom(\overline{X}_*(R\widetilde{A}_{(K,N)}), \mathbb{C})$ can naturally be identified with the subcomplex of $X_\alpha^*(A) = \lim_{\leftarrow K} \lim_{\to U} Hom_{DG}(\overline{X}_*(\Omega R \widetilde{A}_{(K,N)}), \mathcal{E}(U))$ of functionals which are constant on U (and vanish in positive degrees).

For the bivariant complexes recall that there is a natural projection

$$Hom^*_{DG}(X_*(\Omega RA), X_*(\Omega RB)) \to Hom^*(X_*(RA), X_*(RB))$$

which gives a diagram of maps of complexes

$$X^*_\epsilon(A, B) \leftarrow \varprojlim_{\leftarrow \mathcal{K}} \varinjlim_{\to \mathcal{K}'} Hom_{DG}(\overline{X}_*(\Omega R\widetilde{A}_{(K,N)}), \overline{X}_*(\Omega R\widetilde{B}_{(K',N')})) \to X^*_\alpha(A, B)$$

Recall that the projections

$$Hom(\overline{X}_* R\widetilde{A}_{(K,N)}, \overline{X}_* R\widetilde{B}_{(K',N')}) \leftarrow Hom_{DG}(\overline{X}_*(\Omega R\widetilde{A}_{(K,N)}), \overline{X}_*(\Omega R\widetilde{B}_{(K',N')}))$$

are quasiisomorphisms because they split naturally as a map of vector spaces and there exists a natural contracting homotopy on their kernel. By naturality these sections and homotopies fit together so that the projection

$$X^*_\epsilon(A, B) \leftarrow \varprojlim_{\leftarrow \mathcal{K}} \varinjlim_{\to \mathcal{K}'} Hom_{DG}(\overline{X}_*(\Omega R\widetilde{A}_{(K,N)}), \overline{X}_*(\Omega R\widetilde{B}_{(K',N')}))$$

also splits linearly and has a contractible kernel and is therefore a quasiisomorphism, too. The map

$$\varprojlim_{\leftarrow \mathcal{K}} \varinjlim_{\to \mathcal{K}'} Hom_{DG}(\overline{X}_*(\Omega R\widetilde{A}_{(K,N)}), \overline{X}_*(\Omega R\widetilde{B}_{(K',N')})) \to X^*_\alpha(A, B)$$

identifies $\varprojlim_{\leftarrow \mathcal{K}} \varinjlim_{\to \mathcal{K}'} Hom_{DG}(\overline{X}_*(\Omega R\widetilde{A}_{(K,N)}), \overline{X}_*(\Omega R\widetilde{B}_{(K',N')}))$ with the functionals in $X^*_\alpha(A, B)$ that are constant on U.

The induced maps on cohomology provide the desired transformation.

b)

Recall (4.10) that the underlying map of vector spaces of the projection

$$Hom^*_{DG}(X_*(\Omega RA), X_*(\Omega RB) \otimes V.) \to Hom^*_{DG}(X_*(\Omega RA), X_*(RB) \otimes V.)$$

splits naturally and that the kernel of this map of complexes is naturally contractible so that the map

$$X^*_\alpha(A, \mathbb{C}) \to \varprojlim_{\leftarrow \mathcal{K}} \varinjlim_{\to \mathcal{U}} Hom_{DG}(\overline{X}_*(\Omega R\widetilde{A}_{(K,N)}), \overline{X}_*(R\widetilde{\mathbb{C}}) \otimes^{sh}_\pi \mathcal{E}(U))$$

becomes a quasiisomorphism.

The projection (5.17)

$$X_*\pi : \overline{X}_*(R\widetilde{\mathbb{C}}) \to \overline{X}_*(\widetilde{\mathbb{C}})$$

contracting $\overline{X}_*(R\widetilde{\mathbb{C}})$ to $\overline{X}_*\widetilde{\mathbb{C}} \simeq \mathbb{C}$ provides then a homotopy equivalence between the complexes

$$\varprojlim_{\leftarrow \mathcal{K}} \varinjlim_{\to \mathcal{U}} Hom_{DG}(\overline{X}_*(\Omega R\widetilde{A}_{(K,N)}), \overline{X}_*(R\widetilde{\mathbb{C}}) \otimes^{sh}_\pi \mathcal{E}(U))$$

and

$$\varprojlim_{\leftarrow \mathcal{K}} \varinjlim_{\to \mathcal{U}} Hom_{DG}(\overline{X}_*(\Omega R\widetilde{A}_{(K,N)}), \overline{X}_*\widetilde{\mathbb{C}} \otimes_\pi \mathcal{E}(U))$$

$$= \varprojlim_{\leftarrow \mathcal{K}} \varinjlim_{\rightarrow \mathcal{U}} Hom_{DG}(\overline{X}_*(\Omega R\widetilde{A}_{(K,N)}), \mathcal{E}(U))$$

$\square$

The next result shows that asymptotic cohomology is a nontrivial theory.

Theorem 6.9:

Analytic and asymptotic homology coincide, i.e. the transformation of Proposition 6.8 defines a natural equivalence

$$HC_\epsilon^*(\mathbb{C}, -) \xrightarrow{\simeq} HC_\alpha^*(\mathbb{C}, -)$$

Corollary 6.10:

$$HC_\alpha^*(\mathbb{C}, \mathbb{C}) \simeq \begin{cases} \mathbb{C} & * = 0 \\ 0 & * = 1 \end{cases}$$

$\square$

Proof of Theorem 6.9:

We divide the demonstration into several steps.

1) Recall that $\widetilde{\mathbb{C}}$ is a deformation retract of the Ind-Fréchet algebra $\mathcal{R}\widetilde{\mathbb{C}}$ via the homomorphisms $\pi : \mathcal{R}\widetilde{\mathbb{C}} \to \widetilde{\mathbb{C}}$, $s : \widetilde{\mathbb{C}} \to \mathcal{R}\widetilde{\mathbb{C}}$ (see 5.17). The argument in (5.17.c)) extends to the differential graded setting and shows that

$$\overline{X}_*(\Omega\mathcal{R}\widetilde{\mathbb{C}}) \xrightarrow{X_*\Omega\pi} \overline{X}_*(\Omega\widetilde{\mathbb{C}})$$

is a deformation retraction with section

$$\overline{X}_*(\Omega\widetilde{\mathbb{C}}) \xrightarrow{X_*\Omega s} \overline{X}_*(\Omega\mathcal{R}\widetilde{\mathbb{C}})$$

Consequently the number operator N acts trivially on the cohomology of $\overline{X}_*(\Omega\widetilde{\mathbb{C}})$ because the Cartan homotopy formula for the action of N on $\overline{X}_*(\Omega\mathcal{R}\widetilde{\mathbb{C}})$ (see 4.3.) carries over to a homotopy formula for the action of N on $\overline{X}_*(\Omega\widetilde{\mathbb{C}})$.

2) According to 1) and Theorem 4.10. the maps

$$X_\alpha^*(\mathbb{C}, A) = Hom_{DG}^*(\overline{X}_*(\Omega\mathcal{R}\widetilde{\mathbb{C}}), \overline{X}_*(\Omega\mathcal{R}\widetilde{A}) \otimes_\pi^{sh} \mathcal{E}(U)) \to$$

$$\to Hom_{DG}^*(\overline{X}_*(\Omega\mathcal{R}\widetilde{\mathbb{C}}), \overline{X}_*(\mathcal{R}\widetilde{A}) \otimes_\pi^{sh} \mathcal{E}(U)) \xrightarrow{-\circ X_*\Omega s}$$

$$\xrightarrow{-\circ X_*\Omega s} Hom_{DG}^*(\overline{X}_*(\Omega\widetilde{\mathbb{C}}), \overline{X}_*(\mathcal{R}\widetilde{A}) \otimes_\pi^{sh} \mathcal{E}(U)) =$$

$$= \varinjlim_{\rightarrow \mathcal{U}} Hom_{DG}^*(\overline{X}_*(\Omega\widetilde{\mathbb{C}}), \overline{X}_*(\mathcal{R}\widetilde{A}) \otimes_\pi^{sh} \mathcal{E}(U))$$

are quasiisomorphisms. Here we may suppose that the open subsets U of $\mathcal{R}_+^\infty$ involved in the inductive limit are convex (Lemma 1.2.).

3) Fix such a punctured, convex neighbourhood U of ∞ and choose a point $x_0 \in U$. Let Y be the vector field on U generating the linear contraction

$$
\begin{aligned}
[0,1] \times U &\to U \\
(t, x) &\to (1-t)x + tx_0
\end{aligned}
$$

Then i_Y is a derivation of degree -1 on $\mathcal{E}(U)$. Theorem 4.11. can now be applied to conclude that the action of $\mathcal{L}_Y$ on the complex $Hom^*_{DG}(\overline{X}_*(\Omega\widetilde{\mathbb{C}}), \overline{X}_*(\mathcal{R}\widetilde{A}) \otimes^{sh}_\pi \mathcal{E}(U))$ is nullhomotopic. No difficulties due to the sheafification arise because the total orbit of a compact neighbourhood of a point in U under the contracting homotopy remains a compact subset of U. By integrating this action we learn that the constant selfmap of U with image x_0 induces a chain map homotopic to the identity on the complex above. Consequently

$$
Hom^*_{DG}(\overline{X}_*(\Omega\widetilde{\mathbb{C}}), \overline{X}_*(\mathcal{R}\widetilde{A}) \otimes^{sh}_\pi \mathcal{E}(U)) \xrightarrow{(id \otimes i^*_{x_0}) \circ -}
$$

$$
\xrightarrow{(id \otimes i^*_{x_0}) \circ -} Hom^*_{DG}(\overline{X}_*(\Omega\widetilde{\mathbb{C}}), \overline{X}_*(\mathcal{R}\widetilde{A}) \otimes^{sh}_\pi \mathcal{E}(x_0)) =
$$

$$
= Hom^*_{DG}(\overline{X}_*(\Omega\widetilde{\mathbb{C}}), \overline{X}_*(\mathcal{R}\widetilde{A})) = Hom^*(\overline{X}_*(\widetilde{\mathbb{C}}), \overline{X}_*(\mathcal{R}\widetilde{A})) = X^\epsilon_*(A)
$$

is a quasiisomorphism.

4) As the inductive limit is an exact functor on the category of directed systems of vector spaces the quasiisomorphisms

$$
X^\epsilon_*(A) \xleftarrow{qis} Hom^*(\overline{X}_*(\widetilde{\mathbb{C}}), \overline{X}_*(\mathcal{R}\widetilde{A})) \xrightarrow{qis} Hom^*_{DG}(\overline{X}_*(\Omega\widetilde{\mathbb{C}}), \overline{X}_*(\mathcal{R}\widetilde{A}) \otimes^{sh}_\pi \mathcal{E}(U))
$$

of 2) yield a quasiisomorphism

$$
X^*_\epsilon(\mathbb{C}, A) \xrightarrow{qis} X^\epsilon_*(A) = \varinjlim_{\to \mathcal{U}} X^\epsilon_*(A) \xleftarrow{qis} \varinjlim_{\to \mathcal{U}} Hom^*(\overline{X}_*(\widetilde{\mathbb{C}}), \overline{X}_*(\mathcal{R}\widetilde{A}))
$$

$$
\xrightarrow{qis} \varinjlim_{\to \mathcal{U}} Hom^*_{DG}(\overline{X}_*(\Omega\widetilde{\mathbb{C}}), \overline{X}_*(\mathcal{R}\widetilde{A}) \otimes^{sh}_\pi \mathcal{E}(U)) \xrightarrow{qis}
$$

$$
\xrightarrow{qis} Hom^*_{DG}(\overline{X}_*(\Omega\mathcal{R}\widetilde{\mathbb{C}}), \overline{X}_*(\Omega\mathcal{R}\widetilde{A}) \otimes^{sh}_\pi \mathcal{E}(\mathcal{U})) = X^*_\alpha(\mathbb{C}, A)
$$

under taking the inductive limit over the ordered set of convex, punctured neighbourhoods of $\infty \in \mathcal{R}^\infty_+$ which coincides with the transformation of Theorem 6.8.

$\square$

Remark:

The contracting homotopies of $\mathcal{E}(U)$ do not fit together for different U (one cannot find a common basepoint for all U) and do not provide thus a global homotopy formula for the whole asymptotic X-complex. Therefore the exactness of $\varinjlim$ is essential in the demonstration above. As the projective limit functor $\varprojlim$ is very far from being exact, the asymptotic cohomology groups can in fact be quite different from the analytic ones.

$\square$

6-3 Functorial properties of the asymptotic X-complex

Now we will see that the asymptotic X-complex achieves the goal set out in the beginning: it extends the known cyclic theories and is functorial under asymptotic morphisms. This may be viewed as the central result of the paper.

Theorem 6.11:

Let A, B be admissible Fréchet algebras. Then any asymptotic morphism

$$f : A \to B$$

defines an even cocycle in the bivariant, asymptotic X-complex of (A, B):

$$f_* \in X^0_\alpha(A, B)$$

Algebraically f_* is given by the DG-map of complexes (see Chapters 2 and 4)

$$\overline{X}_*(\Omega R\widetilde{A}) \xrightarrow{X_*(\Omega Rf)} \overline{X}_*(\Omega R(\widetilde{B} \otimes_\pi C^\infty(\mathcal{R}^\infty_+))) \xrightarrow{X_*(\Omega m)}$$

$$\xrightarrow{X_*(\Omega m)} \overline{X}_*\Omega(R(\widetilde{B}) \otimes_\pi RC^\infty(\mathcal{R}^\infty_+)) \to \overline{X}_*(\Omega(R\widetilde{B})\widehat{\otimes}_\pi\Omega RC^\infty(\mathcal{R}^\infty_+)) \to$$

$$\to \overline{X}_*(\Omega(R\widetilde{B})\widehat{\otimes}_\pi\Omega C^\infty(\mathcal{R}^\infty_+)) \to \overline{X}_*(\Omega(R\widetilde{B})\widehat{\otimes}_\pi\mathcal{E}(\mathcal{R}^\infty_+)) \xrightarrow{\psi}$$

$$\xrightarrow{\psi} \overline{X}_*(\Omega(R\widetilde{B}))\widehat{\otimes}_\pi\mathcal{E}(\mathcal{R}^\infty_+)$$

This correspondence makes the cohomological (bivariant) X-complex into a contravariant (bivariant) functor on the linear, asymptotic category of admissible Fréchet algebras.

Proof:

Let $f : A \to C^\infty(\mathcal{R}^m_+, B) \cap C_b(\mathbb{R}^m_+, B)$ be an asymptotic morphism from A to B. Fix a "small" open neighbourhood $U \subset B$ of zero. Let $(K, N) \in \mathcal{K}(A, U)$ and choose punctured open neighbourhoods $V_{(K,N)}$ of $\infty \in \mathcal{R}^m_+$ for any $(K, N) \in \mathcal{K}(A)$ such that

$$\{ 8N \omega_f(a, a')(x) \,|\, a, a' \in K^\infty, x \in V_{(K,N)} \} \subset U$$

We claim that

$$X_*(\Omega Rf) \in Hom^0_{DG}(\overline{X}_*(\Omega R\widetilde{A}_{(K,N)}), \overline{X}_*(\Omega R\widetilde{B}) \otimes^{sh}_\pi \mathcal{E}(V_{(K,N)}))$$

Let $W \subset V_{(K,N)}$ be a relatively compact, open subset. Then f provides a map

$$\widetilde{A} \to C^\infty(\mathbb{R}^m_+, \widetilde{B}) \simeq C^\infty(\mathbb{R}^m_+) \otimes_\pi \widetilde{B} \to C^\infty_b(W) \otimes_\pi \widetilde{B}$$

where $C^\infty_b(W)$ is the admissible Fréchet algebra of smooth functions with bounded derivatives on W. Moreover

$$8N\omega_f(K^\infty \times K^\infty) \subset \{ h \in C^\infty_b(W) \otimes_\pi \widetilde{B} \,|\, h(x) \in U, \forall x \in W \}$$

which happens to be a "small" open set in $C_b^\infty(W) \otimes_\pi \tilde{B}$. Proposition 5.12 implies then that f induces a continuous morphism

$$R\tilde{A}_{(K,N)} \xrightarrow{Rf} R(C_b^\infty(W) \otimes_\pi \tilde{B})$$

By Proposition 8.10. the canonical morphism of topological I-adic completions

$$\Omega R(C_b^\infty(W) \otimes_\pi \tilde{B}) \to \Omega R\tilde{B}\hat{\otimes}_\pi \Omega R C_b^\infty(W) \to \Omega R\tilde{B}\hat{\otimes}_\pi \Omega C_b^\infty(W) \to \Omega R\tilde{B}\hat{\otimes}_\pi \mathcal{E}(W)$$

is bounded, too. Consequently the linear maps

$$\overline{X}_*(\Omega R\tilde{A}_{(K,N)}) \to \overline{X}_*(\Omega R\tilde{B}\hat{\otimes}_\pi \mathcal{E}(W)) \xrightarrow{\psi} \overline{X}_*(\Omega R\tilde{B}) \otimes_\pi \mathcal{E}(W)$$

are continuous and natural so that they provide an element

$$X_*(\Omega Rf) \in \varprojlim_{\leftarrow W} Hom_{DG}^0(\overline{X}_*(\Omega R\tilde{A}_{(K,N)}), \overline{X}_*(\Omega R\tilde{B}) \otimes_\pi \mathcal{E}(W))$$

$$= Hom_{DG}^0(\overline{X}_*(\Omega R\tilde{A}_{(K,N)}), \overline{X}_*(\Omega R\tilde{B}) \otimes_\pi^{sh} \mathcal{E}(V_{(K,N)}))$$

which yields the claim. The naturality and compatibility of the construction with the composition product is obvious.

$\square$

6-4 Homotopy properties of the asymptotic X-complex

Theorem 6.12:

Let

$$\delta : A \to A$$

be a bounded derivation on the admissible Fréchet algebra A. Then the Cartan homotopy operators of Theorem 4.10 carry over to the asymptotic setting and provide operators

$$\mathcal{L}_\delta \in X_\alpha^0(A, A) \quad h_\delta \in X_\alpha^1(A, A)$$

that satisfy

$$\partial h_\delta = \mathcal{L}_\delta$$

and consequently the Cartan homotopy formula

$$\mathcal{L}_\delta = \partial_{X_*} \circ h_\delta + h_\delta \circ \partial_{X_*}$$

holds.

$\square$

Corollary 6.13:

Bounded derivations act trivially on (bivariant) asymptotic cohomology.

$\square$

Proof:

In light of Theorems 4.10, 5.21 it will be sufficient to show the continuity of

$$X_*(j) : X_*(\Omega \mathcal{R} A) \to X_*(\mathcal{R}(\Omega A))$$

$$X_*(k) : X_*(\mathcal{R}(\Omega A)) \to X_*(\Omega \mathcal{R} A)$$

So the theorem follows from the

Lemma 6.14:

The morphisms

$$j : \Omega R A_{(K,N)} \to R(\Omega A)_{(K,N)} \quad k : R(\Omega A)_{(K,N)} \to \Omega R A_{(K,N)}$$

are continuous. They give therefore rise to morphisms

$$j : \Omega \mathcal{R} A \to \mathcal{R}(\Omega A) \quad k : \mathcal{R}(\Omega A) \to \Omega \mathcal{R} A$$

of differential graded topologcal I-adic completions.

Proof:

a) Let $\varrho \omega^{n_0}, \ldots, \varrho \omega^{n_k} \in RA_K$. Then

$$j(\varrho \omega^{n_0} \partial(\varrho \omega^{n_1}) \ldots \partial(\varrho \omega^{n_k})) = \sum_{\leq (2n_1+1)\ldots(2n_k+1) \, terms} \varrho' \omega'^{n_0} \varrho' \omega'^{n_1} \ldots \varrho' \omega'^{n_k}$$

with entries in $K^\infty \cup \partial(K^\infty)$. Therefore

$$\| j(\varrho \omega^{n_0} \partial(\varrho \omega^{n_1}) \ldots \partial(\varrho \omega^{n_k})) \|_{N,m}^K$$

$$\leq \sum_{\leq (2n_1+1)\ldots(2n_k+1) \, terms} \| \varrho' \omega'^{n_0} \varrho' \omega'^{n_1} \ldots \varrho' \omega'^{n_k} \|_{N,m}^K$$

$$\leq \sum_{\leq (2n_1+1)\ldots(2n_k+1) \, terms} (2^{m+1})^k \| \varrho' \omega'^{n_0} \|_{N,m+1}^K \cdots \| \varrho' \omega'^{n_k} \|_{N,m+1}^K$$

$$\leq (2n_1 + 1) \ldots (2n_k + 1)(2^{m+1})^k (1 + n_0)^{m+1} N^{-n_0} \ldots (1 + n_k)^{m+1} N^{-n_k}$$

$$\leq (2^{m+2})^k (1 + n_0)^{m+2} N^{-n_0} \ldots (1 + n_k)^{m+2} N^{-n_k}$$

so that the estimate

$$\| j(\varrho \omega^{n_0} \partial(\varrho \omega^{n_1}) \ldots \partial(\varrho \omega^{n_k})) \|_{N,m}^K \leq (2^{m+2})^k \| a^0 \partial a^1 \ldots \partial a^k \|_{N,m+2}^K$$

follows for $a^0, \ldots, a^k \in RA_K$ which proves the continuity of
$j : \Omega^k RA_{(K,N)} \to R(\Omega A)_{(K,N)}^{(k)}$

b) Let $a^0, \ldots, a^j, b^0, \ldots, b^{j'} \in K^\infty$. Then

$$k(\omega(a^0 \partial a^1 \ldots \partial a^j, b^0 \partial b^1 \ldots \partial b^{j'}))$$

$$= \varrho(a^0)\partial\varrho(a^1)\ldots\partial\varrho(a^{j-1})\partial\omega(a^j, b^0)\partial\varrho(b^1)\ldots\partial\varrho(b^{j'}) +$$

$$+ \sum_1^{j-1}(-1)^{i+j}\varrho(a^0)\partial\varrho(a^1)\ldots\partial\omega(a^i, a^{i+1})\ldots\partial\varrho(a^j)\partial\varrho(b^0)\ldots\partial\varrho(b^{j'}) +$$

$$+ (-1)^j\omega(a^0, a^1)\partial\varrho(a^2)\ldots\partial\varrho(a^j)\partial\varrho(b^0)\ldots\partial\varrho(b^{j'})$$

as an easy calculation shows. Therefore

$$\| k(\omega(a^0\partial a^1 \ldots \partial a^j, b^0\partial b^1 \ldots \partial b^{j'})) \|_{N,m}^K \le (j+1)\, 2^m\, N^{-1}$$

To proceed further note that the inequality

$$\| ab \|_{N,m}^K \le 2^{m+1} \| a \|_{N,m+1}^K \| b \|_{N,m+1}^K$$

on $RA_{(K,N)}$ implies

$$\| (c^0\partial c^1 \ldots \partial c^i)(c'^0\partial c'^1 \ldots \partial c'^j) \|_{N,m}^K \le$$

$$C(i,m) \| c^0\partial c^1 \ldots \partial c^i \|_{N,m+1}^K \| c'^0\partial c'^1 \ldots \partial c'^j \|_{N,m+1}^K$$

on $\Omega RA_{(K,N)}$.

Let now

$$\widetilde{\varrho\omega}^n = \varrho\omega^{n_1}\omega'\omega^{n_2}\omega' \ldots \omega'\omega^{n_l} \in R(\Omega A_K)^{(k)}$$

where we suppose that $\omega^{n_i} \in RA_K$ and at least one entry of the terms ω' has strictly positive degree. Clearly $l \le k + 1$. Then

$$\| k(\widetilde{\varrho\omega}^n) \|_{N,m}^K = \| k(\varrho\omega^{n_1}\omega'\omega^{n_2}\omega' \ldots \omega'\omega^{n_l}) \|_{N,m}^K$$

$$\le C'(k,m) \| \varrho\omega^{n_1} \|_{N,m+2k}^K \prod_2^l \| k(\omega') \|_{N,m+2k}^K \| \varrho\omega^{n_i} \|_{N,m+2k}^K$$

$$\le C'(k,m)(1+n_1)^{m+2k}N^{-n_1} \prod_2^l (k+1)2^{m+2k}(1+n_i)^{m+2k}N^{-n_i}$$

$$\le C''(k,m) \prod_1^l (1+n_i)^{m+2k}N^{-n_i}$$

$$\le C''(k,m)(1+n)^{(k+1)(m+2k)}N^{-(n-k)}$$

$$\le C'''(k,N,m)(1+n)^{(k+1)(m+2k)}N^{-n}$$

so that

$$\| k(a) \|_{N,m+2k}^K \le C'''(k,N,m) \| a \|_{N,(k+1)(m+2k)}^K$$

for $a \in R(\Omega A)_K^{(k)}$

$\square$

As a consequence of the Cartan homotopy formula, one obtains the homotopy invariance of asymptotic cohomology groups.

Theorem 6.15:

Asymptotic (bivariant) cyclic cohomology is a bifunctor on the linear, asymptotic homotopy category. In other words, if

$$f, g : A \to B$$

'are continuously homotopic, smooth, asymptotic morphisms between admissible Fréchet algebras A, B, then

$$[f_*] = [g_*] \in HC^0_\alpha(A, B)$$

$\square$

This result is in sharp contrast to what is known for other cyclic theories because in the asymptotic case even continuous and not only smooth homotopies are allowed.

Proof:

If f and g are homotopic there is a factorization

$$A \xrightarrow{F} B[0,1] \xrightarrow[i_1]{i_0} B$$

where F is a smooth asymptotic morphism such that $f = i_0 \circ F$, $g = i_1 \circ F$ and $i_{0,1}$ is given by the evaluation at the endpoints of the unit interval. Here $B[0,1]$ is admissible if B is. So it has to be shown that

$$[i_{0*}] = [i_{1*}] \in HC^0_\alpha(B[0,1], B)$$

The inclusion $\mathcal{C}^\infty([0,1], B) \hookrightarrow B[0,1]$ being an asymptotic HC-equivalence by the derivation lemma (see 7.7.) it suffices to prove

$$[i_{0*}] = [i_{1*}] \in HC^0_\alpha(\mathcal{C}^\infty([0,1], B), B)$$

From now on the arguments are valid for the analytic X-complex as well and provide a demonstration of Theorem 5.21.

The time derivative $\frac{\partial}{\partial t}$ acts as bounded derivation on $\mathcal{C}^\infty([0,1], B)$ and yields an action $\mathcal{L}_{\frac{\partial}{\partial t}}$ on $X_*(\mathcal{R}\mathcal{C}^\infty([0,1], B))$. The evaluation maps define a homomorphism

$$eval : R(\mathcal{C}^\infty([0,1], B)) \to \mathcal{C}^\infty([0,1], RB)$$

which extends to a continuous homomorphism

$$eval : \mathcal{R}(\mathcal{C}^\infty([0,1], B)) \to \mathcal{R}B \otimes_\pi \mathcal{C}^\infty([0,1])$$

as the latter map can be written as composition

$$\mathcal{R}(\mathcal{C}^\infty([0,1], B)) = \mathcal{R}(\mathcal{C}^\infty([0,1]) \otimes_\pi B) \xrightarrow{m}$$

$$\xrightarrow{m} \mathcal{R}B \otimes_\pi \mathcal{R}\mathcal{C}^\infty([0,1]) \xrightarrow{id\otimes\pi} \mathcal{R}B \otimes_\pi \mathcal{C}^\infty([0,1])$$

which is continuous by Proposition 8.10.

We obtain thus a commutative diagram

$$
\begin{array}{ccc}
\mathcal{R}(\mathcal{C}^\infty([0,1],B)) & \xrightarrow{\;\mathcal{L}_{\frac{\partial}{\partial t}}\;} & \mathcal{R}(\mathcal{C}^\infty([0,1],B)) \\
{\scriptstyle eval}\downarrow & & \downarrow{\scriptstyle eval} \\
\mathcal{R}B \otimes_\pi \mathcal{C}^\infty([0,1]) & \xrightarrow[\;\mathcal{L}_{id\otimes\frac{\partial}{\partial t}}\;]{} & \mathcal{R}B \otimes_\pi \mathcal{C}^\infty([0,1])
\end{array}
$$

Recall that the action of $\mathcal{L}_{\frac{\partial}{\partial t}}$ on $X_*(\mathcal{R}(\mathcal{C}^\infty([0,1],B)))$ satisfies a Cartan homotopy formula

$$\mathcal{L}_{\frac{\partial}{\partial t}} = h\partial_{X_*} + \partial_{X_*}h$$

for some operator $h \in X_e^1(\mathcal{C}^\infty([0,1],B),\mathcal{C}^\infty([0,1],B))$

Define now $\chi \in X_e^1(\mathcal{C}^\infty([0,1],B),B)$ as the composition

$$\chi : X_*(\mathcal{R}(\mathcal{C}^\infty([0,1],B))) \xrightarrow{h} X_{*-1}(\mathcal{R}(\mathcal{C}^\infty([0,1],B))) \xrightarrow{X_*eval}$$

$$\xrightarrow{X_*eval} X_{*-1}(\mathcal{R}B \otimes_\pi \mathcal{C}^\infty([0,1])) \xrightarrow{\psi} X_{*-1}(\mathcal{R}B) \otimes_\pi \mathcal{C}^\infty([0,1]) \rightarrow$$

$$\xrightarrow{id\otimes\int_0^1 -dt} X_{*-1}(\mathcal{R}B)$$

(For the definition of ψ see 2.7.) Then

$$\partial\chi = \chi\circ\partial_{X_*} + \partial_{X_*}\circ\chi = \int_0^1 (\psi\circ eval\circ(h\circ\partial + \partial\circ h))dt$$

$$= \int_0^1 \psi\circ eval\circ\mathcal{L}_{\frac{\partial}{\partial t}}dt = \int_0^1 \psi\circ\mathcal{L}_{id\otimes\frac{\partial}{\partial t}}\circ eval\,dt$$

$$= \int_0^1 \frac{\partial}{\partial t}(\psi\circ eval)\,dt = (\psi\circ eval)_1 - (\psi\circ eval)_0 = i_{1*} - i_{0*}$$

The asymptotic case is similar.

$\square$

Corollary 6.16:

Let A,B,C,D be admissible Fréchet algebras. Then

$$HC_\alpha^*(A,C) \oplus HC_\alpha^*(B,C) \xrightarrow{\cong} HC_\alpha^*(A\oplus B,C)$$

$$HC_\alpha^*(D,A\oplus B) \xrightarrow{\cong} HC_\alpha^*(D,A) \oplus HC_\alpha^*(D,B)$$

Proof:

This follows from Proposition 5.22, the Cartan homotopy formula 6.12 and the fact that if $U \subset \mathcal{R}_+^n$, $V \subset \mathcal{R}_+^m$ are neighbourhoods of ∞ then so is $U \times V \subset \mathcal{R}_+^{n+m}$ (1.1).

$\square$

6-5 Pairing of asymptotic cohomology with K-theory

Definition and Proposition 6.17:(Chern Character) [CO$_2$],[CQ]

Let A be a unital admissible Fréchet algebra.
The **Chern Character**

$$ch : K_*(-) \to HC_\alpha^*(\mathbb{C}, -) = HC_*^\alpha(-)$$

is defined by

$$ch_0 : \quad K_0(A) = [\mathbb{C}[e], M_\infty(A)]_\alpha \quad \to \quad HC_\alpha^0(\mathbb{C}, M_\infty(A)) \xrightarrow{\times Tr} HC_\alpha^0(\mathbb{C}, A)$$

$$[f] \qquad \to \qquad [f_*(ch(e))]$$

$$ch_1 : \quad K_1(A) = [\mathbb{C}[u, u^{-1}], M_\infty(A)]_\alpha \quad \to \quad HC_\alpha^1(\mathbb{C}, M_\infty(A)) \xrightarrow{\times Tr} HC_\alpha^1(\mathbb{C}, A)$$

$$[f] \qquad \to \qquad [f_*(ch(u))]$$

The Chern character is a natural transformation of asymptotic homotopy functors. This means that if $f : A \to B$ is an asymptotic morphism, the diagram

$$
\begin{array}{ccc}
K_* A & \xrightarrow{\;\;f_*\;\;} & K_* B \\
\downarrow{ch} & & \downarrow{ch} \\
HC_\alpha^*(\mathbb{C}, A) & \xrightarrow[f_*]{} & HC_\alpha^*(\mathbb{C}, A)
\end{array}
$$

commutes and depends only on the homotopy class of f.

Proof:

The proof is similar to that of Proposition 5.29.

$\square$

Corollary 6.18:

There exists a natural pairing

$$\langle -, - \rangle : K_* \otimes HC^*_\alpha \xrightarrow{ch \otimes id} HC^\alpha_* \otimes HC^*_\alpha \to \mathbb{C}$$

of functors on the linear asymptotic homotopy category. Consequently,

for $x \in K_*(A)$, $\varphi \in HC^*_\alpha(B)$ and any asymptotic morphism $f : A \to B$, the equality

$$\langle ch(f_*(x)), \varphi \rangle = \langle ch(x), f^*(\varphi) \rangle$$

holds. Moreover the two quantities in the equality depend only on the homotopy class of f.

$\square$

The value of the pairing can be calculated by an explicit formula. In fact not a single but a whole class of formulas will be derived. This additional flexibility will turn out to be useful in later chapters, for example when we check the compatibility of the Chern character with products.

Theorem 6.19:[CO$_2$],[CQ] (Explicit formula for the Chern character)

Let A be a unital admissible Fréchet algebra.

a) Let $e = e^2$ be an idempotent in A. Suppose that

$$a \in \lim_{\to \mathcal{K}} RA_{(K,N)}$$

lifts e:

$$\pi(a) = e$$

Then for any $\varphi \in Z^0_\alpha(A)$

$$\langle ch([e]), [\varphi] \rangle = \varphi(a) + \sum_{k=1}^{\infty} \binom{2k}{k} \varphi((a - \frac{1}{2})(a - a^2)^k)$$

where the sum converges to a constant function on any sufficiently small neighbourhood of $\infty \in \mathcal{R}^m_+$.

b) Let $u : uu^{-1} = u^{-1}u = 1$ be invertible in A. Suppose that

$$v, w \in \lim_{\to \mathcal{K}} RA_{(K,N)}$$

lift u, u^{-1}:

$$\pi(v) = u, \quad \pi(w) = u^{-1}$$

Then for any $\psi \in Z^1_\alpha(A)$

$$\langle ch([u]), [\psi] \rangle = \sum_{k=0}^{\infty} \psi(w(1 - vw)^k \, dv)$$

where the sum converges to a constant function on any sufficiently small neighbourhood of $\infty \in \mathcal{R}_+^m$.

c) The choices $a = \varrho(e)$, $v = \varrho(u)$ $w = \varrho(u^{-1})$ yield the well known formulas

$$\langle ch([e]), [\varphi] \rangle = \varphi_0(e) + \sum_{k=1}^{\infty} \binom{2k}{k} \varphi_{2k}(e - \frac{1}{2}, e, \ldots, e)$$

$$\langle ch([u]), [\psi] \rangle = \sum_{k=0}^{\infty} \psi_{2k+1}(u^{-1}, u, \ldots, u^{-1}, u)$$

where the sum converges to a constant function on any neighbourhood U of $\infty \in \mathcal{R}_+^m$ such that

$$\varphi \in Hom_{DG}^0(\overline{X}_*(\Omega R\widetilde{A}_{(\{1,e\},4+\epsilon)}), \mathcal{E}(U))$$

$$\psi \in Hom_{DG}^1(\overline{X}_*(\Omega R\widetilde{A}_{(\{u,u^{-1}\},1+\epsilon)}), \mathcal{E}(U))$$

Remark:

For a given "parameter" $t \in \mathcal{R}_+^{\infty}$, the pairing with a fixed asymptotic cocycle will in general not be defined on the whole K-group of A but only for finitely many classes. It will even not be defined for all idempotent (invertible) matrices representing a given class in K-theory. (Consider for example the asymptotic cocycle (φ^t) on $\mathcal{K}(\mathcal{H})$ constructed in the proof of 7.10: For any given $t \in \mathbb{R}_+ - \mathbb{N}$ one can find easily rank one projections in $\mathcal{K}(\mathcal{H})$ for which the sum defining the coupling with the cocycle diverges.) However the domains of definition of (φ^t) grow as the asymptotic parameter t approaches infinity so that the desired coupling is eventually obtained "at infinity". This explains why the argument in the introduction showing the triviality of the pairing between K-theory and the classical cyclic theories for stable C^*-algebras does not apply to the pairing with asymptotic cohomology. Indeed we will exhibit later a class of C^*-algebras for which this pairing is nondegenerate.

$\square$

Proof:

c): The formulas for the Chern character of an idempotent and an invertible element are immediate from the definitions. The assertions about the domain of convergence follow from the fact that

$$ch(e) \in \overline{X}_0(R\widetilde{\mathbb{C}}_{(\{1,e\},4+\epsilon)}) \quad \text{and} \quad ch(u) \in \overline{X}_1(R\widetilde{\mathbb{C}[u, u^{-1}]}_{(\{u,u^{-1}\},1+\epsilon)})$$

(see Lemma 5.17). The functions $\langle ch(e), \varphi \rangle$, $\langle ch(u), \psi \rangle$ are constant on the claimed neighbourhoods of infinity by 2.9, 2.10.

We are now going to show a), the demonstration of b) being similar. Let $a \in RA_{(K,N)}$ lift $e \in A$ (suppose that $\lambda e \in K$ for some $\lambda > 0$). Then

$$s \to a(s) := (1-s)\,\varrho(e) + s\,a \in RA_{(K,N)}, \; s \in [0,1]$$

is a family of elements lifting $e \in A$ and

$$\pi(a(s) - a(s)^2) = e - e^2 = 0$$

so that $a(s) - a(s)^2 \in \widehat{IA} \cap RA_{(K,N)} \; \forall s \in [0,1]$. Recall that if an element a of an admissible Fréchet algebra is close to an idempotent, i.e. if $4(a - a^2)$ is "small", then a true idempotent can be obtained from a by applying functional calculus with

$$F(x) := x + \sum_{k=1}^{\infty} \binom{2k}{k}\left(x - \frac{1}{2}\right)(x - x^2)^k = x + \left(x - \frac{1}{2}\right)G(x - x^2) \; .$$

(see Lemma 1.20). The power series $G(y)$ has radius of convergence $R = \frac{1}{4}$. Lemma 5.10 allows to apply functional calculus also to the family $a(s)$:

For any M satisfying

$$M > \sup_s 2N \, exp(8 \parallel a(s) - a(s)^2 \parallel_{N,0}^K)$$

$$M > \sup_s 2N \, exp(8 \parallel \frac{\partial}{\partial s}(a(s) - a(s)^2) \parallel_{N,0}^K)$$

one has

$$G(a(s) - a(s)^2) \in RA_{(K,M)} \quad \tfrac{\partial}{\partial s} G(a(s) - a(s)^2) \in RA_{(K,M)}$$

and consequently $F(a(s)) \in RA_{(K,M)}$ is a continuously differentiable one parameter family of idempotents in $RA_{(K,M)}$.

Let now $U \subset \mathcal{R}_+^m$ be a punctured neighbourhood of ∞ such that the asymptotic cocycle φ provides a map

$$\varphi \in Hom_{DG}^0(\overline{X}_*(\Omega R\widetilde{A}_{(K,8M)}), \mathcal{E}(U))$$

Then, for given s, $\langle F(a(s)), \varphi \rangle$ is a constant function on U: for any vector field Y on U

$$Y\langle F(a(s)), \varphi \rangle = \langle F(a(s)), \mathcal{L}_Y \varphi \rangle = \langle F(a(s)), (\partial h_Y + h_Y \partial)\varphi \rangle$$

by the second Cartan homotopy formula (4.11, 5.21)

$$= \langle F(a(s)), \partial(h_Y \varphi) \rangle = \langle \partial F(a(s)), h_Y \varphi \rangle = \langle 0, h_Y \varphi \rangle = 0$$

as $\partial F(a(s))$ is a sum of commutators (see the following argument or 2.9, 2.10).

The value of the pairing is also independent of $s \in [0, 1]$ as the equality

$$\dot{F}(a(s)) = [\dot{F}(a(s))F(a(s)), F(a(s))] + [F(a(s)), F(a(s))\dot{F}(a(s))]$$

and the fact that φ is a trace on $RA_{(K,M)}$ show. Evaluation at $s = 0, 1$ gives finally the equation

$$\langle [ch(e)], [\varphi] \rangle = \varphi(a) + \sum_{k=1}^{\infty} \binom{2k}{k} \varphi((a - \frac{1}{2})(a - a^2)^k)$$

The independence of the value of the pairing from the choice of the idempotent (invertible) matrix over A has already been shown in 5.29, respectively follows from the Cartan homotopy formula applied to the class of e, u in $[\mathbb{C}[e], M_\infty A]_\alpha$, $[\mathbb{C}[u, u^{-1}], M_\infty A]_\alpha$.

$\square$

Chapter 7: Asymptotic cyclic cohomology of dense subalgebras

The phenomenon we are going to consider in this chapter lies at the heart of asymptotic cyclic cohomology and accounts for most of the properties that distinguish asymptotic theory from the cyclic theories known so far. It concerns the comparison between the cohomology of a fixed algebra and of dense subalgebras of "smooth" elements: if there is a regularization procedure to approximate any element in the full algebra by "smooth" elements of the dense subalgebra, then the two algebras are asymptotically cohomology equivalent.

7-1 The Derivation Lemma

Theorem 7.1: (Derivation Lemma)

Let $i : \mathcal{A} \hookrightarrow A$ be an inclusion of admissible Fréchet algebras with dense image. Suppose that the following two conditions are satisfied:

1) There exists a neighbourhood U of 0 in A such that $i^{-1}U$ is "small" in $\mathcal{A}$.

2) There exists a smooth family

$$f_t : \mathbb{R}_+ \to \mathcal{L}(A, \mathcal{A})$$

of bounded linear maps such that

$$\lim_{t \to \infty} i \circ f_t = Id_A \qquad \lim_{t \to \infty} f_t \circ i = Id_\mathcal{A}$$

pointwise on A resp. $\mathcal{A}$.

Then the inclusion i induces an asymptotic HC-equivalence:

$$[i_*] \in HC_\alpha^0(\mathcal{A}, A)$$

$\square$

Let us illustrate the first condition above by noting the following implication

Lemma 7.2:

Suppose that condition 7.1.1) holds for the inclusion $\mathcal{A} \subset A$ of admissible Fréchet algebras with dense image. Then the subalgebra $\mathcal{A}$ is closed under holomorphic functional calculus in A. The same is true for the inclusions $M_n(\mathcal{A}) \subset M_n(A)$ $n > 0$.

Proof:

It is clear by Lemma 1.15 that condition 7.1.1) holds for $\mathcal{A} \subset A$ iff it holds for the inclusion $\tilde{\mathcal{A}} \subset \tilde{A}$ obtained by adjoining units. So one may suppose that $\mathcal{A} \subset A$ is a unital inclusion. It suffices to prove that the spectra of $x \in \mathcal{A}$ in the algebras $\mathcal{A}$ and A coincide. For this one has to verify that x is invertible in $\mathcal{A}$ iff it is invertible in A. So let $x^{-1} \in A$ be an inverse of x. As $\mathcal{A}$ is dense in A one can find y, $y' \in \mathcal{A}$ close to x^{-1} in A such that $xy \in 1 + i^{-1}(U)$, $y'x \in 1 + i^{-1}(U)$. The demonstration of 1.16.1) however shows that $1 + i^{-1}(U) \subset \mathcal{A}$ consists of elements invertible in $\mathcal{A}$. The conclusion follows. For the last assertion note that condition 7.1.1) holds for the inclusion $M_n(\mathcal{A}) \subset M_n(A)$ $n > 0$, too, provided it holds for $\mathcal{A} \subset A$. $\qquad \square$

Proof of Theorem 7.1:

By Lemma 1.15 the inclusion $\tilde{\mathcal{A}} \subset \tilde{A}$ obtained by adjoining units also satisfies the conditions of the theorem. The family of regularizations

$$f : \tilde{A} \to C^\infty(\mathbb{R}_+, \tilde{\mathcal{A}})$$

does not define an asymptotic morphism in general but the induced element

$$f_* \in Hom_{DG}(\overline{X}_*(\Omega R\tilde{A}), \overline{X}_*(\Omega R\tilde{\mathcal{A}}) \otimes_\pi C^\infty(\mathbb{R}_+))$$

belongs nevertheless to the asymptotic X-complex:

$$f_* \in X_\alpha^0(A, \mathcal{A})$$

The class $[f_*] \in HC_\alpha^0(A, \mathcal{A})$ is an asymptotic HC-inverse to $[i_*]$ because $f_t \circ i$ and $i \circ f_t$ are asymptotic morphisms (the families $f_t \circ i$ and $i \circ f_t$ of continuous linear maps are bounded by the theorem of Banach-Steinhaus) smoothly homotopic to the identity and thus

$$[f_*] \circ [i_*] = [(f \circ i)_*] = [id_*^{\mathcal{A}}] \in HC_\alpha^0(\mathcal{A}, \mathcal{A})$$

$$[i_*] \circ [f_*] = [(i \circ f)_*] = [id_*^A] \in HC_\alpha^0(A, A)$$

In fact, the assertion follows from the

Lemma 7.3 :

Let $\mathcal{U}$ be the ordered set of punctured neighbourhoods of ∞ in $\mathbb{R}_+ \cup \{\infty\}$. Then under the conditions of Theorem 7.1.1

$$Rf \in \lim_{\leftarrow \mathcal{K}} \lim_{\to \mathcal{U}} Hom(R\tilde{A}_{(K,N)}, R\tilde{\mathcal{A}} \otimes_\pi^{sh} C^\infty(U))$$

Proof:

Let $(K, N) \in \mathcal{K}(\tilde{A})$ and choose $U =]t_0, \infty[\subset \mathbb{R}_+$ ·such that

$$\{ 8N\omega_{(i \circ f_t)}(a, a') \mid a, a' \in K^\infty, t \in U \}$$

is contained in a ball W in $\tilde{A}$ satisfying the hypothesis 1) of Theorem 7.1.

This is possible because the family $i \circ f_t$ is bounded by the theorem of Banach-Steinhaus so that its curvature decays uniformly on compact sets. (Lemma 1.6)

Consequently

$$8N\omega_{f_t}(a, a') \in i^{-1}(W) \ \forall a, a' \in K^\infty, t \in U$$

which happens to be a "small" ball in $\mathcal{A}$.

As in the proof of Theorem 6.11. one obtains then that

$$Rf : R\tilde{A}_{(K,N)} \to \mathcal{R}\tilde{A} \otimes_\pi^{sh} \mathcal{C}^\infty(U)$$

is continuous. The claimed result follows now from the naturality of the construction.

□

The name "Derivation Lemma" stems from the following observation

Lemma 7.4:

Let A be a Fréchet algebra and let $\{\delta_i, i \in I\}$ be an at most countable set of unbounded derivations on A. Suppose that there is a common dense domain $\mathcal{A}$ of all compositions $\prod_j \delta_{i_j}$. Then every at most countable set of graph seminorms

$$\| a \|_{k,f,m} := \sum_{J \subset \{1,\dots,k\}} \| (\prod_{j \in J} \delta_{i_{f(j)}})a \|_m$$

defines the structure of a Fréchet algebra on $\mathcal{A}$, where $\| - \|_m$ ranges over a set of seminorms defining the topology of A, J runs over the ordered subsets of $\{1, \dots, k\}$ and f is a map from the finite set $\{1, \dots, k\}$ to the index set I.

If A happens to be admissible, then the inclusion $\mathcal{A} \hookrightarrow A$ satisfies condition 7.1.1). Especially $\mathcal{A}$ is admissible, too.

Proof:

We treat for simplicity the case k=1, the reasoning in the general case being similar. Therefore the topology on $\mathcal{A}$ is defined by the seminorms

$$\| a \|'_m := \| \partial a \|_m + \| a \|_m \quad m \in \mathbb{N}$$

Let $U \subset A$ be "small". We claim that $U' := i^{-1}(U)$ will be "small" in $\mathcal{A}$.

Let $K \subset U'$ be compact and choose $\lambda > 1$ such that $\lambda K \subset U'$ which is possible by the compactness of K. One finds for $a_j \in K$

$$\| \prod_1^n a_j \|'_m = \| \sum_{i=1}^n a^1 \ldots \partial(a_i) \ldots a_n \|_m + \| \prod_1^n a_j \|_m$$

$$\leq \sum_{i=1}^n \lambda^{1-n} \| (\lambda a_1) \ldots \partial(a_i) \ldots (\lambda a_n) \|_m + \lambda^{-n} \| \prod_{j=1}^n (\lambda a_j) \|_m$$

By hypothesis $i(\lambda K) \subset U$ has relatively compact multiplicative closure in A. Moreover $\partial(K) \subset A$ is compact. An estimation of the sum above yields therefore

$$\| \prod_{j=1}^n a_j \|'_m \leq (\lambda n C_0 + C_1) \lambda^{-n}$$

If one treats the case $k > 1$ one sees that the number of summands after differentiating the product k times equals n^k which is of subexponential growth in n so that the assertion holds then as well.

□

7-2 Applications

Theorem 7.5:

Let $\alpha : \mathbb{R} \to Aut(A)$ be a continuous (in the topology of pointwise convergence) one-parameter group of automorphisms of the admissible Fréchet-algebra A. Let

$$\mathcal{A}^\infty := \bigcap_{k=0}^\infty \mathcal{A}^k$$

be the subalgebra of smooth elements under α where $\mathcal{A}^k$ denotes the algebra of k-fold continuously differentiable elements with respect to α.

Then the inclusions

$$\mathcal{A}^\infty \hookrightarrow \mathcal{A}^k \hookrightarrow A$$

induce asymptotic HC-equivalences for any $k \in \mathbb{N}$.

Theorem 7.6:

Let M be a smooth, compact manifold without boundary and A an admissible Fréchet algebra.

Then the inclusions of admissible Fréchet algebras

$$C^\infty(M, A) \hookrightarrow C^k(M, A) \hookrightarrow C^0(M, A)$$

induce asymptotic HC-equivalences.

Proof:

Apply the derivation lemma where you choose the following regularization maps:

For Theorem 7.5 take the convolution with a smooth family of smooth, positive functions ν_t on $\mathbb{R}$ with support near 0 which approximate the δ-distribution at 0:

$$f_t: \quad a \to \int_{-\infty}^{+\infty} \nu_t(s)\, \alpha_s(a)\, ds$$

For Theorem 7.6 take convolution with a smooth family of smooth, positive functions on $M \times M$ which approximate the δ-distribution along the diagonal $\Delta \subset M \times M$

$\qquad\qquad\qquad\qquad\qquad\qquad\qquad\qquad\qquad\qquad\qquad\qquad$ $\square$

Proposition 7.7:

Let $(M, \partial M)$ be a compact manifold with boundary and let $\widetilde{M} := M \cup_{\partial M} \partial M \times [0, \infty[$ be the noncompact manifold obtained by adding a collar. Let $p : \widetilde{M} \to M$ be the map which equals the identity on M and projects the collar to the boundary. Define

$$C^\infty(M, \partial M) := \{f \in C^\infty(\widetilde{M}) \mid f \text{ is locally constant outside } M\}$$

$$C^\infty(M) := C^\infty(\widetilde{M})/(g,\ g|_M = 0)$$

Then for any admissible Fréchet algebra A the natural inclusions

$$C^\infty(M, \partial M, A) \hookrightarrow C^\infty(M, A) \hookrightarrow C(M, A)$$

are asymptotic HC-equivalences.

Proof:

Analogous to 7.6. it is easily shown that the inclusions

$$\mathcal{C}^\infty(M, \partial M, A) \hookrightarrow C(M, A) \hookleftarrow \mathcal{C}^\infty(M, A)$$

satisfy condition 7.1.1) and that all three algebras are admissible if A is. To obtain the needed regularization maps identify a neighbourhood of $\partial M \subset M$ with $M \times]-\infty, 0]$ so that a neighbourhood of the collar $M \times [0, \infty[$ can be identified with $M \times \mathbb{R}$.

Let $\nu_t \in \mathcal{C}_c^\infty(]-\frac{1}{t}, \frac{1}{t}[)$, $t \in \mathbb{R}_+$ be a smooth family of smooth kernels with support concentrated near 0 and approaching the δ-distribution at 0.

Let furthermore Φ_t, $t \geq 1$ be a smooth family of diffeomorphisms of $\mathbb{R}$ with compact support that equal the translation $L_{\frac{1}{t}} : x \to x + \frac{1}{t}$ on a large compact interval around 0 and such that $\lim_{t \to \infty} \Phi_t = Id$ pointwise as operators on $\mathcal{C}_c^\infty(\mathbb{R})$. For example one may take $\Phi_t(x) = x + \frac{1}{t} - \phi(x)\frac{1}{t}$ where $\phi(x) = \phi(-x)$ is smooth, vanishes on $[-1, 1]$, equals 1 outside a large interval $[-C, C]$ and satisfies $|\frac{\partial}{\partial x}\phi(x)| < \frac{1}{2}$ on $\mathbb{R}$.

Then we define

$$\chi_t : \quad C(M, A) \quad \hookrightarrow \quad C(\widetilde{M}, A) \quad \to \quad \mathcal{C}^\infty(M, \partial M, A)$$

$$f \quad \to \quad p^* f \quad \to \quad \nu_t * (p \circ \Phi_{\frac{1}{t}})^* f$$

On the other hand

$$\chi_t' : \quad C(\widetilde{M}, A) \quad \to \quad \mathcal{C}^\infty(\widetilde{M}, A)$$

$$g \quad \to \quad \nu_t * (\Phi_{-\frac{1}{t}})^* g$$

preserves the ideal of functions vanishing on $M \subset \widetilde{M}$ and descends thus to a family of maps

$$\chi_t' : C(M, A) \to \mathcal{C}^\infty(M, A)$$

It is easily shown that χ_t, χ_t' are regularization maps for the inclusions

$$\mathcal{C}^\infty(M, \partial M, A) \hookrightarrow C(M, A), \; \mathcal{C}^\infty(M, A) \hookrightarrow C(M, A)$$

So the derivation lemma may be applied to them and yields the claim.

□

Proposition 7.8:

Let $\mathcal{SC}$ denote the algebra of smooth functions on the closed unit interval which vanish at the endpoints. For any admissible Fréchet algebra A the canonical map

$$SA := \mathcal{SC} \otimes_\pi A \to SA$$

induces an asymptotic HC-equivalence.

Proof:

This follows from the previous proposition because the tensor product algebra SA can easily be identified with the algebra $C_0^\infty([0,1], A)$. One only has to modify the regularization maps of 7.7 so as to preserve the ideals of functions vanishing on the endpoints of the unit interval.

$\square$

There is still another situation where the derivation lemma can be applied.

Theorem 7.9:

Let A be a separable C^*-algebra and let τ be an (unbounded), densely defined, positive trace on A.

Let $l^1(A, \tau)$ be the domain of τ. It is a twosided ideal in A which becomes a Banach algebra under the graph norm

$$\| y \|_1 := \sup_{z \in A, \|z\| \leq 1} |\tau(yz)| + \| y \|$$

Then the inclusion

$$l^1(A, \tau) \hookrightarrow A$$

induces an asymptotic HC-equivalence.

Proof:

It is well known that the domain of a positive trace on a C^*-algebra is a twosided, dense ideal ([D]) which is complete in the norm $\| - \|_1$ described above.

The inequality

$$\| yz \|_1 \leq \| y \|_1 \| z \|_A \quad \forall y \in l^1(A, \tau); z \in A$$

is obvious from the definitions and shows that hypothesis 1) of the derivation lemma is satisfied for the inclusion $i : l^1(A, \tau) \hookrightarrow A$.

To find a regularization map $f_t : A \to l^1(A, \tau)$ one observes that, as A is separable, there exists a smooth one-parameter, positive, bounded approximate unit $(u_t)_{t \in \mathbb{R}} \subset A$ which consists of elements of the ideal $l^1(A, \tau)$. The family

$$f_t := \quad A \to l^1(A, \tau)$$

$$z \to u_t z$$

yields then a regularization map with the desired properties:

Clearly

$$\lim_{t \to \infty} i \circ f_t(z) = z \quad \forall z \in A$$

because (u_t) is an approximate unit for A.

On the other hand, for $y \in l^1(A, \tau)$ one finds

$$\| u_t y - y \|_1 \leq |\tau(u_t yz - yz)| + \epsilon + \| u_t y - y \|$$

for some

$$z = \sum_{finite} a_j b_j \in A; \| a_j \| \leq 1, \| b_j \| \leq 1$$

(every element of a C^*-algebra is a finite linear combination of positive elements, i.e. of squares).

So

$$|\tau(u_t yz - yz)| \leq \left|\tau\left(\sum(b_j u_t - b_j) ya_j\right)\right| \leq$$

$$\leq \sum \| (b_j u_t - b_j)y \|_1 \leq \sum \| b_j u_t - b_j \| \| y \|_1$$

which tends to 0 as t tends to infinity. Therefore

$$\lim_{t \to \infty} f_t \circ i(y) = y \ \forall y \in l^1(A, \tau)$$

so that i satisfies the second hypothesis of the derivation lemma, too.

◻

Theorem 7.10:

For any C^*-algebra A the inclusion

$$A \otimes_\pi l^1(\mathcal{H}) \hookrightarrow A \otimes_{C^*} \mathcal{K}(\mathcal{H})$$

induces an asymptotic HC-equivalence. Here the tensor products are supposed to be the projective Banach tensor product on the left and the C^*-tensor product on the right hand side respectively.

Proof:

Take $\mathcal{H} = l^2(I\!\!N)$ and let

$$M_n(\mathbb{C}) = End(\langle e_1, \ldots, e_n \rangle) \subset \mathcal{L}(\mathcal{H})$$

Suppose that F is a densely defined, positive, unbounded operator on $\mathcal{H}$ with $F^{-1} \in \mathcal{K}(\mathcal{H}) \| F^{-1} \| \leq 1$ such that $\{ e_k; k \in I\!\!N \}$ forms a complete system of eigenvectors of F.

If A is faithfully represented on $\mathcal{H}'$ then $A \otimes_{C^*} \mathcal{K}(\mathcal{H})$ is faithfully represented on $\mathcal{H}' \hat{\otimes} \mathcal{H}$

Lemma 7.11:

1)
$$a \to \| (Id \otimes F)a(Id \otimes F) \|_{\mathcal{H}' \hat{\otimes} \mathcal{H}}$$

defines a norm on $M_\infty(A) \subset A \otimes_{C^*} \mathcal{K}(\mathcal{H})$. Denote the completion by $\mathcal{A}$.

2)
$$\| a^0 \dots a^n \|_{\mathcal{A}} \le \| a^0 \|_{\mathcal{A}} \left(\prod_1^{n-1} \| a^j \|_{A \otimes_{C^*} \mathcal{K}(\mathcal{H})} \right) \| a^n \|_{\mathcal{A}}$$

3) If F^{-1} is of trace class, one has a commutative diagram of inclusions

$$
\begin{array}{ccc}
\mathcal{A} & \longrightarrow & A \otimes_\pi l^1(\mathcal{H}) \\
\| & & \downarrow \\
\mathcal{A} & \longrightarrow & A \otimes_{C^*} \mathcal{K}(\mathcal{H})
\end{array}
$$

$\square$

Continuation of the proof of Theorem 7.10:

Let $(u_t) \in l^1(\mathcal{H})$ be a bounded approximate unit for $\mathcal{K}(\mathcal{H})$. Suppose that (in the notations of the lemma above) F and u_t commute, that $Fu_t = u_t F$ extends to a bounded operator on $\mathcal{H}$ and that $t \to Fu_t : \mathbb{R}_+ \to \mathcal{L}(\mathcal{H})$ is smooth.

For example one may take

$u_N := P_N$ = the orthogonal projection onto $\langle e_1, \dots, e_N \rangle$

and suitable convex combinations for intermediate values of t.

Define

$$\chi_t : \quad A \otimes_{C^*} \mathcal{K}(\mathcal{H}) \quad \to \quad \mathcal{A}$$

$$a \qquad \to \quad (1 \otimes u_t)a(1 \otimes u_t) \quad 1 \otimes u_t \in \mathcal{M}(A) \otimes_{C^*} \mathcal{K}$$

The maps χ_t are obviously bounded.

Because of Lemma 7.11,2) the derivation lemma may be applied to the inclusions

$$\mathcal{A} \hookrightarrow A \otimes_\pi l^1(\mathcal{H}) \quad \mathcal{A} \hookrightarrow A \otimes_{C^*} \mathcal{K}(\mathcal{H})$$

This shows that two of the three inclusions in diagram 7.11,3) induce asymptotic HC-equivalences and therefore also the third inclusion

$$A \otimes_\pi l^1(\mathcal{H}) \hookrightarrow A \otimes_{C^*} \mathcal{K}(\mathcal{H})$$

$\square$

Chapter 8: Products

The next basic step after having defined the various cyclic theories consists in developing operations and especially product operations for them. The first one, the exterior product corresponds to the cup product on cyclic cohomology.

To define it it is necessary to construct a natural chain map

$$\times : \widehat{X}_* R(A \otimes B) \to \widehat{X}_* RA \widehat{\otimes} \widehat{X}_* RB$$

on the algebraic, resp.

$$\times : \widehat{X}_* \mathcal{R}(A \otimes_\pi B) \to \widehat{X}_* \mathcal{R}A \widehat{\otimes}_\pi \widehat{X}_* \mathcal{R}B$$

on the topological and

$$\times : \widehat{X}_* \Omega\mathcal{R}(A \otimes_\pi B) \to \widehat{X}_* \Omega\mathcal{R}A \widehat{\otimes}_\pi \widehat{X}_* \Omega\mathcal{R}B$$

on the differential graded level. The construction is again based on the interplay between the cyclic bicomplex (resp. periodic de Rham complex of A) and the X-complex of the tensor algebra RA presented in Chapter 3. It is not difficult to derive, starting from the natural homomorphism

$$\Omega(A \otimes B) \to \Omega A \widehat{\otimes} \Omega B$$

of enveloping differential graded algebras a chain map

$$\Omega_*^{PdR}(A \otimes B) \to X_* A \widehat{\otimes} X_* B$$

which induces products on cyclic cohomology of degree less or equal to one. Passing to tensor algebras one can derive a chain map

$$X_* R(A \otimes B) \to X_* RA \widehat{\otimes} X_* RB$$

which preserves I-adic filtrations and yields by taking the associated graded complex a product on (periodic) cyclic cohomology which coincides in degrees less or equal to one with the product we started from. Moreover the product carries over to the analytic and the asymptotic X-complex.

The product thus obtained is not associative on the chain level but associative up to homotopy by an explicit chain homotopy involving only a fixed finite number of multilinear algebraic operations in the entries of the tensors under discussion. This allows to carry over the chain homotopy to the topological and differential graded setting.

To get some information about the nature of the exterior product on cyclic cohomology obtained in this way we compare it with the product on K-theory via the Chern character. The formula finally obtained is

$$\langle ch(a \times b), \varphi \times \psi \rangle = \frac{1}{(2\pi i)^{ij}} \langle ch(a), \varphi \rangle \langle ch(b), \psi \rangle$$

for

$$a \in K_i(A),\, b \in K_j(B),\, \varphi \in HC^i_{\epsilon,\alpha}(A),\, \psi \in HC^j_{\epsilon,\alpha}(B)\,;\, i,j \in \{0,1\}.$$

which shows that both products correspond to each other up to the factor $2\pi i$ appearing when all classes involved are of odd dimension.

The period factor $2\pi i$ will necessarily come up in comparing any kind of multiplicative structure on K-theory and periodic (analytic, asymptotic) cyclic cohomology for the following reason. The cyclic theories involved are a priori defined by $\mathbb{Z}/2\mathbb{Z}$ graded chain complexes whereas any product of two classes in K_1 will a priori lie in K_2 and can only a posteriori be identified via Bott periodicity with a class in K_0. It is the fact that Bott periodicity (a deep transcendental result) is involved which is responsible for the "period factor" $2\pi i$. It is also not possible to get rid of the constant $2\pi i$ by changing the Chern character by introducing normalization constants: the naturality of the Chern character under asymptotic morphisms implies that the only freedom of choice one has is to multiply the global formulas for ch_0 resp. ch_1 by a constant. The introduction of any constant in front of ch_0 would destroy the multiplicativity of the Chern character in even degrees whereas the only reasonable modification in odd dimensions would be to replace ch_1 by $ch'_1 = \frac{1}{2\pi i} ch_1$ which destroys the purely algebraic character of the definition of ch_1 but makes the character of the fundamental class in $K_1(\mathcal{C}^\infty(S^1))$ integral. This would however change the character formula for a product only to

$$\langle ch(a \times b), \varphi \times \psi \rangle = (2\pi i)^{ij} \langle ch(a), \varphi \rangle \langle ch(b), \psi \rangle$$

which is similar to the one obtained originally. The last remaining possibility is to change the chain map $\times$ in order to make the Chern character strictly multiplicative. Experience shows however that it seems to be difficult to construct an explicit chain map

$$X^* RA \widehat{\otimes} X^* RB \to X^* R(A \otimes B)$$

which has the same effect on cohomology as $\times$ if at least one factor is evendimensional but differs from $\times$ if both factors are of odd dimension. So we leave the product as it stands.

From the Eilenberg-Zilber theorem in periodic cyclic cohomology it is known that

$$\times : \widehat{X}_* R(A \otimes B) \to \widehat{X}_* RA \widehat{\otimes} \widehat{X}_* RB$$

is a quasiisomorphism so that there has to exist a chain map

$$\widehat{X}_* RA \widehat{\otimes} \widehat{X}_* RB \to \widehat{X}_* R(A \otimes B)$$

providing an inverse up to homotopy of $\times$. Such an inverse would yield an exterior product for the bivariant analytic and asymptotic cohomology theories. Here we treat however only a simple consequence, namely the existence of a "slant" product

$$\backslash : K_* A \otimes HC^*_{\epsilon,\alpha}(A \otimes_\pi B) \to HC^*_{\epsilon,\alpha}(B)$$

which can easily be established directly. It is useful for checking the injectivity of the exterior product with cyclic cohomology classes that lie in the image of the Chern character.

As an application of the exterior product operation we show the stable Morita invariance of asymptotic cyclic cohomology: For any C^*-algebra A the inclusion

$$A \to A \otimes_{C^*} \mathcal{K}(\mathcal{H})$$

is an asymptotic HC-equivalence. This is in sharp contrast to the behaviour of periodic or entire cyclic cohomology as it provides nontrivial cocycles on stable C^*-algebras.

It should be noted that in the meantime a natural homotopy inverse to the chain map $\times$ has been constructed. It can be used to show that there is an Eilenberg-Zilber quasiisomorphism

$$X_* \mathcal{R}(A \otimes_\pi B) \xrightarrow{qis} X_* \mathcal{R}A \otimes_\pi X_* \mathcal{R}B$$

and to prove the existence of an associative exterior product

$$HC^*_{\epsilon,\alpha}(A, B) \otimes HC^*_{\epsilon,\alpha}(C, D) \to HC^*_{\epsilon,\alpha}(A \otimes_\pi C, B \otimes_\pi D)$$

on bivariant analytic, resp. asymptotic cyclic cohomology. (See [P]).

8-1 Exterior products

We work at first on a purely algebraic level and begin by considering the effect of the desired chain maps on cohomology. Algebras are supposed to be unital throughout.

Recall the following remark:

Lemma 8.1:

Let C_*, D_* be chain complexes (bounded from below) of vector spaces. Then a map of complexes $\Phi : C_* \to D_*$ is determined, up to chain homotopy, by its effect on homology: $\Phi_* : h(C_*) \to h(D_*)$. Conversely, any homomorphism from the homology of C_* to the homology of D_* arises in this way.

Proof:

A morphism of complexes (of degree d) $\Phi : C_* \to D_*$ is the same thing as a cocycle (of degree d) in the Hom-complex $Hom_*(C., D.)$ Moreover, two cocycles in the Hom-complex are homologuous if and only if the associated maps of chain complexes are chain homotopic. The assertion follows then from the fact that the universal coefficient spectral sequence collapses for complexes of vector spaces yielding

$$h(Hom_*(C., D.)) \simeq Hom(h(C.), h(D.))$$

$\square$

The lemma shows immediately that there is in general no reasonable map

$$X^*(A) \otimes X^*(B) \to X^*(A \otimes B)$$

because the X-complex takes care only of the cyclic cohomology up to degree 1 (see 2.2) and the product of two classes of degree 1 ought have degree 2. However there have to exist reasonable product maps if one considers better approximations of the periodic de Rham complex than the very crude X-complex:

The homology of the complex

$$Hom\left(\Omega_*^{PdR}(A \otimes B)/F^3\Omega_*^{PdR}(A \otimes B), \mathbb{C}\right)$$

equals

$$h(Hom\left(\Omega_*^{PdR}(A \otimes B)/F^3\Omega_*^{PdR}(A \otimes B), \mathbb{C})\right) = \begin{cases} HC^2(A \otimes B) & * = 0 \\ HC^1(A \otimes B)/kerS & * = 1 \end{cases}$$

whereas the homology of the tensor product of complexes $X^*(A) \otimes X^*(B)$ equals

$$h(X^*A \otimes X^*B) = \begin{cases} HC^0(A)/kerS \otimes HC^0(B)/kerS \oplus HC^1(A) \otimes HC^1(B) & * = 0 \\ HC^0(A)/kerS \otimes HC^1(B) \oplus HC^1(A) \otimes HC^0(B)/kerS & * = 1 \end{cases}$$

as Cuntz and Quillen show [CQ]. The preceding lemma yields therefore

Proposition 8.2:

a) There exists a unique homotopy class of morphisms of complexes

$$\sigma : \Omega_*^{PdR}(A \otimes B)/F^3\Omega_*^{PdR}(A \otimes B) \to X_*A \otimes X_*B$$

inducing a product on the cohomology of the dual complexes that coincides up to stabilization by S with the Yoneda product [CO] of the corresponding Hochschild cohomology groups. This means that it is given by the following table where $\sharp$ denotes Connes's product [CO]

$$
\begin{array}{ccc}
HC^0/kerS \otimes HC^0/kerS & \xrightarrow{S \circ \sharp} & HC^2 \\
HC^1 \otimes HC^1 & \xrightarrow{\frac{1}{2}\sharp} & HC^2 \\
HC^0/kerS \otimes HC^1 & \xrightarrow{\sharp} & HC^1/kerS \\
HC^1 \otimes HC^0/kerS & \xrightarrow{\sharp} & HC^1/kerS
\end{array}
$$

b) There is a map of complexes representing the homotopy class described in a) defined by making commutative the following diagram of $\mathbb{Z}/2\mathbb{Z}$-graded vector spaces

$$\Omega_*^{PdR}(A \otimes B) \xrightarrow{\Psi} X_*(R(A \otimes B)) \simeq \Omega(A \otimes B)$$

$$\downarrow \qquad\qquad\qquad \downarrow \chi$$

$$\qquad\qquad\qquad\qquad \Omega A \widehat{\otimes} \Omega B$$

$$\qquad\qquad\qquad\qquad \downarrow \pi$$

$$\Omega_*^{PdR}(A \otimes B)/F^3\Omega_*^{PdR}(A \otimes B) \quad\to\quad X_*A \otimes X_*B$$

The map Ψ and the isomorphism of the upper line are those of Theorem 3.11. The map $\chi : \Omega(A \otimes B) \to \Omega A \widehat{\otimes} \Omega B$ is the identity in degree 0 and the composition

$$\Omega(A \otimes B) \xrightarrow{N^{-1}} \Omega(A \otimes B) \xrightarrow{\nu} \Omega A \widehat{\otimes} \Omega B$$

where N is the number operator and ν is the morphism of differential graded algebras which corresponds to the inclusion $A \otimes B \to \Omega A \widehat{\otimes} \Omega B$ via the adjunction

$$Hom_{Alg}(A, B_0) = Hom_{DG}(\Omega A, B)$$

$\square$

Proof:

The proof is lengthy but straightforward.

Let us translate the preceding proposition into terms of X-complexes:

Corollary 8.3:

a) By composing the canonical map $X_*R(A \otimes B) \to \Omega_*^{PdR}(A \otimes B)$ with the canonical projection and the map above one obtains a map of complexes

$$\mu : X_*R(A \otimes B) \to X_*A \widehat{\otimes} X_*B$$

b) An explicit representative of the homotopy class of this map is given by $\mu :=$ $\mu_0 \oplus \mu_1$

$$\mu_0 : X_0R(A \otimes B) \qquad\to\qquad X_0A\widehat{\otimes}X_0B \oplus X_1A\widehat{\otimes}X_1B$$

$$\varrho(a^0 \otimes b^0) \qquad\to\qquad a^0 \otimes b^0$$

$$\varrho(a^0 \otimes b^0)\omega(a^1 \otimes b^1, a^2 \otimes b^2) \;\to\; -\tfrac{1}{2}(a^0da^1a^2 \otimes b^0b^1db^2 - a^0a^1da^2 \otimes b^0db^1b^2)$$

$$\varrho\omega^n \qquad\to\qquad 0 \ (n > 1)$$

$$\mu_1 : X_1R(A \otimes B) \qquad\to\qquad X_0A\widehat{\otimes}X_1B \oplus X_1A\widehat{\otimes}X_0B$$

$$\varrho(a^0 \otimes b^0)d\varrho(a^1 \otimes b^1) \;\to\; \tfrac{1}{2}(a^0a^1 + a^1a^0) \otimes b^0db^1 + a^0da^1 \otimes \tfrac{1}{2}(b^0b^1 + b^1b^0)$$

$$\varrho\omega^m d\varrho \qquad\to\qquad 0 \ (m > 0)$$

The same is true in the case of graded algebras and the graded periodic de Rham complex.

The graded tensor product $\widehat{\otimes}$ indicates that switching the factors in the product yields the commutative diagram

$$
\begin{array}{ccc}
X_* R(A \otimes B) & \xrightarrow{\mu} & X_* A \widehat{\otimes} X_* B \\
\downarrow{\scriptstyle X_*(sw)} & & \downarrow{\scriptstyle \widehat{sw}} \\
X_* R(B \otimes A) & \xrightarrow{\mu} & X_* B \widehat{\otimes} X_* A
\end{array}
$$

where the graded switch map is given by

$$
\widehat{sw}(x \widehat{\otimes} y) = (-1)^{deg(x)deg(y)} (y \widehat{\otimes} x)
$$

$\square$

Proof:

Elementary.

Whereas our construction so far does not yield anything interesting for general algebras because cyclic cohomology above degree one is ignored, it already provides a product on the chain level for the cyclic complexes of tensor algebras which are of cohomological dimension one respectively a product on the chain level for the quasiisomorphic X-complexes of tensor algebras.

Recall that the exterior product of asymptotic morphisms resp. based linear maps is well defined and yields via the adjunction

$$
\begin{array}{ccc}
Hom_C(A \otimes A', RA \otimes RA') & = & Hom_{Alg}(R(A \otimes A'), RA \otimes RA') \\
\varrho_A \otimes \varrho_{A'} & \leftrightarrow & m
\end{array}
$$

a homomorphism of algebras

$$
m : R(A \otimes A') \to RA \otimes RA'
$$

Lemma 8.4:

There exists a natural map of complexes

$$
\mu' : X_* R(A \otimes B) \to X_* RA \widehat{\otimes} X_* RB
$$

which is defined as the composition

$$
X_* R(A \otimes B) \xrightarrow{X_*(i)} X_* R(R(A \otimes B)) \xrightarrow{X_*(Rm)} X_* R(RA \otimes RB) \xrightarrow{\mu} X_* RA \widehat{\otimes} X_* RB
$$

If A, B happen to be differential graded algebras and $A \otimes B$ is the graded tensor product, viewed as differential graded algebra in the obvious way, then the above map of complexes becomes a DG-map.

$\square$

The product μ' is not associative on the level of chain complexes, but associative up to a canonical chain homotopy.

Proposition 8.5:

The following two maps of chain complexes

$$\Phi_0 : X_*R(A\otimes B\otimes C) \to X_*R(R(A\otimes B)\otimes C) \to X_*R(A\otimes B)\widehat{\otimes}X_*C \to X_*A\widehat{\otimes}X_*B\widehat{\otimes}X_*C$$

$$\Phi_1 : X_*R(A\otimes B\otimes C) \to X_*R(A\otimes R(B\otimes C)) \to X_*A\widehat{\otimes}X_*R(B\otimes C) \to X_*A\widehat{\otimes}X_*B\widehat{\otimes}X_*C$$

$$\Phi_0 = (\mu\otimes Id)\circ\mu\circ X_*R(\varrho_{A\otimes B}\otimes Id_C) \quad \Phi_1 = (Id\otimes\mu)\circ\mu\circ X_*R(Id_A\otimes\varrho_{B\otimes C})$$

are chain homotopic. An explicit homotopy is provided by the degree one map

$$\Theta : X_*R(A\otimes B\otimes C) \to X_*A\widehat{\otimes}X_*B\widehat{\otimes}X_*C$$

$$\varrho w^k \to 0 \ \ k\neq 1 \quad \varrho w^j d\varrho \to 0 \ \ j\neq 0$$

$$\varrho(a^0\otimes b^0\otimes c^0)d\varrho(a^1\otimes b^1\otimes c^1)$$

$$\downarrow$$

$$\tfrac{1}{4}\left(a^0da^1\widehat{\otimes}[b^0,b^1]\widehat{\otimes}c^0dc^1\right)$$

$$\varrho(a^0\otimes b^0\otimes c^0)\omega(a^1\otimes b^1\otimes c^1, a^2\otimes b^2\otimes c^2)$$

$$\downarrow$$

$$\tfrac{1}{4}(a^2a^0da^1\widehat{\otimes}b^0b^1db^2\widehat{\otimes}c^2c^0dc^1 + a^2a^0da^1\widehat{\otimes}b^0b^1db^2\widehat{\otimes}c^1c^2dc^0$$

$$+a^1a^2da^0\widehat{\otimes}b^0b^1db^2\widehat{\otimes}c^2c^0dc^1 + a^0a^1da^2\widehat{\otimes}b^2b^0db^1\widehat{\otimes}c^0c^1dc^2$$

$$-a^0a^1da^2\widehat{\otimes}b^2b^0db^1\widehat{\otimes}c^1c^2dc^0 - a^1a^2da^0\widehat{\otimes}b^2b^0db^1\widehat{\otimes}c^0c^1dc^2)$$

□

Proof:

Lengthy but elementary.

□

Lemma 8.6:

The diagrams

$$X_*R(A \otimes B \otimes C) \xrightarrow{\mu'} X_*R(A \otimes B)\widehat{\otimes}X_*RC \xrightarrow{(\mu',1)} X_*RA\widehat{\otimes}X_*RB\widehat{\otimes}X_*RC$$

$$\downarrow \qquad\qquad\qquad\qquad\qquad\qquad\qquad\qquad\qquad \|$$

$$X_*R(RA \otimes RB \otimes RC) \xrightarrow[\Phi_0]{} X_*RA\widehat{\otimes}X_*RB\widehat{\otimes}X_*RC$$

$$X_*R(A \otimes B \otimes C) \xrightarrow{\mu'} X_*RA\widehat{\otimes}X_*R(B \otimes C) \xrightarrow{(1,\mu')} X_*RA\widehat{\otimes}X_*RB\widehat{\otimes}X_*RC$$

$$\downarrow \qquad\qquad\qquad\qquad\qquad\qquad\qquad\qquad\qquad \|$$

$$X_*R(RA \otimes RB \otimes RC) \xrightarrow[\Phi_1]{} X_*RA\widehat{\otimes}X_*RB\widehat{\otimes}X_*RC$$

commute.

Proof:

This follows by combining the commutative diagrams

$$X_*R(A \otimes B \otimes C) \xrightarrow{X_*R(\varrho_{A\otimes B}\otimes\varrho_C)} X_*R(R(A \otimes B) \otimes RC)$$

$$X_*R(\varrho_A\otimes\varrho_B\otimes\varrho_C)\downarrow \qquad\qquad\qquad\qquad \downarrow X_*R(R(\varrho_A\otimes\varrho_B)\otimes id_C)$$

$$X_*R(RA \otimes RB \otimes RC) \xrightarrow{X_*R(\varrho_{RA\otimes RB}\otimes Id)} X_*R(R(RA \otimes RB) \otimes RC)$$

and

$$X_*R(R(A \otimes B) \otimes RC) \xrightarrow{\mu} X_*R(A \otimes B)\widehat{\otimes}X_*RC$$

$$\downarrow \qquad\qquad\qquad\qquad\qquad\qquad \downarrow$$

$$X_*R(R(RA \otimes RB) \otimes RC) \xrightarrow{\mu} X_*R(RA \otimes RB)\widehat{\otimes}X_*RC$$

$$\downarrow \mu \otimes Id$$

$$X_*RA\widehat{\otimes}X_*RB\widehat{\otimes}X_*RC$$

obtaining thus the first diagram of the lemma. The commutativity of the second diagram is shown similarly.

□

It is important that only multilinear algebraic operations are involved in the chain homotopy. This will enable one to carry the homotopy-associativity over to the topologized setting.

In the algebraic case we obtain by taking I-adic filtrations into account the

Theorem 8.7:

Let $X^*_{fin}R$ be the complex of linear functionals on X_*R that vanish on elements of high I-adic valuation (3.12.) Then the chain map μ' provides a map of complexes

$$\mu' : X^*_{fin}RA \, \widehat{\otimes} \, X^*_{fin}RB \to X^*_{fin}R(A \otimes B)$$

which is homotopy associative and induces thus an associative "**exterior product**"

$$\times : PHC^*(A) \widehat{\otimes} PHC^*(B) \to PHC^*(A \otimes B)$$

on its cohomology groups.

Proof:

Among the maps used in defining the chain map μ' (8.4) $m : R(A \otimes B) \to RA \otimes RB$ preserves I-adic filtrations if the right hand side is given the product filtration as the commutative diagram

$$
\begin{array}{ccc}
R(A \otimes B) & \xrightarrow{\quad m \quad} & RA \otimes RB \\
\pi \downarrow & & \downarrow \pi \otimes \pi \\
A \otimes B & \xrightarrow{\quad = \quad} & A \otimes B
\end{array}
$$

shows. The map μ (8.3) vanishes on elements of I-adic valuation bigger than one. Taking this into account, we conclude with Lemma 5.1 (where the effect of $i : RA \to RRA$ is investigated) that $\mu' : X_*R(A \otimes B) \to X_*RA \otimes X_*RB$ shifts I-adic valuations by at most 1. The conclusion follows. The homotopy-associativity is shown by a similar analysis of the chain maps of 8.5, 8.6.

□

The exterior product carries over to the differential graded setting:

Theorem 8.8:

a) The composition of maps of complexes (see chapter 4 for the definitions)

$$\times: \quad X_*(\Omega R(A \otimes B)) \xrightarrow{X_*j} \quad X_*R(\Omega(A \otimes B)) \xrightarrow{X_*(R\nu)} \quad X_*R(\Omega A \widehat{\otimes} \Omega B)$$

$$X_*R(\Omega A \widehat{\otimes} \Omega B) \xrightarrow{\mu'_{\Omega A \widehat{\otimes} \Omega B}} X_*R(\Omega A) \widehat{\otimes} X_*R(\Omega B) \xrightarrow{X_*k \widehat{\otimes} X_*k} X_*\Omega RA \widehat{\otimes} X_*\Omega RB$$

induces a natural map of differential graded X-complexes

$$\times: \quad X_{DG}^\bullet(RA) \widehat{\otimes} X_{DG}^\bullet(RB) \to X_{DG}^\bullet(R(A \otimes B))$$

b) The maps

$$X_{DG}^\bullet RA \widehat{\otimes} X_{DG}^\bullet RB \widehat{\otimes} X_{DG}^\bullet RC \xrightarrow{(\times, id)} X_{DG}^\bullet R(A \otimes_\pi B) \widehat{\otimes} X_{DG}^\bullet RC \xrightarrow{\times} X_{DG}^\bullet R(A \otimes_\pi B \otimes_\pi C)$$

and

$$X_{DG}^\bullet RA \widehat{\otimes} X_{DG}^\bullet RB \widehat{\otimes} X_{DG}^\bullet RC \xrightarrow{(id, \times)} X_{DG}^\bullet RA \widehat{\otimes} X_{DG}^\bullet R(B \otimes_\pi C) \xrightarrow{\times} X_{DG}^\bullet R(A \otimes_\pi B \otimes_\pi C)$$

are naturally chain homotopic.

□

Proof:

a): is obvious from the definitions and by the multilinearity of μ which turns μ' automatically into a DG-map if the involved algebras are differential graded. b): we divide the proof into several steps:

Lemma 8.9:

The diagrams

$$
\begin{array}{ccc}
X_*\Omega R(A \otimes B \otimes C) & \xrightarrow{\times} & X_*\Omega R(A \otimes B) \widehat{\otimes} X_*\Omega RC \\
X_*j \quad \downarrow & & \downarrow \quad X_*j \widehat{\otimes} X_*j \\
X_*R(\Omega(A \otimes B \otimes C)) & \to & X_*R(\Omega(A \otimes B)) \widehat{\otimes} X_*R\Omega C \\
\downarrow & & \downarrow \\
X_*R(\Omega A \widehat{\otimes} \Omega B \widehat{\otimes} \Omega C) & \xrightarrow[\mu']{} & X_*R(\Omega A \widehat{\otimes} \Omega B) \widehat{\otimes} X_*R\Omega C
\end{array}
$$

$$X_*\Omega R(A \otimes B)\widehat{\otimes}X_*\Omega RC \xrightarrow{(\times,1)} X_*\Omega RA\widehat{\otimes}X_*\Omega RB\widehat{\otimes}X_*\Omega RC$$

$$X_*j\widehat{\otimes}X_*j \qquad \downarrow \qquad\qquad\qquad\qquad \downarrow \qquad X_*j$$

$$X_*R(\Omega(A \otimes B))\widehat{\otimes}X_*R\Omega C \;\to\; X_*R\Omega A\widehat{\otimes}X_*R\Omega B\widehat{\otimes}X_*R\Omega C$$
$$\downarrow \qquad\qquad\qquad\qquad\qquad \downarrow$$
$$X_*R(\Omega A\widehat{\otimes}\Omega B)\widehat{\otimes}X_*R\Omega C \xrightarrow[(\mu'.1)]{} X_*R\Omega A\widehat{\otimes}X_*R\Omega B\widehat{\otimes}X_*R\Omega C$$

commute up to homotopy.

There is a similar diagram showing that

$$X_*\Omega R(A \otimes B \otimes C) \;\to\; X_*\Omega RA\widehat{\otimes}X_*\Omega R(B \otimes C) \;\to\; X_*\Omega RA\widehat{\otimes}X_*\Omega RB\widehat{\otimes}X_*\Omega RC$$

$$\downarrow \qquad\qquad\qquad\qquad \downarrow \qquad\qquad\qquad\qquad \downarrow$$

$$X_*R(\Omega A\widehat{\otimes}\Omega B\widehat{\otimes}\Omega C) \;\to\; X_*R\Omega A\widehat{\otimes}X_*R(\Omega B\widehat{\otimes}\Omega C) \;\to\; X_*R\Omega A\widehat{\otimes}X_*R\Omega B\widehat{\otimes}X_*R\Omega C$$

commutes up to homotopy.

Proof:

While the lower squares commute strictly, one finds for the upper ones:

$$(X_*j\widehat{\otimes}X_*j)\circ\times = (X_*j\widehat{\otimes}X_*j)\circ(X_*k\widehat{\otimes}X_*k)\circ\mu\circ X_*j = (X_*jk\widehat{\otimes}X_*jk)\circ\mu\circ X_*j \sim \mu\circ X_*j$$

because $X_*(jk)$ is chain homotopic to the identity. The chain homotopy even preserves I-adic filtrations because j and k do so (Lemma 4.9) (See also 8.16).

Proof of Theorem 8.8,b): Combining the preceding lemmas yields the following diagrams which commute up to homotopy:

$$X_*(\Omega R(A \otimes B \otimes C) \xrightarrow{(\times,1)\circ\times} X_*(\Omega RA)\widehat{\otimes}X_*(\Omega RB)\widehat{\otimes}X_*(\Omega RC)$$

$$X_*(\varrho^{\otimes^3} \circ j) \downarrow \qquad\qquad\qquad\qquad \downarrow X_*j^{\otimes^3}$$

$$X_*R(R\Omega A\widehat{\otimes}R\Omega B\widehat{\otimes}R\Omega C) \xrightarrow[\Phi_0]{} X_*R\Omega A\widehat{\otimes}X_*R\Omega B\widehat{\otimes}X_*R\Omega C$$

$$\downarrow X_*k^{\otimes^3}$$

$$X_*(\Omega RA)\widehat{\otimes}X_*(\Omega RB)\widehat{\otimes}X_*(\Omega RC)$$

$$X_*(\Omega R(A \otimes B \otimes C)) \xrightarrow{(1,\times)\circ\times} X_*(\Omega RA)\widehat{\otimes}X_*(\Omega RB)\widehat{\otimes}X_*(\Omega RC)$$

$$X_*(\varrho^{\otimes^3} \circ j) \downarrow \qquad\qquad\qquad\qquad \downarrow X_*j^{\otimes^3}$$

$$X_*R(R\Omega A\widehat{\otimes}R\Omega B\widehat{\otimes}R\Omega C) \xrightarrow[\Phi_1]{} X_*R\Omega A\widehat{\otimes}X_*R\Omega B\widehat{\otimes}X_*R\Omega C$$

$$\downarrow X_*k^{\otimes^3}$$

$$X_*(\Omega RA)\widehat{\otimes}X_*(\Omega RB)\widehat{\otimes}X_*(\Omega RC)$$

Following the diagrams one way, one obtains $(\times,1)\circ\times$, resp.$((1,\times)\circ\times)$ because $k\circ j = Id$. Therefore the explicit chain homotopy Θ between Φ_0 and Φ_1, constructed in Proposition 8.5 yields an explicit chain homotopy between $(\times,1)\circ\times$ and $(1,\times)\circ\times$. Because the homotopy operator Θ is multilinear, it is compatible with gradings and derivations and provides therefore a homotopy operator

$$\Theta'_{DG} : X^*_{DG}R(A \otimes B \otimes C) \to X^*_{DG}RA \widehat{\otimes} X^*_{DG}RB \widehat{\otimes} X^*_{DG}RC[-1]$$

□

The algebraic construction of the exterior product being achieved, topologies can be taken into account.

Proposition 8.10:

Let A, B be admissible Fréchet algebras. The natural homomorphism adjoint to the product of the universal based linear maps

$$m : R(A \otimes B) \to RA \otimes RB$$

induces continuous morphisms

$$m \in \varprojlim_{\leftarrow \mathcal{K}} \varinjlim_{\to \mathcal{K}' \times \mathcal{K}''} Hom(R(A \otimes_\pi B)_{(K,N)}, RA_{(K',N')} \otimes_\pi RB_{(K'',N'')})$$

of Fréchet algebras, i.e. a homomorphisms

$$m : \mathcal{R}(A \otimes_\pi B) \to \mathcal{R}A \otimes_\pi \mathcal{R}B$$

of topological I-adic completions.

□

In order to prove the proposition we show first the

Lemma 8.11:

Let A, B be admissible Fréchet algebras and suppose that K is a multiplicatively closed compact subset of a "small" open ball U in $A \otimes_\pi B$. Then there exist multiplicatively closed compact sets $K' \subset U' \subset A$, $K'' \subset U'' \subset B$ such that, with

$$K' \otimes K'' := \{a \otimes b \mid a \in K', b \in K''\}$$

the following holds: For any $N \geq 1$ there exists $M > 0$ such that the identity on $R(A \otimes B)$ induces a continuous map (see 5.6)

$$R(A \otimes_\pi B)_{(K,N)} \to R(A \otimes_\pi B)_{(K' \otimes K'', M)}$$

Consequently

$$"\lim_{K}" R(A \otimes_\pi B)_{(K,N)} \simeq "\lim_{K' \otimes K''}" R(A \otimes_\pi B)_{(K' \otimes K'', M)}$$

where on the left hand side the limit is taken over all compact subsets of $U \subset A$.

Proof:

We may assume that K is a nullsequence in $A \otimes_\pi B$ contained in the algebraic tensor product $A \otimes B$ by Lemma 5.6. Choose increasing sequences of seminorms $\| \|_A^j$, $\| \|_B^{j'}$ defining the topologies of the admissible Fréchet algebras A, B such that the open unit balls $U_{\| \|_A^j}$, $V_{\| \|_B^{j'}}$ are "small" for all $j, j' \in I\!N$. Denote by $\| \|_{A \otimes_\pi B}^j$ the projective cross norm associated to $\| \|_A^j$, $\| \|_B^j$ on $A \otimes_\pi B$.

Put $\widetilde{K} := \bigcup_n \frac{1}{n} K$. Then $\widetilde{K}$ is a nullsequence in the algebraic tensor product $A \otimes B$, too, and because we work with the projective tensor product, $\widetilde{K}$ may be written (after exclusion of finitely many elements) as

$$\widetilde{K} := \{\gamma^j = \sum_{k=0}^{n_j} a_k^j \otimes b_k^j \mid \sum_{k=0}^{n_j} \| a_k^j \|_A^{\phi(j)} \| b_k^j \|_B^{\phi(j)} \leq 2 \| \sum_{k=0}^{n_j} a_k^j \otimes b_k^j \|_{A \otimes_\pi B}^{\phi(j)} \}$$

$$\| \sum_{k=0}^{n_j} a_k^j \otimes b_k^j \|_{A \otimes_\pi B}^{\phi(j)} < \frac{1}{2} \quad \lim_{j \to \infty} \| \sum_{k=0}^{n_j} a_k^j \otimes b_k^j \|_{A \otimes_\pi B}^{\phi(j)} = 0$$

where $\phi(n)$ tends to ∞ with n.

Then the sets

$$K_0' := \{\alpha_k^j := (2 \| \sum_{k=0}^{n_j} a_k^j \otimes b_k^j \|_{A \otimes_\pi B}^{\phi(j)})^{\frac{1}{2}} \frac{a_k^j}{\| a_k^j \|_A^{\phi(j)}} \mid j \in I\!N, 0 \leq k \leq n_j\} \subset A$$

$$K_1' := \{\beta_k^j := (2 \| \sum_{k=0}^{n_j} a_k^j \otimes b_k^j \|_{A \otimes_\pi B}^{\phi(j)})^{\frac{1}{2}} \frac{b_k^j}{\| b_k^j \|_B^{\phi(j)}} \mid j \in I\!N, 0 \leq k \leq n_j\} \subset B$$

are nullsequences contained in the open unit balls $U_{\| \|_A^1}$, (resp. $V_{\| \|_B^1}$). These being "small", it follows that

$$\widetilde{K}' := \overline{\text{mult. closure of } K_0'} \subset A$$

$$\tilde{K}'' := \overline{\text{mult. closure of } K_1'} \subset B$$

are compact, too. For some $C \geq 1$, the cones $K'(K'')$ over $\frac{1}{C}\tilde{K}'(\frac{1}{C}\tilde{K}'')$ with vertex 0 will be multiplicatively closed and contained in "small" balls $U \subset A$, $U' \subset B$.

As any element of K is in the linear span of $K' \otimes K''$ there is a natural inclusion of algebras

$$R(A \otimes B)_K \subset R(A \otimes B)_{K' \otimes K''} \subset R(A \otimes B)$$

Let

$$x = \sum_\gamma \lambda_\gamma \varrho_\gamma w^{k_\gamma} \in R(A \otimes B)_K = R(A \otimes B)_{\tilde{K}}$$

be such that

$$\sum_\gamma |\lambda_\gamma| (1 + k_\gamma)^m N^{-k_\gamma} \leq \| x \|_{N,m}^{\tilde{K}} + \epsilon$$

Now

$$\varrho w^n(\gamma^0, \ldots, \gamma^{2n}) = \varrho w^n(\ldots, \sum_k a_k^j \otimes b_k^j, \ldots)$$

$$= \sum_k \frac{\| a_k^j \|_A^{\phi(j)} \| b_k^j \|_B^{\phi(j)}}{2 \| \sum_{k'=0}^{n_j} a_{k'}^j \otimes b_{k'}^j \|_{A \otimes_\pi B}^{\phi(j)}} \varrho w^n(\ldots, \sum_k \alpha_k^j \otimes \beta_k^j, \ldots)$$

so that

$$\| x \|_{C^2 N, m}^{K' \otimes K''}$$

$$\leq \sum_\gamma |\lambda_\gamma| \sum_{k_0, \ldots, k_{2k_\gamma}} (\prod_0^{2k_\gamma} \frac{\| a_k^j \|_A^{\phi(j)} \| b_k^j \|_B^{\phi(j)}}{2 \| \sum_{k'=0}^{n_j} a_{k'}^j \otimes b_{k'}^j \|_{A \otimes_\pi B}^{\phi(j)}}) \| \varrho w^{k_\gamma}(\ldots, \alpha_k^j \otimes \beta_k^j, \ldots) \|_{C^2 N, m}^{K' \otimes K''}$$

$$\leq \sum_\gamma |\lambda_\gamma| C^{2k_\gamma + 1} (1 + k_\gamma)^m (C^2 N)^{-k_\gamma} \leq C(\| x \|_{N,m}^{\tilde{K}} + \epsilon)$$

As there exists $C' \geq 1$ such that

$$R(A \otimes_\pi B)_{(K,N)} \to R(A \otimes_\pi B)_{(\tilde{K}, C'N)}$$

is continuous, the conclusion follows.

□

The demonstration of Proposition 8.10. can now be achieved by

Lemma 8.12:

Let $K' \subset A$, $K'' \subset B$ be multiplicatively closed "small" compact subsets of admissible Fréchet algebras A, B. Suppose that $K' \otimes K''$ is "small" in $A \otimes_\pi B$. Then the canonical map

$$m : R(A \otimes B) \to RA \otimes RB$$

induces a continuous morphism

$$R(A \otimes_\pi B)_{(K' \otimes K'', N)} \to RA_{(K', M)} \otimes_\pi RB_{(K'', M)}$$

for any N, if M is choosen large enough. Consequently one obtains

$$m : \ "\lim_{K' \otimes K''}" R(A \otimes_\pi B)_{(K' \otimes K'', N)} \to \mathcal{R}A \otimes_\pi \mathcal{R}B$$

Proof:

Under the canonical morphism $m : R(A \otimes B) \to RA \otimes RB$

$$\varrho(a \otimes b) \qquad \to \qquad \varrho(a) \otimes \varrho(b)$$

$$\omega(a \otimes b, a' \otimes b') \quad \to \quad \omega(a, a') \otimes \varrho(bb') + \varrho(aa') \otimes \omega(b, b') - \omega(a, a') \otimes \omega(b, b')$$

Thus

$$m : R(A \otimes_\pi B)_{K' \otimes K''} \to RA_{K'} \otimes RB_{K''}$$

algebraically and

$$\varrho\omega^n(a^0 \otimes b^0, \ldots, a^{2n} \otimes b^{2n}) \quad a^i \in K', \ b^j \in K''$$

maps to

$$\sum_{3^n \, terms} \varrho^{i_0} \omega^{i_1} \varrho^{i_2} \ldots \omega^{i_l}(\ldots, a^i, \ldots) \otimes \varrho^{j_0} \omega^{j_1} \varrho^{j_2} \ldots \omega^{j_m}(\ldots, b^j, \ldots)$$

$$= \sum_{\leq (3.8^2)^n \, terms} \varrho\omega^k(\ldots, a', \ldots) \otimes \varrho\omega^{k'}(\ldots, b', \ldots)$$

by Lemma 5.1. where $k \leq n$, $k' \leq n$, $k + k' \geq n$. and the entries of a' (b') belong to K'. (K'')
Therefore

$$\| m(x) \|_{\| \|_{192N,m'}^{K'} \otimes \| \|_{192N,m''}^{K''}} \leq C \ \| x \|_{N,m'+m''}^{K' \otimes K''}$$

and the lemma is proved.

$\square$

Proposition 8.13:

Let A, B be unital, admissible Fréchet algebras. The chain map

$$\mu' : X_* R(A \otimes B) \to X_* RA \widehat{\otimes} X_* RB$$

extends to a continuous map of X-complexes of topological I-adic completions

$$\mu' : X_* \mathcal{R}(A \otimes_\pi B) \to X_* \mathcal{R}A \widehat{\otimes}_\pi X_* \mathcal{R}B$$

Proof:

Recall (8.4.) that μ' was defined as the composition

$$X_*R(A{\otimes}B) \xrightarrow{X_*(i)} X_*R(R(A{\otimes}B)) \xrightarrow{X_*(Rm)} X_*R(RA{\otimes}RB) \xrightarrow{\mu} X_*(RA)\widehat{\otimes}X_*(RB)$$

The morphism $X_*(i)$ induces a map of Ind-objects

$$X_*(i) : X_*\mathcal{R}(A \otimes_\pi B) \to X_*\widehat{R}(\mathcal{R}(A \otimes_\pi B))$$

by Proposition 5.11. The universal homomorphism m yields morphisms of Ind-objects

$$m : \mathcal{R}(A \otimes_\pi B) \to \mathcal{R}A \otimes_\pi \mathcal{R}B$$

by Proposition 8.10. and thus a map of complexes

$$X_*Rm : X_*\widehat{R}(\mathcal{R}(A \otimes_\pi B)) \to X_*\widehat{R}(\mathcal{R}A \otimes_\pi \mathcal{R}B)$$

The map

$$\mu : X_*\widehat{R}(A \otimes B) \to X_*A \widehat{\otimes} X_*B$$

involves only multiplication and summation in A and B and vanishes on elements of I-adic valuation > 1 so that it also yields a morphism of formal inductive limits

$$\mu : X_*\widehat{R}(\mathcal{R}A \otimes_\pi \mathcal{R}B) \to X_*\mathcal{R}A \widehat{\otimes}_\pi X_*\mathcal{R}B$$

Composing all these maps provides finally the morphism of formal inductive limit complexes

$$\mu' : X_*\mathcal{R}(A \otimes_\pi B) \to X_*\mathcal{R}A \widehat{\otimes}_\pi X_*\mathcal{R}B$$

$\square$

The aim of this paragraph, the construction of an exterior product for analytic and asymptotic cohomology can be achieved now. The involved algebras are not supposed to be unital anymore.

Theorem 8.14:

a) The map

$$\mu' : \overline{X}_*R(A\widetilde{\otimes}B) \to \overline{X}_*R\widetilde{A} \widehat{\otimes} \overline{X}_*R\widetilde{B}$$

induces natural **chain maps of analytic X-complexes**

$$\times : X^*_{\epsilon,V}(A) \widehat{\otimes} X^*_{\epsilon,W}(B) \to X^*_{\epsilon,V\otimes_\pi W}(A \otimes_\pi B)$$

$$\times : X^*_\epsilon(A) \widehat{\otimes} X^*_\epsilon(B,C) \to X^*_\epsilon(A \otimes_\pi B, C)$$

b) The maps

$$X_\epsilon^*(A) \widehat{\otimes} X_\epsilon^*(B) \widehat{\otimes} X_\epsilon^*(C) \xrightarrow{(\times,1)} X_\epsilon^*(A \otimes_\pi B) \widehat{\otimes} X_\epsilon^*(C) \xrightarrow{\times} X_\epsilon^*(A \otimes_\pi B \otimes_\pi C)$$

and

$$X_\epsilon^*(A) \widehat{\otimes} X_\epsilon^*(B) \widehat{\otimes} X_\epsilon^*(C) \xrightarrow{(1,\times)} X_\epsilon^*(A) \widehat{\otimes} X_\epsilon^*(B \otimes_\pi C) \xrightarrow{\times} X_\epsilon^*(A \otimes_\pi B \otimes_\pi C)$$

are naturally chain homotopic.

c) A similar statement holds if one of the complexes involved on the left is a bivariant one.

d) The chain maps $\times$ define associative **"exterior products"**

$$\times : HC_\epsilon^*(A) \widehat{\otimes} HC_\epsilon^*(B) \to HC_\epsilon^*(A \otimes_\pi B)$$

$$\times : HC_\epsilon^*(A) \widehat{\otimes} HC_\epsilon^*(B,C) \to HC_\epsilon^*(A \otimes_\pi B, C)$$

e) Naturality means that for any algebra homomorphisms $f : A \to A'$, $g : B \to B'$ the square

$$
\begin{array}{ccc}
X_\epsilon^*(A') \widehat{\otimes} X_\epsilon^*(B') & \longrightarrow & X_\epsilon^*(A' \otimes B') \\
{\scriptstyle f^* \otimes g^*} \downarrow & & \downarrow {\scriptstyle (f \otimes g)^*} \\
X_\epsilon^*(A) \widehat{\otimes} X_\epsilon^*(B) & \longrightarrow & X_\epsilon^*(A \otimes B)
\end{array}
$$

commutes.

Proof:

a): Follows from Proposition 8.13.

b) The morphisms in the diagrams of Lemma 8.6 extend to morphisms of the corresponding Ind-objects, where one has to take $X_* R(\mathcal{R}A \otimes_\pi \mathcal{R}B \otimes_\pi \mathcal{R}C)$ in the lower left corner. The maps Φ_0, Φ_1 and the chain homotopies of Proposition 8.5 vanish on elements of high I-adic filtration and involve only a fixed finite number of additions and multiplications and extend therefore also to the corresponding formal inductive limits.

$\square$

Theorem 8.15:

a) The map

$$\times : \overline{X}_{DG}^* R\widetilde{A} \widehat{\otimes} \overline{X}_{DG}^* R\widetilde{B} \to \overline{X}_{DG}^* R(\widetilde{A \otimes B})$$

of Theorem 8.8 induces **chain maps of asymptotic X-complexes**

$$\times : X_\alpha^*(A) \widehat{\otimes} X_\alpha^*(B) \to X_\alpha^*(A \otimes_\pi B)$$

$$\times : X_\alpha^*(A) \widehat{\otimes} X_\alpha^*(B,C) \to X_\alpha^*(A \otimes_\pi B, C)$$

b) The maps

$$X_\alpha^*(A) \mathbin{\widehat{\otimes}} X_\alpha^*(B) \mathbin{\widehat{\otimes}} X_\alpha^*(C) \xrightarrow{(\times,1)} X_\alpha^*(A \otimes_\pi B) \mathbin{\widehat{\otimes}} X_\alpha^*(C) \xrightarrow{\times} X_\alpha^*(A \otimes_\pi B \otimes_\pi C)$$

and

$$X_\alpha^*(A) \mathbin{\widehat{\otimes}} X_\alpha^*(B) \mathbin{\widehat{\otimes}} X_\alpha^*(C) \xrightarrow{(1,\times)} X_\alpha^*(A) \mathbin{\widehat{\otimes}} X_\alpha^*(B \otimes_\pi C) \xrightarrow{\times} X_\alpha^*(A \otimes_\pi B \otimes_\pi C)$$

are chain homotopic.

c) A similar statement holds if one of the complexes involved on the left is a bivariant one.

d) The chain maps $\times$ define associative **"exterior products"**

$$\times : HC_\alpha^*(A) \mathbin{\widehat{\otimes}} HC_\alpha^*(B) \to HC_\alpha^*(A \otimes_\pi B)$$

$$\times : HC_\alpha^*(A) \mathbin{\widehat{\otimes}} HC_\alpha^*(B,C) \to HC_\alpha^*(A \otimes_\pi B, C)$$

e) The exterior product is a natural transformation of linear asymptotic homotopy functors, i.e. if $[f] \in [A, A']_\alpha$ $[g] \in [B, B']_\alpha$ are asymptotic morphisms, the diagram

$$
\begin{array}{ccc}
HC_\alpha^*(A') \widehat{\otimes} HC_\alpha^*(B') & \xrightarrow{f^* \otimes g^*} & HC_\alpha^*(A) \widehat{\otimes} HC_\alpha^*(B) \\
\times \downarrow & & \downarrow \times \\
HC_\alpha^*(A' \otimes_\pi B') & \xrightarrow{(f \otimes g)^*} & HC_\alpha^*(A \otimes_\pi B)
\end{array}
$$

commutes.

Proof:

a): All maps in the definition of $\times$ (8.8) are continuous: $X_* j$ by (6.14), $X_* R\nu$ by definition of the topology on ΩA, μ' by (8.13) and $X_* k$ by (6.14). Therefore $\times$ induces maps

$$X_\alpha^*(A) \mathbin{\widehat{\otimes}} X_\alpha^*(B) \to \varprojlim_{\mathcal{K}} \varinjlim_{U \times V} Hom_{DG}^*(\overline{X}_*(\Omega R(\widetilde{A \otimes_\pi B})_{(K,N)}), \mathcal{E}(U) \widehat{\otimes}_\pi \mathcal{E}(V))$$

$$= \varprojlim_{\mathcal{K}} \varinjlim_{U \times V} Hom_{DG}^*(\overline{X}_*(\Omega R(\widetilde{A \otimes_\pi B})_{(K,N)}), \mathcal{E}(U \times V))$$

$$X_\alpha^*(A) \otimes X_\alpha^*(B,C) \to \varprojlim_{\mathcal{K}} \varinjlim_{U \times V} Hom_{DG}^*(\overline{X}_*(\Omega R(\widetilde{A \otimes_\pi B})_{(K,N)}), \overline{X}_*(\Omega R\widetilde{C}) \otimes_\pi^{sh} \mathcal{E}(U \times V))$$

As $U \times V$ is a neighbourhood of $\infty \in \mathcal{R}_+^{n+m}$ if U and V are so in $\mathcal{R}_+^n$, $\mathcal{R}_+^m$ (1.1), we are done.

b): The demonstration is similar to that of the preceding Theorem 8.14 b) with the single difference that diagram 8.9 has to be used in addition. Its maps are continuous and yield morphisms of the associated formal inductive limits as well, but it commutes only up to homotopy. The proof is thus completed by the following

Lemma 8.16:

Let A be a unital admissible Fréchet algebra and consider the natural morphisms

$$j : \Omega\mathcal{R}A \to \mathcal{R}(\Omega A), \quad k : \mathcal{R}(\Omega A) \to \Omega\mathcal{R}A$$

of differential graded algebras (see 6.14). Then $j \circ k$ is smoothly homotopic to the identity $id_{\mathcal{R}(\Omega A)}$. Consequently $(j \circ k)_*$ is chain homotopic to the identity on $X_* \mathcal{R}(\Omega A)$.

Proof:

The homomorphism $j \circ k$ is given by

$$j \circ k : \qquad R(\Omega A) \qquad\qquad \to \qquad\qquad R(\Omega A)$$

$$\varrho(a^0 \partial a^1 \dots \partial a^n) \quad \to \quad \varrho(a^0)\varrho(\partial a^1) \dots \varrho(\partial a^n)$$

Note that

$$j \circ k |_{RA} = Id_{RA}$$

so that the curvature of $j \circ k$ vanishes on $A \times A \subset \Omega A \times \Omega A$. This implies that the homotopy

$$R(\Omega A) \xrightarrow{i} RR(\Omega A) \xrightarrow{RF} R(\Omega A)[0,1]$$

induced by the linear homotopy

$$F := (1-t)\, j \circ k + t\, Id : R(\Omega A) \to R(\Omega A)[0,1]$$

factorizes, when restricted to $R(\Omega A)^{(k)}$ as

$$R(\Omega A) \xrightarrow{i} RR(\Omega A) \to RR(\Omega A)/I^k R(\Omega A) \xrightarrow{RF} R(\Omega A)[0,1]$$

Taking topologies into account, one knows from Lemma 6.14 that F extends to a continuous linear map

$$F : \mathcal{R}(\Omega A) \to \mathcal{R}(\Omega A)[0,1]$$

Thus we obtain with the help of Lemma 5.11 that the composition

$$\mathcal{R}(\Omega A)^{(k)} \xrightarrow{i} \widehat{R}\mathcal{R}(\Omega A) \to \widehat{R}\mathcal{R}(\Omega A)/\widehat{I}^k \mathcal{R}(\Omega A) =$$

$$= \bigoplus_0^k \mathcal{R}(\Omega A) \otimes \overline{\mathcal{R}(\Omega A)}^{\otimes 2j} \xrightarrow{RF} \mathcal{R}(\Omega A)[0,1]$$

provides a continuous homotopy F' connecting $j \circ k$ with the identity. Moreover it is clear that the restriction of F' on $\mathcal{R}(\Omega A)^{(k)}$ is a polynomial function in t of degree at most k. It follows that $\mathcal{R}(\Omega A)^{(k)}$ decomposes into a topological finite direct sum

$$\mathcal{R}(\Omega A)^{(k)} = \bigoplus_0^k \mathcal{R}_j$$

of weight spaces

$$\mathcal{R}_j := \{x \in \mathcal{R}(\Omega A)^{(k)},\ F'(x,t) = t^j x, t \in \mathbb{C}\}$$

As the homotopy F' is evidently smooth on each weight space $\mathcal{R}_j$ we are done.

Parts c) and d) of Theorem 8.15 are clear and a) follows readily from the definition of the exterior product. Finally e) can be verified by a lengthy diagram chase.

□

8-2 Stable Morita invariance of asymptotic cohomology

Proposition 8.17:

For any admissible Fréchet algebra the natural inclusion

$$i: A \to M_n(A)$$

induces both an analytic and asymptotic HC-equivalence.

Proof:

See the next theorem, where the analogous "infinite dimensional" statement is treated.

□

The following stability result for asymptotic cohomology is in sharp contrast to the behaviour of ordinary (resp. entire) cyclic cohomology.

Theorem 8.18:

For any C^*-algebra A the natural inclusion

$$i: A \hookrightarrow A \otimes_{C^*} \mathcal{K}(\mathcal{H})$$

of C^*-algebras defines an asymptotic HC-equivalence.

Proof:

The inclusion i_A factors via

$$A \xrightarrow{i'_A} A \otimes_\pi l^1(\mathcal{H}) \xrightarrow{i''_A} A \otimes_{C^\bullet} \mathcal{K}(\mathcal{H})$$

where i''_A is an asymptotic HC-equivalence by the derivation lemma (Theorem 7.1, 7.10). We claim that an asymptotic HC-inverse to i^A_* is provided by

$$j^A := (Id_{A*} \times Tr) \circ [i''_{A*}]^{-1} \in HC^0_\alpha(A \otimes_{C^\bullet} \mathcal{K}(\mathcal{H}), A)$$

In fact

$$j^A \circ i^A_* = (Id^A_* \times Tr) \circ (i^{A''}_*)^{-1} \circ (i^{A''}_*) \circ (i^{A'}_*)$$

$$= (Id^A_* \times Tr) \circ (i^{A'}_*) = (Id^A_* \times Tr) \circ (Id^A \otimes_\pi i'_{\mathbb{C}*})$$

$$= Id^A_* \times (Tr \circ i'_{\mathbb{C}*})$$

by the naturality of the exterior product

$$= Id^A_* \times (i'^*_{\mathbb{C}} Tr) = Id^A_* \times \tau_0 = Id^A_*$$

with $\tau_0 \in HC^0(\mathbb{C})$ the normalized trace on $\mathbb{C}$.

To show that

$$i^A_* \circ j^A = Id^{A \otimes_{C^\bullet} \mathcal{K}(\mathcal{H})}_*$$

we use Atiyah's "rotation trick":
The two inclusions

$$i^{A \otimes_{C^\bullet} \mathcal{K}(\mathcal{H})} : A \otimes_{C^\bullet} \mathcal{K}(\mathcal{H}) \simeq A \otimes_{C^\bullet} \mathcal{K}(\mathcal{H}) \otimes_{C^\bullet} \mathbb{C} \to A \otimes_{C^\bullet} \mathcal{K}(\mathcal{H}) \otimes_{C^\bullet} \mathcal{K}(\mathcal{H})$$

$$\bar{i}^{A \otimes_{C^\bullet} \mathcal{K}(\mathcal{H})} : A \otimes_{C^\bullet} \mathcal{K}(\mathcal{H}) \simeq A \otimes_{C^\bullet} \mathbb{C} \otimes_{C^\bullet} \mathcal{K}(\mathcal{H}) \to A \otimes_{C^\bullet} \mathcal{K}(\mathcal{H}) \otimes_{C^\bullet} \mathcal{K}(\mathcal{H})$$

are canonically homotopic.
By the naturality of the exterior product (8.14) one has

$$i^A_* \circ j^A = j^{A \otimes_{C^\bullet} \mathcal{K}(\mathcal{H})} \circ \bar{i}^{A \otimes_{C^\bullet} \mathcal{K}(\mathcal{H})}_*$$

so that

$$i^A_* \circ j^A = j^{A \otimes_{C^\bullet} \mathcal{K}(\mathcal{H})} \circ \bar{i}^{A \otimes_{C^\bullet} \mathcal{K}(\mathcal{H})}_* = j^{A \otimes_{C^\bullet} \mathcal{K}(\mathcal{H})} \circ i^{A \otimes_{C^\bullet} \mathcal{K}(\mathcal{H})}_*$$

by homotopy invariance

$$= Id^{A \otimes_{C^\bullet} \mathcal{K}(\mathcal{H})}_*$$

by what we just proved.

$\square$

Corollary 8.19:

$$HC_\alpha^*(\mathcal{K}(\mathcal{H})) \simeq \begin{cases} \mathbb{C} & * = 0 \\ 0 & * = 1 \end{cases}$$

$\square$

There are many ways to construct nontrivial cocycles on $\mathcal{K}(\mathcal{H})$. One possibility was described in the proof of Theorem 7.10.

8-3 Compatibility of the Chern character with exterior products

In this paragraph the behaviour of the pairing

$$K_* \otimes HC_{\epsilon,\alpha}^* \to \mathbb{C}$$

with respect to exterior products is investigated.

The final result (Theorem 8.22) is not completely obvious because the X-complex is a priori $\mathbb{Z}/2$-graded whereas K-theory becomes two periodic only by the Bott periodicity theorem. So although hidden in the statements, the periodicity theorem shows up in the demonstrations.

All algebras involved are assumed to be unital.

Consider first of all a special case: the compatibility of the canonical isomorphism

$$\Phi : K_1 A \xrightarrow{\simeq} K_0 SA = K_0(SA)$$

with the pairing with asymptotic cohomology. (For the definition of SA see 7.8.)

Recall that the isomorphism Φ is constructed as follows:

Let $u \in GL_n(A)$ and choose a smooth path $v : [0,1] \to GL_{2n}(A)$ connecting

$$v(0) = \begin{pmatrix} u & 0 \\ 0 & u^{-1} \end{pmatrix} \quad \text{and} \quad v(1) = \begin{pmatrix} 1_n & 0 \\ 0 & 1_n \end{pmatrix}$$

As an example, one may take

$$v := w \begin{pmatrix} u & 0 \\ 0 & 1 \end{pmatrix} w^{-1} \begin{pmatrix} 1 & 0 \\ 0 & u^{-1} \end{pmatrix}$$

with

$$w(t) := \begin{pmatrix} \cos\frac{\pi}{2}t & -\sin\frac{\pi}{2}t \\ \sin\frac{\pi}{2}t & \cos\frac{\pi}{2}t \end{pmatrix}$$

Then the class $\Phi[(u)] \in K_0 SA$ can be represented by the formal difference of the idempotents

$$\Phi(u) := [e_0] - [e_1] = \left[v \begin{pmatrix} 1_n & 0 \\ 0 & 0 \end{pmatrix} v^{-1} \right] - \left[\begin{pmatrix} 1_n & 0 \\ 0 & 0 \end{pmatrix} \right]$$

Lemma 8.20:

Let $\psi \in X^1 A \subset X^1 RA$ be a cyclic 1-cocycle on A and let

$$\tau_1 \in X^1(S\mathbb{C}) \quad \tau_1((fdg)_\natural) = \int_0^1 f dg$$

be the fundamental class of the circle. Recall the map $\mu : X_0 R(A \otimes B) \to X_1 A \otimes X_1 B$ (Proposition 8.3).

Then for any $u \in GL(A)$

$$\langle ch(\Phi(u)), \mu^*(\psi \otimes \tau_1) \rangle = \langle ch(u), \psi \rangle = \langle u^{-1} du, \psi \rangle$$

where $\Phi(u)$ is supposed to be given by the element constructed explicitly above.

Proof:

Lengthy but straightforward.

$\square$

Proposition 8.21:

Let A be a unital admissible Fréchet algebra and let $\psi \in HC^1_{\epsilon,\alpha}(A)$. Let

$$Tr \times \psi \times \tau_1 \in HC^0_{\epsilon,\alpha}(M_\infty SA)$$

be the exterior product of $Tr \times \psi$ with the fundamental class of the circle. Then, under the isomorphism $\Phi : K_1 A \xrightarrow{\simeq} K_0(SA)$

$$\langle ch(\Phi(u)), Tr \times \psi \times \tau_1 \rangle = \langle ch(u), Tr \times \psi \rangle$$

for all $u \in K_1 A$.

Proof:

Let $[u] \in K_1 A$ be represented by $u \in GL_n(A)$ and let

$$v := w \begin{pmatrix} u & 0 \\ 0 & 1 \end{pmatrix} w^{-1} \begin{pmatrix} 1 & 0 \\ 0 & u^{-1} \end{pmatrix} \quad e := v \begin{pmatrix} 1_n & 0 \\ 0 & 0 \end{pmatrix} v^{-1}$$

be as above. Then

$$\langle ch(\Phi(u)), Tr \times \psi \times \tau_1 \rangle = \langle F(\varrho(e_0)), Tr \times \psi \times \tau_1 \rangle - \langle F(\varrho(e_1)), Tr \times \psi \times \tau_1 \rangle$$

$$= \langle F(\varrho(e_0)), Tr \times \psi \times \tau_1 \rangle$$

with

$$F(x) = x + \sum_{k=1}^{\infty} \binom{2k}{k} (x - \frac{1}{2}) (x - x^2)^k$$

(See Lemma 1.20) Remember that the exterior product on cyclic cohomology is induced by the map of complexes

$$X_*R(A \otimes B) \xrightarrow{X_*(Rm \circ i)} X_*R(RA \otimes RB) \xrightarrow{\mu} X_*RA \otimes X_*RB$$

(Lemma 8.4) so that

$$\langle F(\varrho(e_0)), Tr \times \psi \times \tau_1 \rangle = \langle (Rm \circ i) F(\varrho(e_0)), \mu^*(Tr \times \psi \otimes \pi^* \tau_1) \rangle$$

$$= \langle (R(Id \otimes \pi) \circ Rm \circ i) F(\varrho(e_0)), \mu^*(Tr \times \psi \otimes \tau_1) \rangle$$

by the naturality of μ

$$= \langle F(R(Id \otimes \pi) \circ Rm \circ i(\varrho(e_0))), \mu^*(Tr \times \psi \otimes \tau_1) \rangle = \langle F(a), \mu^*(Tr \times \psi \otimes \tau_1) \rangle$$

with

$$a := v_1 \begin{pmatrix} 1_n & 0 \\ 0 & 0 \end{pmatrix} v_1^{-1} \qquad v_1 := w \begin{pmatrix} \varrho(u) & 0 \\ 0 & 1 \end{pmatrix} w^{-1} \begin{pmatrix} 1 & 0 \\ 0 & \varrho(u^{-1}) \end{pmatrix}$$

where a exists and the sum defining $F(a)$ converges in $SRA_{(K,N)}$ provided that $(K,N) \in \mathcal{K}$ is large enough.

On the other hand, $\varrho(u)$ becomes invertible in $RA_{(K,N)}$ for (K,N) large and the equality

$$\langle ch(u), \psi \rangle = \langle \varrho(u)^{-1} d\varrho(u), Tr \times \psi \rangle = \langle ch(\Phi(\varrho(u))), \mu^*(Tr \times \psi \otimes \tau_1) \rangle$$

becomes valid by Lemma 8.20.

By definition

$$\langle ch(\Phi(\varrho(u))), \mu^*(Tr \times \psi \otimes \tau_1) \rangle = \langle F(b), \mu^*(Tr \times \psi \otimes \tau_1) \rangle$$

with

$$b := v_2 \begin{pmatrix} 1_n & 0 \\ 0 & 0 \end{pmatrix} v_2^{-1} \qquad v_2 := w \begin{pmatrix} \varrho(u) & 0 \\ 0 & 1 \end{pmatrix} w^{-1} \begin{pmatrix} 1 & 0 \\ 0 & \varrho(u)^{-1} \end{pmatrix}$$

$$b \in SRA_{(K,N)}$$

The projections

$$(\pi \otimes Id)(a) = e = (\pi \otimes Id)(b)$$

being equal to the idempotent $e = e^2$ representing $\Phi(u)$, we find by using the linear homotopy connecting a and b and the Cartan homotopy formula as in the proof of Theorem 6.19 that

$$\langle F(a), \mu^*(Tr \times \psi \otimes \tau_1) \rangle = \langle F(b), \mu^*(Tr \times \psi \otimes \tau_1) \rangle$$

which proves the proposition.

$\square$

The main result of this paragraph is

Theorem 8.22:

Let A, B be unital admissible Fréchet algebras. Let

$$a \in K_i(A),\, b \in K_j(B),\, \varphi \in HC^i_{\epsilon,\alpha}(A),\, \psi \in HC^j_{\epsilon,\alpha}(B)\,;\, i,j \in \{0,1\}.$$

Then

$$\langle ch(a \times b), \varphi \times \psi \rangle = \frac{1}{(2\pi i)^{ij}} \langle ch(a), \varphi \rangle \langle ch(b), \psi \rangle$$

Remark:

The appearance of the factor $2\pi i$ has a conceptual reason: While the product of two one-dimensional cohomology classes is automatically a zero-dimensional class because the X-complexes are $\mathbb{Z}/2$-graded, the product of two classes in K_1 lies a priori in K_2 and can only be identified via the Bott periodicity theorem with a class in K_0. The fact that the periodicity theorem (a deep transcendental result) is involved is responsible for the "period factor" $2\pi i$.

$\square$

Remark:

The theorem also justifies finally our choice of constants in the definition of the exterior product on cyclic cohomology.

$\square$

Proof:

After replacing all algebras by matrix algebras over them one may suppose all classes in K-theory to be represented by idempotent (invertible) elements rather than by idempotent (invertible) matrices. Also the trace Tr may be suppressed from the notation (This uses $(Tr \times \varphi) \times (Tr \times \psi) = Tr \times (\varphi \times \psi)$ which is true on the level of cochains).

First case: $i = j = 0$

Let $a = [e_1] \in K_0(A),\, b = [e_2] \in K_0(B)$ Then $a \times b = [e_1 \otimes e_2]$ and

$$\varphi \times \psi = (X_*(\pi \circ Rm \circ i))^*(\varphi \otimes \psi)$$

where $X_*(\pi \circ Rm \circ i)$ is given by

$$X_0 R(A \otimes B) \xrightarrow{X_*(Rm \circ i)} X_0 R(RA \otimes RB) \xrightarrow{X_* \pi} X_0(RA \otimes RB) = X_0(RA) \otimes X_0(RB)$$

Therefore

$$\langle ch(a \times b), \varphi \times \psi \rangle = \langle F(\varrho(e_1 \otimes e_2)), (\pi \circ Rm \circ i)^*(\varphi \otimes \psi) \rangle$$

$$= \langle ((\pi \circ Rm \circ i) \, F(\varrho(e_1 \otimes e_2)), \varphi \otimes \psi \rangle = \langle F((\pi \circ Rm \circ i)(\varrho(e_1 \otimes e_2))), \varphi \otimes \psi \rangle$$

$$= \langle F((\varrho(e_1)) \otimes \varrho(e_2)), \varphi \otimes \psi \rangle$$

The sum defining $F((\varrho(e_1)) \otimes \varrho(e_2))$ converges in $RA_{(K,N)} \otimes_\pi RB_{(K',N')}$ for $(K, N) \in \mathcal{K}(A)$, $(K', N') \in \mathcal{K}(B)$ large enough. On the other hand

$$\langle ch(a), \varphi \rangle \langle ch(b), \psi \rangle = \langle F(\varrho(e_1)), \varphi \rangle \langle F(\varrho(e_2)), \psi \rangle$$

$$= \langle F(\varrho(e_1)) \otimes F(\varrho(e_2)), \varphi \otimes \psi \rangle$$

As

$$\pi_A \otimes \pi_B(F((\varrho(e_1)) \otimes \varrho(e_2))) = e_1 \otimes e_2 = \pi_A \otimes \pi_B(F(\varrho(e_1)) \otimes F(\varrho(e_2)))$$

we may conclude with the help of Lemma 5.10 and by applying the Cartan homotopy formula as in the proof of Theorem 6.19.

Second case: $i = 0, j = 1$

Let $a \in K_0(A)$, $b \in K_1(B) \xrightarrow[\Phi]{\approx} K_0(\mathcal{S}B)$. Then

$$\langle ch(a \times b), \varphi \times \psi \rangle = \langle ch(\Phi(a \times b)), (\varphi \times \psi) \times \tau_1 \rangle$$

by Proposition 8.21

$$= \langle ch(a \times \Phi(b)), \varphi \times (\psi \times \tau_1) \rangle$$

by definition of Φ and the associativity of the exterior product on cohomology

$$= \langle ch(a), \varphi \rangle \langle ch(\Phi(b)), \psi \times \tau_1 \rangle$$

by the first case treated above

$$= \langle ch(a), \varphi \rangle \langle ch(b), \psi \rangle$$

by Proposition 8.21 again.

Third case: $i = j = 1$

Let $a \in K_1(A)$, $b \in K_1(B)$. We find

$$\langle ch(a), \varphi \rangle \langle ch(b), \psi \rangle = \langle ch(\Phi(a)), \varphi \times \tau_1 \rangle \langle ch(\Phi(b)), \psi \times \tau_1 \rangle$$

by Proposition 8.21

$$= \langle ch(\Phi(a) \times \Phi(b)), (\varphi \times \tau_1) \times (\psi \times \tau_1) \rangle$$

by the first case treated above

$$= \langle ch((a \times b) \times Bott), (\varphi \times \psi) \times (\tau_1 \times \tau_1) \rangle$$

Here the isomorphism

$$A \otimes_\pi S\mathbb{C} \otimes_\pi B \otimes_\pi S\mathbb{C} \simeq A \otimes_\pi B \otimes_\pi S^2\mathbb{C}$$

has been applied. Furthermore, the associativity of the exterior product and the identity

$$\Phi(a) \times \Phi(b) = (a \times b) \times Bott \in K_0(S^2(A \otimes_\pi B))$$

have been used. The last expression equals

$$\langle ch(a \times b), \varphi \times \psi \rangle \langle ch(Bott), \tau_1 \times \tau_1 \rangle$$

by the first case above so that it remains to calculate

$$\langle ch(Bott), \tau_1 \times \tau_1 \rangle$$

Let $\pi : [0,1] \to S^1 = \mathbb{R}/\mathbb{Z}$ be the projection map and let $\pi \times \pi : [0,1] \times [0,1] \to T^2$ be its square. Both the Bott-element and the cyclic cocycle $\tau_1 \times \tau_1$ on $S^2\mathbb{C}$ are pullbacks of corresponding elements on T^2:

$$Bott = (\pi \times \pi)_* \beta' \qquad \tau_1 \times \tau_1 = (\pi \times \pi)^*(\tau_1' \times \tau_1')$$

where $\tau_1' \times \tau_1' \in HC^2(C^\infty(T^2)) \simeq H_2(T^2, \mathbb{C})$ equals the fundamental class. To evaluate the pairing

$$\langle ch(Bott), \tau_1 \times \tau_1 \rangle = \langle ch(\beta'), \tau_1' \times \tau_1' \rangle$$

note that β' is the pullback of the canonical Bott element

$$\beta \in K^0(S^2, pt) = K_0(C_0(\mathbb{R}^2)) = K_0(SC^\infty(S^1))$$

under a smooth degree one map $f : T^2 \to S^2$. As the homological fundamental classes correspond to each other under such a map the above pairing can also be evaluated on S^2. On S^2 it is known finally that the Bott element is the image of the fundamental class

$$[u] = [t \to exp(2\pi it)] \in K_1(C^\infty(S^1))$$

under

$$\Phi : K_1(C^\infty(S^1)) \to K_0(SC^\infty(S^1))$$

whereas the homological fundamental class of $C^\infty(S^2)$ equals again $\tau_1 \times \tau_1$ when restricted to $SC^\infty(S^1)$.

So we conclude by Proposition 8.21 that

$$\langle ch(Bott), \tau_1 \times \tau_1 \rangle = \langle ch(u), \tau_1 \rangle = \tau_1(u^{-1}du)$$

$$= \int_0^1 exp(-2\pi it) d\, exp(2\pi it) = 2\pi i$$

This terminates the proof of Theorem 8.22.

□

8-4 Slant products

From the Künneth isomorphism in periodic cyclic homology one knows that for unital algebras A, B the chain map

$$\times : X_* \widehat{R}(A \otimes B) \to X_* \widehat{R}A \widehat{\otimes} X_* \widehat{R}B$$

constructed in 8.7, which defines the exterior product, has a homotopy inverse, i.e. there exists a chain map

$$X_* \widehat{R}A \widehat{\otimes} X_* \widehat{R}B \to X_* \widehat{R}(A \otimes B)$$

that provides an inverse up to homotopy of $\times$. (To verify this claim strictly we should first of all identify our product with Connes's product up to suitable normalization constants, which we have not done, but which seems probable in light of the compatibility of both products with the product in K-theory.)

Such a chain map could be used to construct an exterior product on the bivariant X-complexes generalizing the composition product and the exterior product.

As a special case we would then obtain a pairing

$$\backslash : X_* RA \widehat{\otimes} X^* R(A \otimes B) \to X^* RB$$

satisfying

$$\alpha \backslash (\beta \times \gamma) = \langle \alpha, \beta \rangle \gamma$$

for $\alpha \in X_j RA,\ \beta \in X^j RA,\ \gamma \in X^k RB$

and by composing with the Chern character a pairing

$$\backslash : K_*(A) \widehat{\otimes} HC^*(A \otimes B) \to HC^*(B)$$

While we have not constructed a homotopy inverse of the exterior product yet, the slant product with K-theory can be defined in a simple and direct manner.

See also the remark at the end of the introduction of this chapter.

Theorem 8.23:

Let A, B be admissible Fréchet algebras and suppose that B is unital.

a) There exists a natural homomorphism

$$\backslash : K_*(B) \to HC^*_{\epsilon,\alpha}(A, A \otimes_\pi B)$$

called the "**slant product**".

b) The naturality can be expressed as follows:

If $f \in [A, A']_{smooth}$ $g \in [B, B']_{smooth}$ (resp. $f \in [A, A']_\alpha$ $g \in [B, B']_\alpha$) are smooth (unital) homotopy classes of (asymptotic) morphisms, then, for $x \in K_*(B)$ the cohomology classes

$$x\backslash \in HC^*_{\epsilon,\alpha}(A, A \otimes_\pi B), \ (f \otimes g)_* \in HC^0_{\epsilon,\alpha}(A \otimes_\pi B, A' \otimes_\pi B')$$

$$f_* \in HC^0_{\epsilon,\alpha}(A, A') \text{ and } g_*(x)\backslash \in HC^*_{\epsilon,\alpha}(A', A' \otimes_\pi B')$$

satisfy

$$(f \otimes g)_* \circ (x\backslash) = (g_*(x)\backslash) \circ f_*$$

in $HC^*_{\epsilon,\alpha}(A, A' \otimes_\pi B')$, i.e. the diagram

$$
\begin{array}{ccc}
"A" & \xrightarrow{x\backslash} & "A \otimes_\pi B" \\
f_* \downarrow & & \downarrow (f \otimes g)_* \\
"A'" & \xrightarrow{g_*(x)\backslash} & "A' \otimes_\pi B'"
\end{array}
$$

commutes.

Proof:

Construction of the slant product:

Even case: For unital B the group $K_0(B)$ can be represented by formal differences of (piecewise smooth) homotopy classes of idempotent matrices over B. An idempotent matrix $e \in M_n(B)$ defines a homomrphism of algebras

$$
\begin{array}{cccc}
e_* : & A & \to & A \otimes M_n(B) \\
& a & \to & a \otimes e
\end{array}
$$

Put

$$
\begin{array}{cccc}
\backslash : & K_0(B) & \to & HC^0_{\epsilon,\alpha}(A, A \otimes_\pi B) \\
\\
& e & \to & [Id_*^{A \otimes B} \times Tr_{M_n \mathbb{C}}] \circ [e_*]
\end{array}
$$

It is clear that the map is well defined because $HC^*_{\epsilon,\alpha}$ are smooth homotopy functors.

Odd case:

Let $\tau_1 \in HC^1(S\mathbb{C})$ the fundamental class of the circle and denote by Φ the isomorphism $\Phi : K_1 A \to K_0 SA$. Define then $\backslash$ to be the composition

$$K_1(B) \xrightarrow{\Phi} K_0 SB \xrightarrow{\backslash} HC^0_{\epsilon,\alpha}(A, SA \otimes_\pi B) \xrightarrow{(\tau_1 \times id_*)} HC^1_{\epsilon,\alpha}(A, A \otimes_\pi B)$$

Naturality of the slant product:

The naturality of the construction under continuous morphisms of admissible Fréchet algebras is obvious. With respect to asymptotic morphisms, we consider the cases $f = id$ and $g = id$ separately from which the general case follows. The last one poses no problems. So let $g : B \to B'$ be a smooth asymptotic morphism and let $e = e^2 \in A$ be an idempotent in A. Then

$$g_*([e]) \in K_0 B'$$

can be represented by the idempotent

$$e' := F(g(e)) \in B'$$

where F is the power series of 1.20 (We suppress asymptotic parameters and suppose them always to be chosen such that the curvature of f at (e, e) is sufficiently small.)

The slant product

$$g_*([e]) \backslash \in HC_\alpha^0(A, A \otimes_\pi B')$$

is then induced by the morphism $a \to a \otimes F(g(e))$ whereas

$$g_* \circ (e\backslash) \in HC_{\epsilon,\alpha}^0(A, A \otimes_\pi B')$$

is induced by $a \to a \otimes g(e)$. As these asymptotic morphisms are homotopic by a linear homotopy the conclusion follows.

$\square$

Theorem 8.24:

Let A, B, C be admissible Fréchet algebras and suppose A to be unital.

There exist natural bilinear pairings

$$K_*A \,\widehat{\otimes}\, HC_{\epsilon,\alpha}^*(A \otimes_\pi B) \to HC_{\epsilon,\alpha}^*(B)$$

$$K_*A \,\widehat{\otimes}\, HC_{\epsilon,\alpha}^*(A \otimes_\pi B, C) \to HC_{\epsilon,\alpha}^*(B, C)$$

$$HC_{\epsilon,\alpha}^*(B, C) \,\widehat{\otimes}\, K_*A \to HC_{\epsilon,\alpha}^*(B, A \otimes_\pi C)$$

which are given by applying slant- and composition-products.

Let $x \in K_i(A)$, $\varphi \in HC_{\epsilon,\alpha}^i(A)$, $\psi \in HC_{\epsilon,\alpha}^j(B)$, (resp. $\psi \in HC_{\epsilon,\alpha}^j(B, C)$). Then the equality

$$x \backslash (\varphi \times \psi) = \langle ch(x), \varphi \rangle \, \psi$$

holds.

Proof:

The existence of the pairings is clear as well as several naturality properties which we do not state explicitly but which áre easily derived. It remains to check the compatibility of the slant- and the exterior product claimed in b). By definition of the exterior product the chain map $\times \circ e\backslash$ may be obtained by taking either of the two possible paths around the commutative diagram

$$
\begin{array}{ccc}
\overline{X}_* \mathcal{R} \widetilde{A} & \xrightarrow{e\backslash} & \overline{X}_* \mathcal{R}(A \widetilde{\otimes}_\pi B) \\
\downarrow & & \downarrow \\
\overline{X}_* \mathcal{R}(\widetilde{A} \otimes \widetilde{\mathbb{C}}) & \rightarrow & \overline{X}_* \mathcal{R}(\widetilde{A} \otimes \widetilde{B}) \\
\mu' \downarrow & & \downarrow \mu' \\
\overline{X}_* \mathcal{R}(\widetilde{A}) \widehat{\otimes}_\pi \overline{X}_* \mathcal{R}(\widetilde{\mathbb{C}}) & \xrightarrow{Id \otimes e\backslash} & \overline{X}_* \mathcal{R}(\widetilde{A}) \widehat{\otimes}_\pi \overline{X}_* \mathcal{R}(\widetilde{B})
\end{array}
$$

However the identity map on $\overline{X}_* \mathcal{R}(\widetilde{\mathbb{C}})$ is chain homotopic to the composition

$$
\overline{X}_* \mathcal{R}(\widetilde{\mathbb{C}}) \xrightarrow{qis} X_* \mathcal{R}\mathbb{C} \xrightarrow{X_* \pi} X_* \mathbb{C} = \mathbb{C} \rightarrow \overline{X}_* \mathcal{R}(\widetilde{\mathbb{C}})
$$
$$
1 \rightarrow ch(e)
$$

so that $\times \circ e\backslash$ may be represented as well by the chain map

$$
\overline{X}_* \mathcal{R}(\widetilde{A}) \rightarrow \overline{X}_* \mathcal{R}(\widetilde{A}) \otimes \mathbb{C} \rightarrow \overline{X}_* \mathcal{R}(\widetilde{A}) \widehat{\otimes} \overline{X}_* \mathcal{R}(\widetilde{B})
$$

$$
\alpha \rightarrow \alpha \otimes 1 \rightarrow \alpha \widehat{\otimes} ch(e)
$$

from which the conclusion

$$
e\backslash(\varphi \times \psi) = (\varphi \widehat{\otimes} \psi) \circ \times \circ e\backslash = \varphi \langle ch(e), \psi \rangle
$$

readily follows.

In odd dimensions

$$
u\backslash(\varphi \times \psi) = \Phi(u)\backslash(\varphi \times \psi) \times \tau_1
$$

by definition of the slant product

$$
= \Phi(u)\backslash\varphi \times (\psi \times \tau_1) = \varphi \langle ch(\Phi(u)), \psi \times \tau_1 \rangle
$$

by the even case just treated

$$
= \varphi \langle ch(u), \psi \rangle
$$

by Proposition 8.21.

$\square$

Chapter 9: Exact sequences

The possibility of calculating asymptotic cohomology groups relies essentially on its excision properties. The main drawback of the theory as it stands is that we cannot say anything about this question if we regard the category of all admissible Fréchet algebras. It is only after suspension that we get positive "stable" results, but it remains desirable to obtain unstable excision theorems because suspension destroys the purely algebraic-analytic character of the theory.

The most basic excision problem concerns the extension

$$0 \to SA \to CA \to A \to 0$$

where a corresponding exact sequence for bivariant cohomology would provide natural HC-equivalences

$$\beta \in HC_\alpha^1(A, SA) \quad \alpha \in HC_\alpha^1(SA, A)$$

This would imply a cohomological periodicity theorem

$$HC_\alpha^*(A, B) \simeq HC_\alpha^{*+1}(SA, B) \simeq HC_\alpha^{*+1}(A, SB)$$

Whereas a natural candidate for the HC-equivalence $\alpha \in HC_\alpha^1(SA, A)$ is provided by the exterior product with the fundamental class of the circle there seems at the moment to be no reasonable definition of $\beta \in HC_\alpha^1(A, SA)$.

The case $A = \mathbb{C}$ is the only one where α can be shown to be an unstable HC-equivalence, which proves the important fact that the coefficient groups $HC_\alpha^*(S\mathbb{C}, S\mathbb{C})$ of stable asymptotic cohomology are what they ought to be.

The most straightforward approach to cohomological periodicity and the construction of β is to derive them from Bott periodicity in K-theory. The suitable version of Bott-periodicity was developed by Connes and Higson [CH] who construct (nonlinear) asymptotic morphisms

$$\beta_E : SA \otimes_{\mathbb{C}^*} \mathcal{K} \to S^3 A \otimes_{\mathbb{C}^*} \mathcal{K} \quad \alpha_E : S^3 A \otimes_{\mathbb{C}^*} \mathcal{K} \to SA \otimes_{\mathbb{C}^*} \mathcal{K}$$

which are homotopy inverse to each other and induce the Bott-periodicity map in K-theory. These morphisms do not exist unstably because it is essential that the source of an asymptotic morphism describing a Bott element is a "group up to homotopy", which is true for suspensions but not for general algebras.

Using this approach it is possible to define an element

$$\beta' \in HC_\alpha^0(SA, S^3 A)$$

and finally a stable Bott element

$$\beta \in HC_\alpha^1(SA, S^2 A)$$

by taking the exterior product with the fundamental class of the circle. The construction of β' is based on the observation that although nonlinear, the restriction of the Connes-Higson morphism β_E to the "smooth" subalgebra $\mathcal{C}_0^\infty(]0, 1[, A)$ of

SA may be represented by a linear asymptotic morphism. The inclusion of this subalgebra into SA is an asymptotic HC-equivalence by the derivation lemma.

The stable cohomological periodicity theorem

$$HC_\alpha^*(SA, B) \simeq HC_\alpha^{*+1}(S^2A. B) \quad HC_\alpha^*(A, SB) \simeq HC_\alpha^{*+1}(A, S^2B)$$

follows by showing that the stable Bott element is an asymptotic HC-inverse of the Dirac element $\alpha \in HC_\alpha^1(S^2A, SA)$. The demonstration is based on Atiyah's rotation trick but rather cumbersome due to the necessity of descending to "smooth" subalgebras which are not preserved during the constructions in the proof.

The general excision theorems in asymptotic cyclic cohomology can easily be derived from the stable periodicity theorem. A well known argument from algebraic topology showing that fibrations and cofibrations coincide in the stable homotopy category can be used to derive for each homomorphism

$$f : A \to B$$

of admissible Fréchet algebras two six term exact cofibration (Puppe)-sequences relating the stable asymptotic cohomology of $f : A \to B$ to that of the mapping cone C_f of f. If

$$0 \to J \to A \xrightarrow{f} B \to 0$$

is an extension of separable C^*-algebras (i.e. if J possesses a bounded, positive, quasicentral approximate unit), then it was shown by Connes and Higson [CH] that the ideal J is stably asymptotically homotopy equivalent to the mapping cone C_f. By descending to suitable "smooth" subalgebras to gain linearity of the considered asymptotic homotopy equivalences we show that the kernel J of f is stably asymptotically HC-equivalent to the mapping cone of f if f possesses a bounded linear splitting. This provides six term exact sequences for the stable asymptotic cohomology of an extension of separable C^*-algebras with bounded linear section.

Finally some comment on the difficulties in constructing an unstable Bott element $\beta \in HC_\alpha^1(A, SA)$ for arbitrary admissible Fréchet algebras. There are several ways to obtain chain maps

$$\beta : X_*RA \to X_{*+1}R(SA)$$

with the right algebraic properties: one can use either H-unitality and excision for ordinary cyclic homology following Wodzicki or can mimic the procedure of Elliott, Natsume and Nest. They obtain a Bott element from the deformation of the crossed product of the algebra of Schwartz functions on the real line with the group $\mathbb{R}$ acting by rescaling the Pontrjagin dual real line and derive Bott periodicity from Takai duality. Finally there have been attempts of Lott to define a bivariant Bott element by using the superconnection formalism of Quillen.

In all these approaches a typical feature arises: the obtained chain map is such that the image of a tensor with entries belonging to a compact subset of A will be a tensor with entries in the multiplicative closure of the union of the given compact subset of A and the set of elements of a bounded, approximate unit of $S\mathbb{C}$, which does not form a compact set anymore. Therefore a theory allowing the entries of tensors to belong only to compact sets is unlikely to possess a bivariant Bott element. But on the other hand it is not possible to modify the theory by allowing the entries of tensors over a nonunital algebra to belong to the union of a compact set and a bounded, approximate unit because approximate units are not preserved under homomorphisms of algebras. Some further study is necessary to find the way out of this dilemma.

9-1 The stable periodicity theorem

Construction of a stable Bott-element:

The Bott class β generating the infinite cyclic group

$$K_0(S^2\mathbb{C}) := Ker(K_0(C(S^2)) \xrightarrow{eval_\infty} K_0(\mathbb{C}))$$

can be represented by the formal difference

$$\beta = [e_0] - [e_1]$$

of two idempotent matrices

$$e_0 := \frac{1}{1+|z|^2} \begin{pmatrix} |z|^2 & z \\ \bar{z} & 1 \end{pmatrix} \qquad e_1 := \begin{pmatrix} 1 & 0 \\ 0 & 0 \end{pmatrix}$$

in $M_2(C(S^2)) = M_2(C(\mathbb{C} \cup \{\infty\}))$ which coincide at ∞.

Due to Connes-Higson there exists a corresponding class in

$$E(\mathbb{C}, S^2\mathbb{C}) = [[S\mathbb{C}, S^3\mathbb{C} \otimes_{C^\cdot} \mathcal{K}]]$$

the group of homotopy classes of (nonlinear) asymptotic morphisms from $S\mathbb{C}$ to $S^3\mathbb{C} \otimes_{C^\cdot} \mathcal{K}$. Its image under $E(\mathbb{C}, S^2\mathbb{C}) \to E(\mathbb{C}, C(S^2))$ coincides with $(e_0)_* - (e_1)_*$ (where we have identified an idempotent in $C(S^2) \otimes_{C^\cdot} \mathcal{K}$ with a homomorphism of $\mathbb{C}$ to the latter algebra). We will adopt their construction to obtain a Bott class in bivariant asymptotic cohomology.

Definition 9.1:

Let $(u_t)_{t\in\mathbb{R}_+}$, $0 \leq u_t \leq 1$ be a continuous, bounded approximate unit for $S^2\mathbb{C} = C_0(\mathbb{R}^2)$. Then

$$"Bott" : \quad S\mathbb{C} = C_0(]-1,1[) \quad \to \quad C(\mathbb{R}_+, M_4(S^3\mathbb{C}))$$

$$f \quad \to \quad f(D_t)\begin{pmatrix} e_0 & 0 \\ 0 & e_1 \end{pmatrix}$$

$$D_t := \begin{pmatrix} \sin\frac{\pi}{2}x & (\cos\frac{\pi}{2}x)(1-u_t) \\ (\cos\frac{\pi}{2}x)(1-u_t) & -\sin\frac{\pi}{2}x \end{pmatrix}$$

defines a linear, asymptotic morphism $S\mathbb{C} \to M_4(S^3\mathbb{C})$).

Here the third suspension coordinate corresponds to "x" and the first two coordinates correspond to $S^2\mathbb{C}$.

□

For a proof see 10.2.

Lemma 9.2:

The diagram

$$S\mathbb{C} \xrightarrow{\;"Bott"\;} M_4(S^3\mathbb{C}) \quad = \quad M_4(S^3\mathbb{C})$$

$$\downarrow e_0 \otimes id_{S\mathbb{C}} \vee e_1 \otimes \breve{id}_{S\mathbb{C}} \qquad\qquad\qquad\qquad \downarrow i$$

$$(M_2(\mathcal{C}^\infty(S^2)) \otimes_\pi S\mathbb{C})^2 \quad \to \quad (M_4(\mathcal{C}^\infty(S^2)) \otimes_\pi S\mathbb{C} \quad \to \quad M_4(C_0(\mathbb{R} \times S^2))$$

commutes up to homotopy. Here $\breve{id} : S\mathbb{C} \to S\mathbb{C}$ denotes the canonical inversion corresponding to the degree -1 involution of the unit interval and $\vee : S\mathbb{C} \to S\mathbb{C} \oplus S\mathbb{C}$ is the splicing map (see 9.4.).

Proof:

Consider the family of asymptotic morphisms

$$F_\lambda : \quad S\mathbb{C} \quad \to \quad M_4(C_0(\mathbb{R} \times S^2))$$
$$f \quad \to \quad f(D_{t,\lambda}) \begin{pmatrix} e_0 & 0 \\ 0 & e_1 \end{pmatrix}$$

$$D_{t,\lambda} := \begin{pmatrix} sin\frac{\pi}{2}x & (1-\lambda)(cos\frac{\pi}{2}x)(1-u_t) \\ (1-\lambda)(cos\frac{\pi}{2}x)(1-u_t) & -sin\frac{\pi}{2}x \end{pmatrix}$$

($D_{t,\lambda}$ is constructed from D_t by replacing the approximate unit (u_t) of $S^2\mathbb{C}$ by $\lambda 1_{C(S^2)} + (1-\lambda)u_t$). Then $F_0 = i \circ "Bott"$ and

$$F_1 : f \to \begin{pmatrix} f(sin\frac{\pi}{2}x)e_0 & 0 \\ 0 & f(-sin\frac{\pi}{2}x)e_1 \end{pmatrix}$$

is naturally homotopic to $e_0 \otimes id_{S\mathbb{C}} \vee e_1 \otimes \breve{id}_{S\mathbb{C}}$ as $x \to sin\frac{\pi}{2}x$ defines a homeomorphism of $[-1, 1]$ which is isotopic to the identity.

Definition 9.3.a):

Let A be an admissible Fréchet algebra. Let $\tau_1 \in HC^1(S\mathbb{C})$ be the fundamental class of the unit interval. Consider the diagrams

$$SA \xleftarrow{\;k^A\;} SA \xrightarrow{\;"Bott" \otimes_\pi Id^A\;} M_4(S^3\mathbb{C}) \otimes_\pi A$$

$$M_4(S^3\mathbb{C}) \otimes_\pi A \xrightarrow{\;k'\;} M_4(S^3A) \xleftarrow{\;k''\;} M_4(S\mathbb{C}) \otimes_\pi S^2A$$

of (asymptotic) morphisms, where k, k'' are asymptotic HC-equivalences by the derivation lemma (see Chapter 7). Put then

$$\beta_{SA} := [\frac{1}{2\pi i}(Tr \times \tau_1) \times id_*^{S^2A}] \circ [k''_*]^{-1} \circ [k'_*] \circ ["Bott" \otimes_\pi Id^A] \circ [k_*^A]^{-1}$$

$$\beta_{SA} \in HC^1_\alpha(SA, S^2A)$$

is called the **"stable Bott element"**.

□

If $f : A \to B$ is an (asymptotic) morphism the diagram

$$
\begin{array}{ccc}
"SA" & \xrightarrow{\ \beta_{SA}\ } & "S^2A" \\
\scriptstyle{Sf_*}\big\downarrow & & \big\downarrow\scriptstyle{S^2f_*} \\
"SB" & \xrightarrow[\ \beta_{SB}\ ]{} & "S^2B"
\end{array}
$$

commutes as is immediately clear from the definitions.

Definition 9.3.b):

Let A be an admissible Fréchet algebra. Then

$$\alpha_A := [\tau_1 \times id^A_*] \circ [k^A_*]^{-1} \in HC^1_\alpha(SA, A)$$

is called the **"Dirac element"**.

□

The diagram

$$
\begin{array}{ccc}
"SA" & \xrightarrow{\ \alpha_A\ } & "A" \\
\scriptstyle{Sf_*}\big\downarrow & & \big\downarrow\scriptstyle{f_*} \\
"SB" & \xrightarrow[\ \alpha_B\ ]{} & "B"
\end{array}
$$

commutes as for the stable Bott element.

Theorem 9.4: (Stable periodicity theorem)

Let A be an admissible Fréchet algebra. The stable Bott- and Dirac-elements

$$\beta_{SA} \in HC^1_\alpha(SA, S^2A) \quad \alpha_{SA} \in HC^1_\alpha(S^2A, SA)$$

define asymptotic HC-equivalences inverse to each other:

$$\alpha_{SA} \circ \beta_{SA} = id^{SA}_* \in HC^0(SA, SA)$$

$$\beta_{SA} \circ \alpha_{SA} = id^{S^2A}_* \in HC^0(S^2A, S^2A)$$

Proof:

First part:
$$\alpha_{SA} \circ \beta_{SA} = id_*^{SA} \in HC^0(SA, SA)$$

By definition $\alpha_{SA} \circ \beta_{SA}$ equals

$$[\tau_1 \times id_*^{SA}] \circ [k_*^{SA}]^{-1} \circ [\frac{1}{2\pi i}(Tr \times \tau_1) \times id^{S^2 A}] \circ [k_*'']^{-1} \circ [k_*'] \circ [" Bott" \otimes_\pi Id^A] \circ [k_*]^{-1}$$

First of all note that the diagram

$$
\begin{array}{ccccccc}
& & M_4(S^2 SA) & & \xrightarrow{k'''} & M_4(S^3 A) & \\
& & \| & & & \uparrow & k'' \\
M_4(S^2 SA) & = & M_4(S^2 SA) & & \rightarrow & M_4(SS^2 A) & \\
& & \downarrow & [(Tr \times \tau_1)]\times & & \downarrow & [(Tr \times \tau_1)]\times \\
[\tau_2]\times \downarrow & & SSA & & \xrightarrow{k^{SA}} & S^2 A & \\
& & \downarrow & & [\tau_1]\times & & \\
SA & = & SA & & & &
\end{array}
$$

commutes where

$$\tau_2 := \frac{1}{2\pi i}(Tr \times \tau_1 \times \tau_1) \in HC^2(M_4(S^2 \mathbb{C}))$$

(Here we denote by S the smooth and by S the continuous suspensions.)

Therefore $\alpha_{SA} \circ \beta_{SA}$ equals

$$(*) \quad [\tau_2 \times id_*^{SA}] \circ [k_*''']^{-1} \circ [k_*'] \circ [" Bott" \otimes_\pi Id^A] \circ [k_*]^{-1}$$

The cohomology class $\tau_2 \in HC^2(M_4(S^2 \mathbb{C}))$ is known to be induced from a class $\tau_2' \in HC^2(M_4(\mathcal{C}^\infty(S^2)))$ which is in fact the fundamental class of the two sphere. This gives rise by lemma 9.2. to the diagram

$$
\begin{array}{ccc}
& SA & \xrightarrow{\varphi} (M_2(\mathcal{C}^\infty(S^2)) \otimes_\pi SA)^2 \\
& & \downarrow \\
k' \circ (" Bott" \otimes id^A) \quad \downarrow & & M_4(\mathcal{C}^\infty(S^2)) \otimes_\pi SA \\
& & \downarrow \quad k''' \\
M_4(S^3 A) & \rightarrow & M_4(C_0(\mathbb{R} \times S^2, A)) \\
k''' \quad \uparrow & & \uparrow \quad k''' \\
M_4(S^2 \mathbb{C}) \otimes_\pi SA & \rightarrow & M_4(\mathcal{C}^\infty(S^2)) \otimes_\pi SA \\
\tau_2 \times \quad \downarrow & & \downarrow \quad \tau_2' \times
\end{array}
$$

which commutes up to homotopy, where

$$\varphi = (e_0 \otimes id \vee e_1 \otimes \check{id}) \circ k^A$$

From this we derive

$$\alpha_{SA} \circ \beta_{SA} =$$

$$= [\tau_2 \times id_*^{SA}] \circ [k_*''']^{-1} \circ [k_*'] \circ [" Bott" \otimes_\pi Id^A] \circ [k_*]^{-1}$$

$$= [\tau_2' \times id_*^{SA}] \circ [k_*''']^{-1} \circ [k_*'''] \circ [(e_0 \otimes id \vee e_1 \otimes \check{id})_*]$$

$$= [\tau_2' \times id_*^{SA}] \circ [(e_0 \otimes id \vee e_1 \otimes \check{id})_*]$$

$$= [\tau_2' \times id_*^{SA}] \circ [(e_0 \otimes id_{SA})_*] + [\tau_2' \times id_*^{SA}] \circ [(e_1 \otimes \check{id}_{SA})_*]$$

(as will be verified at the end of the demonstration)

$$= e_0 \backslash [\tau_2' \times id_*^{SA}] + e_1 \backslash [\tau_2' \times id_*^{SA}]$$

by definition of the slant product

$$= \langle ch(e_0), \tau_2' \rangle [id_*^{SA}] + \langle ch(e_1), \tau_2' \rangle [\check{id}_*^{SA}]$$

by theorem 8.24

$$= 1[id_*^{SA}] + 0[\check{id}_*^{SA}] = [id_*^{SA}]$$

by the well known integrality properties of the ordinary Chern character (see [CO]).

Second Part:

$$\beta_{SA} \circ \alpha_{SA} = id_*^{S^2 A}$$

By definition $\beta_{SA} \circ \alpha_{SA}$ equals

$$[\frac{1}{2\pi i}(Tr \times \tau_1) \times id^{S^2 A}] \circ [k_*'']^{-1} \circ [k_*'] \circ [(\text{"}Bott\text{"} \otimes_\pi Id^A)_*] \circ [k_*^A]^{-1} \circ [\tau_1 \times id_*^{SA}] \circ [k_*^{SA}]^{-1}$$

Note that all bivariant classes involved are induced by exterior products or (asymptotic) morphisms. The naturality of the exterior product with respect to (asymptotic) morphisms implies then

$$[k_*'']^{-1} \circ [k_*'] \circ [(\text{"}Bott\text{"} \otimes_\pi Id^A)_*] \circ [k_*^A]^{-1} \circ [\tau_1 \times id_*^{SA}] =$$

$$= [\tau_1 \times id_*^{M_4(S\mathbb{C}) \otimes_\pi S^2 A}] \circ [Sk_*'']^{-1} \circ [Sk_*'] \circ [S(\text{"}Bott\text{"} \otimes_\pi Id^A)_*] \circ [Sk_*^A]^{-1}$$

(S still denoting the smooth suspension). Inserting this into the formula for $\beta_{SA} \circ \alpha_{SA}$ yields

$$\beta_{SA} \circ \alpha_{SA} =$$

$$= [\frac{1}{2\pi i}(Tr \times \tau_1) \times id^{S^2 A}] \circ [\tau_1 \times id_*^{M_4(S\mathbb{C}) \otimes_\pi S^2 A}] \circ [Sk_*'']^{-1} \circ$$

$$\circ [Sk_*'] \circ [S(\text{"}Bott\text{"} \otimes_\pi Id^A)_*] \circ [Sk_*^A]^{-1} \circ [k_*^{SA}]^{-1}$$

$$= [\tau_2 \times id^{S^2 A}] \circ [Sk_*'']^{-1} \circ [Sk_*'] \circ [S(\text{"}Bott\text{"} \otimes_\pi Id^A)_*] \circ [Sk_*^A]^{-1} \circ [k_*^{SA}]^{-1}$$

Now there is a commutative diagram

$$
\begin{array}{ccc}
& S^2A & \xrightarrow{sw} & S^2A \\
Sk^A & \uparrow & & \uparrow & k^{SA} \\
& S(SA) & & S(SA) \\
k^{SA} & \uparrow & & \uparrow & S(k^A) \\
& S(SA) & & S(SA) \\
\psi & \downarrow & & \downarrow & \psi' \\
& M_4(S^3) \otimes_\pi SA & & S(M_4(S^3\mathbb{C}) \otimes_\pi A) \\
k' & \downarrow & & \downarrow & Sk' \\
& M_4(S^3SA) & & S(M_4(S^3A)) \\
k'' & \uparrow & & \uparrow & Sk'' \\
& M_4(S\mathbb{C}) \otimes_\pi S^2SA & \xrightarrow[SW]{} & S(M_4(S\mathbb{C}) \otimes_\pi S^2A)
\end{array}
$$

where

$$ \psi = "Bott" \otimes_\pi Id^{SA} \quad \psi' = S("Bott" \otimes_\pi Id^A) $$

and

$$ sw : S^2A \to S^2A $$

$$
\begin{array}{ccc}
SW: & M_4(S\mathbb{C}) \otimes_\pi S^2SA & \xrightarrow{\cong} & S(M_4(S\mathbb{C}) \otimes_\pi S^2A) \\
& \cap & & \cap \\
& M_4(S^4A) & \xrightarrow{\cong} & M_4(S^4A)
\end{array}
$$

are the isomorphisms given in suspension coordinates by the matrices

$$ sw = \begin{pmatrix} 0 & -1 \\ 1 & 0 \end{pmatrix} \quad SW = \begin{pmatrix} 0 & 0 & 0 & -1 \\ 1 & 0 & 0 & 0 \\ 0 & 1 & 0 & 0 \\ 0 & 0 & 1 & 0 \end{pmatrix} $$

respectively. Note that

$$ [sw_*] = [id_*] \in HC^0_\alpha(S^2A, S^2A), \ [SW_*] = [id_*] \in HC^0_\alpha(S^4A, S^4A) $$

as both permutation matrices sw, SW are of positive determinant and thus connected by a continuous path to the identity matrix giving rise to a homotopy of the induced maps on suspensions connecting the switch maps to the identity.

This implies the equality

$$ \beta_{SA} \circ \alpha_{SA} = $$

$$ = [\tau_2 \times id^{S^2A}] \circ [Sk''_*]^{-1} \circ [Sk'_*] \circ [S("Bott" \otimes_\pi Id^A)_*] \circ [Sk^A_*]^{-1} \circ [k^{SA}_*]^{-1} $$

$$ = [\tau_2 \times id^{S^2A}] \circ [Sk''_*]^{-1} \circ [Sk'_*] \circ [S("Bott" \otimes_\pi Id^A)_*] \circ [Sk^A_*]^{-1} \circ [k^{SA}_*]^{-1} \circ [sw_*] $$

$$ = [\tau_2 \times id^{S^2A}] \circ [SW_*] \circ [k''_*]^{-1} \circ [k'_*] \circ [("Bott" \otimes_\pi Id^{SA})_*] \circ [k^{SA}_*]^{-1} \circ [Sk^A_*]^{-1} $$

Consider the commutative diagram

$$M_4(S^3SA) \qquad\qquad S(M_4(S^3A))$$

$$k'' \quad \uparrow \qquad\qquad\qquad \uparrow \qquad\qquad Sk''$$

$$M_4(S\mathbb{C}) \otimes_\pi S^2SA \xrightarrow[SW]{} S(M_4(S\mathbb{C}) \otimes_\pi S^2A)$$

$$i'' \quad \uparrow$$

$$M_4(S^2\mathbb{C}) \otimes_\pi SSA \qquad\qquad \uparrow \qquad\qquad i$$

$$i' \quad \uparrow$$

$$M_4(S^4A) \xrightarrow[SW']{} M_4(S^4A)$$

$$\downarrow \qquad\qquad\qquad \downarrow$$

$$M_4(S^4A) \xrightarrow[SW\sim id]{} M_4(S^4A)$$

where all vertical maps induce asymptotic HC-equivalences by the derivation lemma and where $[SW_*] = [id_*] \in HC_\alpha^0(S^4A, S^4A)$ by what we said above.

Note that

$$k'' \circ i'' = k'''$$

in the notations of the first part of the proof. From this we derive the equality

$$[SW_*] \circ [k''_*]^{-1} = [i_*] \circ [SW'_*] \circ [i'_*]^{-1} \circ [k'''_*]^{-1}$$

$$= [i_*] \circ [i'_*]^{-1} \circ [k'''_*]^{-1} = [(id^{M_4(S^2\mathbb{C})} \otimes_\pi Sk^A)_*] \circ [k'''_*]^{-1}$$

which implies by the naturality of the exterior product

$$[\tau_2 \times id_*^{S^2A}] \circ [SW_*] \circ [k''_*]^{-1} =$$

$$= [\tau_2 \times id_*^{S^2A}] \circ [(id^{M_4(S^2\mathbb{C})} \otimes_\pi Sk^A)_*] \circ [k'''_*]^{-1}$$

$$= [(Sk^A)_*] \circ [\tau_2 \times id_*^{SSA}] \circ [k'''_*]^{-1}$$

Inserting this into the expression for $\beta \circ \alpha$ one finds

$$\beta_{SA} \circ \alpha_{SA} =$$

$$= [\tau_2 \times id^{S^2A}] \circ [SW_*] \circ [k''_*]^{-1} \circ [k'_*] \circ [(\text{"}Bott\text{"} \otimes_\pi Id^{SA})_*] \circ [k^{SA}_*]^{-1} \circ [Sk^A_*]^{-1}$$

$$= [(Sk^A)_*] \circ ([\tau_2 \times id_*^{SSA}] \circ [k'''_*]^{-1} \circ [k'_*] \circ [(\text{"}Bott\text{"} \otimes_\pi Id^{SA})_*] \circ [k^{SA}_*]^{-1}) \circ [Sk^A_*]^{-1}$$

$$= [(Sk^A)_*] \circ (\alpha_{SA} \circ \beta_{SA}) \circ [Sk^A_*]^{-1}$$

by the equality (*) of the first part of the demonstration

$$= [(Sk^A)_*] \circ [id_*^{SSA}] \circ [Sk^A_*]^{-1}$$

by what we proved so far

$$= [id_*^{S^2A}]$$

It remains to verify the assertion about the action of $\tilde{id}^{SA}$ on asymptotic cohomology. It is well known that the suspension of an algebra becomes a group object in the homotopy category of algebras. The "addition up to homotopy" being given by the splicing map

$$\vee: \quad SA \oplus SA \quad \rightarrow \quad SA \qquad (f, g) \quad \rightarrow \quad f \vee g := \begin{cases} f(2t) & 0 \le t \le \frac{1}{2} \\ g(2t - 1) & \frac{1}{2} \le t \le 1 \end{cases}$$

The compositions

$$SA \xrightarrow{(id \oplus 0)} SA \oplus SA \xrightarrow{\vee} SA$$
$$SA \xrightarrow{(0 \oplus id)} SA \oplus SA \xrightarrow{\vee} SA$$

are homotopic to the identity whereas

$$SA \xrightarrow{(id \oplus \tilde{id})} SA \oplus SA \xrightarrow{\vee} SA$$

is nullhomotopic. From this we conclude

$$0 = [\vee_*] \circ [(id \oplus \tilde{id})_*] = [\vee_*] \circ [id_*^{SA \oplus SA}] \circ [(id \oplus \tilde{id})_*]$$

$$= [\vee_*] \circ ([(id \oplus 0)_*] + [(0 \oplus id)_*]) \circ [(id \oplus \tilde{id})_*]$$

by the structure theorem 6.16 about the asymptotic cohomology of a direct sum

$$= [\vee_*] \circ [(id \oplus 0)_*] \circ [id_*] + [\vee_*] \circ [(0 \oplus id)_*] \circ [\tilde{id}_*]$$

$$= [id_*] + [\tilde{id}_*]$$

which proves the claim.

$\square$

Proposition 9.5:

The Dirac element $\alpha_{\mathbb{C}} \in HC_\alpha^1(S\mathbb{C}, \mathbb{C})$ defines an asymptotic HC-equivalence and consequently

$$HC_\alpha^*(S\mathbb{C}, S\mathbb{C}) = \begin{cases} \mathbb{C} & * = 0 \\ 0 & * = 1 \end{cases}$$

Proof:

1) Construction of an HC-inverse $\beta_{\mathbb{C}}$ of $\alpha_{\mathbb{C}}$:

Let $e_0, e_1 \in M_2(C^\infty(S^2))$ be the matrices of 9.2. defining the Bott class and let

$$e_0, e_1 : \tilde{\mathbb{C}} \rightarrow M_2(C^\infty(S^2))$$

be the corresponding unital homomorphisms. Let $f : S^1 \times S^1 \rightarrow S^2$ be a smooth map of degree one and denote as before by τ_1 the cohomological fundamental class of the circle. Put

$$\phi_{0,1} := [\tilde{k}_*] \circ [\frac{1}{2\pi i}(Tr \times \tau_1) \times id_*^{C^\infty(S^1)}] \circ [M_2(f)_*] \circ [e_{0,1 *}]$$

$$\phi_{0,1} \in Hom^1(X_*(R\tilde{\mathbb{C}}), X_*(RC(S^1)))$$

where $\tilde{k} : C^\infty(S^1) \to C(S^1)$ is the inclusion. As both chain maps annihilate $1_{\mathbb{C}} \in X_*(R\tilde{\mathbb{C}})$ they descend to chain maps

$$\phi_{0,1} \in Hom^1(\overline{X}_*(R\tilde{\mathbb{C}}), \overline{X}_*(RC(S^1)))$$

$$= Hom^1(\overline{X}_*(R\tilde{\mathbb{C}}), \overline{X}_*(R\widetilde{S\mathbb{C}})) = X_\epsilon^1(\mathbb{C}, S\mathbb{C})$$

As chain maps both ϕ_0 and ϕ_1 are cocycles in the bivariant analytic X-complex so that one can define

$$\beta_{\mathbb{C}} := [\phi_0] - [\phi_1] \in HC_\epsilon^1(\mathbb{C}, S\mathbb{C}) = HC_\alpha^1(\mathbb{C}, S\mathbb{C})$$

2)

$$\alpha_{\mathbb{C}} \circ \beta_{\mathbb{C}} = id_*^{\mathbb{C}} \in HC_\alpha^*(\mathbb{C}, \mathbb{C})$$

By definition

$$\alpha_{\mathbb{C}} \circ \beta_{\mathbb{C}} =$$

$$= [\tau_1 \times id_*^{\mathbb{C}}] \circ [k_*^{\mathbb{C}}]^{-1} \circ [\tilde{k}_*] \circ [\frac{1}{2\pi i}(Tr \times \tau_1) \times id_*^{C^\infty(S^1)}] \circ [M_2(f)_*] \circ ([e_0{}_*] - [e_1{}_*])$$

$$= [\tau_1 \times id_*^{\mathbb{C}}] \circ [\frac{1}{2\pi i}(Tr \times \tau_1) \times id_*^{C^\infty(S^1)}] \circ [M_2(f)_*] \circ ([e_0{}_*] - [e_1{}_*])$$

by definition of $k^{\mathbb{C}}$ and $\tilde{k}$

$$= [\frac{1}{2\pi i}(Tr \times \tau_1 \times \tau_1) \times id_*^{\mathbb{C}}] \circ [M_2(f)_*] \circ ([e_0{}_*] - [e_1{}_*])$$

$$= [\tau_2' \times id_*^{\mathbb{C}}] \circ ([e_0{}_*] - [e_1{}_*])$$

in the notations of the previous theorem

$$= e_0 \backslash [\tau_2' \times id_*^{\mathbb{C}}] - e_1 \backslash [\tau_2' \times id_*^{\mathbb{C}}]$$

$$= (\langle ch(e_0), \tau_2' \rangle - \langle ch(e_1), \tau_2' \rangle) id_*^{\mathbb{C}} = id_*^{\mathbb{C}}$$

(see 9.4.)

3)

$$\beta_{\mathbb{C}} \circ \alpha_{\mathbb{C}} = [id_*^{S\mathbb{C}}] \in HC^0(S\mathbb{C}S\mathbb{C})$$

By definition

$$\beta_{\mathbb{C}} \circ \alpha_{\mathbb{C}} =$$

$$= [\tilde{k}_*] \circ [\frac{1}{2\pi i}(Tr \times \tau_1) \times id_*^{C^\infty(S^1)}] \circ [M_2(f)_*] \circ ([e_0{}_*] - [e_1{}_*]) \circ [\tau_1 \times id_*^{\mathbb{C}}] \circ [k_*^{\mathbb{C}}]^{-1}$$

$$= [\tilde{k}_*] \circ [\frac{1}{2\pi i}(Tr \times \tau_1) \times id_*^{C^\infty(S^1)}] \circ [M_2(f)_*] \circ [\tau_1 \times id_*^{M_2(C^\infty(S^2))}] \circ ([Se_0{}_*] - [Se_1{}_*]) \circ [k_*^{\mathbb{C}}]^{-1}$$

$$= [\tilde{k}_*] \circ [\frac{1}{2\pi i}(Tr \times \tau_1) \times id_*^{C^\infty(S^1)}] \circ [\tau_1 \times id_*^{M_2(C^\infty(S^1 \times S^1))}] \circ [M_2(Sf)_*] \circ ([Se_0{}_*] - [Se_1{}_*]) \circ [k_*^{\mathbb{C}}]^{-1}$$

$$= [\tilde{k}_*] \circ [\frac{1}{2\pi i}(Tr \times \tau_1 \times \tau_1) \times id_*^{C^\infty(S^1)}] \circ [M_2(i')_*] \circ [M_2(Sf)_*] \circ ([Se_0{}_*] - [Se_1{}_*]) \circ [k_*^{\mathbb{C}}]^{-1}$$

where

$$i' : M_2(SC^\infty(S^1 \times S^1)) \to M_2(C^\infty(S^1 \times S^1 \times S^1))$$

is the inclusion.

Now the diagrams

$$
\begin{array}{ccccc}
 & & S\mathbb{C} & & \\
k & & \uparrow & & \\
 & & S\mathbb{C} & \xrightarrow{SW \circ \text{"Bott"}} & M_4(S^3\mathbb{C}) \\
id^S \otimes e_0 \vee \tilde{id}^S \otimes e_1 & & \downarrow & & \downarrow i \\
 & & SM_2(C^\infty(S^2))^2 & \xrightarrow{k'''} & M_4(C_0(\mathbb{R} \times S^2))
\end{array}
$$

and

$$
\begin{array}{ccccc}
 & M_4(S^3\mathbb{C}) & \xrightarrow{i} & M_4(C_0(\mathbb{R} \times S^2)) & \\
SW^{-1} & \downarrow & & \uparrow & k''' \\
 & M_4(S^3\mathbb{C}) & & M_4(SC^\infty(S^2)) & \\
i & \downarrow & & \downarrow & M_4(Sf) \\
 & M_4(C_0(S^2 \times \mathbb{R})) & & M_4(SC^\infty(S^1 \times S^1)) & \\
k'' & \uparrow & & \downarrow & M_4(i') \\
 & M_4(C^\infty(S^2) \otimes_\pi S) & \xrightarrow{M_4(f \otimes i')} & M_4(C^\infty(S^1 \times S^1 \times S^1)) &
\end{array}
$$

commute up to homotopy as can be seen from 9.2., where $SW : M_4(S^3\mathbb{C}) \to M_4(S^3\mathbb{C})$ corresponds to the map induced by the cyclic permutation ($1\,2\,3$) of the suspension coordinates. SW is homotopic to the identity. As the involution $\tilde{id}^{SA}$ induces -1 on asymptotic cohomology groups (see the proof of theorem 9.4) one may conclude that

$$[M_2(i')_*] \circ [M_2(Sf)_*] \circ ([Se_{0*}] - [Se_{1*}]) =$$
$$= [M_4(f \otimes i')_*] \circ [k'']^{-1} \circ [i_*] \circ [\text{"Bott"}_*]$$

Thus

$$\beta_\mathbb{C} \circ \alpha_\mathbb{C} =$$

$$= [\tilde{k}_*] \circ [\frac{1}{2\pi i}(Tr \times \tau_1 \times \tau_1) \times id_*^{C^\infty(S^1)}] \circ [M_4(f \otimes i')_*] \circ [k'']^{-1} \circ [i_*] \circ [\text{"Bott"}_*] \circ [k_*^\mathbb{C}]^{-1}$$

$$= [\tilde{k}_*] \circ [\tau_2' \times id_*^{C^\infty(S^1)}] \circ [k'']^{-1} \circ [i_*] \circ [\text{"Bott"}_*] \circ [k_*^\mathbb{C}]^{-1}$$

$$= [\tau_2' \times id_*^{C(S^1)}] \circ [k''']^{-1} \circ [k'_*] \circ [\text{"Bott"}_*] \circ [k_*^\mathbb{C}]^{-1}$$

in the notations of the previous theorem

$$= \alpha_{S\mathbb{C}} \circ \beta_{S\mathbb{C}}$$

by equation $(*)$ of 9.4.

$$= [id_*^{S\mathbb{C}}]$$

by the stable periodicity theorem.

$\square$

9-2 Puppe sequences and the first excision theorem

Any asymptotic morphism $f : A \to B$ of admissible Fréchet algebras induces maps of complexes

$$f_* : X_\alpha^*(C, A) \to X_\alpha^*(C, B) \quad f^* : X_\alpha^*(A, D) \leftarrow X_\alpha^*(B, D)$$

Recall that the **cone of a map** $f : X_* \to Y_*$ **of complexes** is given by the complex

$$Cone(f_*, X_*, Y_*) := \begin{cases} Cone_n := X_{n+1} \oplus Y_n \\ \partial_{Cone} := \begin{pmatrix} -\partial_{X_*} & 0 \\ f_* & \partial_{Y_*} \end{pmatrix} \end{cases}$$

The cohomology of the cone is then called the **relative cohomology** of the pair

$$f_* : X_* \to Y_*$$

and is related to the cohomology of X_* and Y_* via a long exact sequence which in fact becomes a six term exact sequence for $\mathbb{Z}/2$-graded complexes.

Our aim in this chapter is to describe the relative asymptotic X-complexes

$$Cone(f_*, X_\alpha^*(C, A), X_\alpha^*(C, B)) \quad Cone(f^*, X_\alpha^*(A, D), X_\alpha^*(B, D))$$

in suitable cases in a more direct manner. These results form the basis for calculations of asymptotic cohomology groups.

If $f : A \to B$ is a morphism of algebras, the **mapping cone** C_f of f is defined by the cartesian square

$$\begin{array}{ccc} C_f & \longrightarrow & CB \\ \downarrow & & \downarrow{\scriptstyle eval} \\ A & \underset{f}{\longrightarrow} & B \end{array}$$

If f is continuous and if A, B are admissible Fréchet algebras, then C_f is admissible, too.

Theorem 9.6: (First Excision Theorem)

Let A, B, C, D be admissible Fréchet algebras and let

$$f : A \to B$$

be a continuous homomorphism. Then there are natural six term exact Puppe sequences

$$HC_\alpha^0(C, SA) \xrightarrow{Sf_*} HC_\alpha^0(C, SB)$$

$$HC_\alpha^0(C, SC_f) \qquad HC_\alpha^1(C, SC_f)$$

$$HC_\alpha^1(C, SB) \xleftarrow{Sf_*} HC_\alpha^1(C, SA)$$

$$HC_\alpha^0(SA, D) \xleftarrow{Sf^*} HC_\alpha^0(SB, D)$$

$$HC_\alpha^0(SC_f, D) \qquad HC_\alpha^1(SC_f, D)$$

$$HC_\alpha^1(SB, D) \xrightarrow{Sf^*} HC_\alpha^1(SA, D)$$

Moreover, the cone of the bivariant X-Complexes under Sf and the bivariant X-Complexes of the mapping cone of Sf are naturally quasiisomorphic:

$$X_\alpha^*(C, SC_f) \xrightarrow{qis} Cone(Sf_*, X_\alpha^*(C, SA), X_\alpha^*(C, SB))[1]$$

$$X_\alpha^*(SC_f, D) \xleftarrow{qis} Cone(Sf^*, X_\alpha^*(SA, D), X_\alpha^*(SB, D))$$

Proof:

We begin with the construction of the six term exact sequences. They are obtained by applying the functors $X_\alpha^*(C, -)$ and $X_\alpha^*(-, D)$ to the cofibre sequence

$$\to S^2C_f \to S^2A \xrightarrow{S^2f} S^2B \xrightarrow{Sj} SC_f \xrightarrow{Si} SA \xrightarrow{Sf} SB$$

and taking cohomology. The stable Bott periodicity theorem for asymptotic cohomology allows then to turn this sequence of cohomology groups into a periodic one. In order to show its exactness it suffices to prove that the cofibre sequence induces long exact sequences in asymptotic homology resp. cohomology. As a long cofibre sequence consists (up to homotopy equivalences) of short cofibre sequences, it remains to show the exactness of

$$HC^*_\alpha(C, SC_f) \xrightarrow{Si_*} HC^*_\alpha(C, SA) \xrightarrow{Sf_*} HC^*_\alpha(C, SB)$$

and

$$HC^*_\alpha(SC_f, D) \xleftarrow{Si^*} HC^*_\alpha(SA, D) \xleftarrow{Sf^*} HC^*_\alpha(SB, D)$$

respectively. The composition of the two maps is zero in both cases because the composition $f \circ i$ is nullhomotopic as the diagram

$$
\begin{array}{ccc}
C_f & \longrightarrow & CB \\
{\scriptstyle i}\downarrow & & \downarrow{\scriptstyle eval} \\
A & \xrightarrow{\ f\ } & B
\end{array}
$$

shows.

First case: The cohomological X-complex

$$HC^*_\alpha(SC_f, D) \xleftarrow{Si^*} HC^*_\alpha(SA, D) \xleftarrow{Sf^*} HC^*_\alpha(SB, D)$$

Let $[\varphi] \in HC^*_\alpha(SA, D)$ be such that $[Si^*\varphi] = 0 \in HC^*_\alpha(SC_f, D)$. So there exists

$$\psi \in X^{*-1}_\alpha(SC_f, D) \text{ with } \partial \psi = i^* \varphi$$

ψ is well defined up to a cocycle. It follows that

$$Sj^* \psi \in X^{*-1}_\alpha(S^2 B, D)$$

is a cocycle:

$$\partial (Sj^* \psi) = Sj^* (\partial \psi) = Sj^* Si^* \varphi = S(i \circ j)^* \varphi = 0$$

as $i \circ j = 0$. Its cohomology class is well defined up to the image of

$$Sj^* : HC^{*-1}_\alpha(SC_f, D) \to HC^{*-1}_\alpha(S^2 B, D)$$

We put

$$\nu := [Sj^* \psi] \circ \beta_{SB} \in HC^*_\alpha(SB, D)$$

and claim

$$Sf^* \nu = [\varphi] \in HC^*_\alpha(SA, D)$$

The demonstration proceeds in several steps.

1) The cohomology class of $Sf^*\nu$ is independent of all choices made: Any element of the indeterminacy group is killed under Sf^*: If $[\chi] \in HC^{*-1}_\alpha(SC_f, D)$ then

$$Sf^* ((Sj^* \chi) \circ \beta_{SB}) = \chi \circ Sj_* \circ \beta_{SB} \circ Sf_* = \chi \circ S(j \circ Sf)_* \circ \beta_{SA} = 0$$

because $j \circ Sf$ is nullhomotopic.

2) The construction of $Sf^*\nu$ is natural with respect to maps of cofibration sequences:

If

$$S^2A \;\rightarrow\; S^2B \;\rightarrow\; SC_f \;\rightarrow\; SA \;\xrightarrow{Sf}\; SB$$

$$\downarrow \qquad\quad \downarrow \qquad\quad \downarrow \qquad\quad \downarrow \qquad\qquad \downarrow$$

$$S^2A' \;\rightarrow\; S^2B' \;\rightarrow\; SC_{f'} \;\rightarrow\; SA' \;\xrightarrow[Sf']{}\; SB'$$

is induced from a commutative square

$$
\begin{array}{ccc}
A & \xrightarrow{\;f\;} & B \\
{\scriptstyle g}\downarrow & & \downarrow{\scriptstyle h} \\
A' & \xrightarrow[f']{} & B'
\end{array}
$$

and

$$\nu \in HC_\alpha^*(SB,D), \quad \nu' \in HC_\alpha^*(SB',D)$$

are classes associated to

$$\varphi' \in HC_\alpha^*(SA',D), \quad \varphi := Sg^*\varphi' \in HC_\alpha^*(SA,D)$$

by the procedure above, then

$$Sf^*\,\nu = Sg^*\,(Sf'^*\,\nu')$$

3)

Applying 2) to the square

$$
\begin{array}{ccc}
A & \xrightarrow{\;Id\;} & A \\
{\scriptstyle Id}\downarrow & & \downarrow{\scriptstyle f} \\
A & \xrightarrow[f]{} & B
\end{array}
$$

shows that only the case of the cofibre sequence

$$S^2A \rightarrow S^2A \xrightarrow{Sj'} CSA \xrightarrow{eval\otimes id} SA \rightarrow SA$$

has to be treated.

4) Consider the exact sequence

$$0 \rightarrow S\mathbb{C} \rightarrow C_0^\infty(]0,1]) := \{f \in C^\infty([0,1]), f(0)=0\} \xrightarrow{eval} \mathbb{C} \rightarrow 0$$

where $eval$ is given by evaluation at 1. Then

$$\tau_0 \in X^0\mathbb{C} \qquad\qquad \tau_0(1) = 1$$

$$\tau_1 \in X^1(C_0^\infty(]0,1])) \qquad \tau_1((f\,dg)_\natural) = \int_0^1 f\,dg$$

are related by

$$\partial \tau_1 = eval^* \tau_0$$

As we saw already,

$$[j'^* \tau_1] \in HC^1(S\mathbb{C})$$

is the fundamental class of the circle.

The fact that the exterior product on cohomology was constructed via a map of complexes shows that for any cocycle $\varphi \in X^*_\alpha(SA, D)$ and $eval \otimes id : C^\infty_0(]0,1]) \otimes_\pi SA \to \mathbb{C} \otimes SA$

$$(eval \otimes id)^* \varphi = (eval \otimes id)^* (\tau_0 \times \varphi) = (eval^* \tau_0) \times \varphi = \partial \tau_1 \times \varphi = \partial (\tau_1 \times \varphi)$$

with $\tau_1 \times \varphi \in X^{*+1}_\alpha(C^\infty_0(]0,1]) \otimes_\pi SA, D)$.

The inclusion

$$k : C^\infty_0(]0,1]) \otimes_\pi SA \to CSA$$

being an asymptotic HC-equivalence (7.8), we see that the cochain

$$(\tau_1 \times \varphi) \circ (k_*)^{-1} \in X^{*+1}_\alpha(CSA, D)$$

satisfies

$$\partial((\tau_1 \times \varphi) \circ (k_*)^{-1}) = \varphi \circ (eval \otimes id)_* \circ k_*^{-1} = \varphi \circ (eval \otimes id)_* = (eval \otimes id)^* \varphi$$

So we can achieve the construction 1) by putting

$$\nu := ((\tau_1 \times \varphi) \circ (k_*)^{-1}) \circ Sj'_* \circ \beta_{SA} \in Z^*_\alpha(SA, D)$$

and find

$$id^* [\nu] = [(\tau_1 \times \varphi) \circ (j' \otimes id)^{SA}_* \circ (k_*)^{-1} \circ \beta_{SA}]$$

$$= [(j'^* \tau_1 \times \varphi) \circ (k_*)^{-1} \circ \beta_{SA}] = [\varphi \circ \alpha_{SA} \circ \beta_{SA}] = [\varphi]$$

by the periodicity theorem.

Second case: the homological X-complex:

$$HC^*_\alpha(C, SC_f) \xrightarrow{Si_*} HC^*_\alpha(C, SA) \xrightarrow{Sf_*} HC^*_\alpha(C, SB)$$

Let $[\varphi] \in HC^*_\alpha(C, SA)$ be such that $Sf_* [\varphi] = 0$. Then

$$[\varphi'] := \beta_{SA} \circ [\varphi] \in HC^{*+1}_\alpha(C, S^2A)$$

satisfies

$$S^2 f_*[\varphi'] = S^2 f_* \circ \beta_{SA} \circ [\varphi] = \beta_{SB} \circ Sf_* [\varphi] = 0$$

So

$$S^2 f_* \varphi' = \partial \psi'$$

for some

$$\psi' \in X^*_\alpha(C, S^2B)$$

which is well defined up to a cocycle again. Then

$$Sj_* \psi' \in X^*_\alpha(C, SC_f)$$

satisfies

$$\partial (Sj_* \psi') = Sj_* (\partial \psi') = S(j \circ Sf)_* \varphi'$$

As $j \circ Sf$ is canonically nullhomotopic, there exists a natural element

$$h \in X_\alpha^1(S^2 A, SC_f) \text{ with } \partial h = S(j \circ Sf)_*$$

given by the Cartan homotopy operator.

We may then conclude that

$$\nu' := -Sj_* \circ \psi' + h \circ \varphi' \in X_\alpha^*(C, SC_f)$$

is a cocycle whose cohomology class is well defined up to the image of

$$Sj_* : HC_\alpha^*(C, S^2 B) \to HC_\alpha^*(C, SC_f)$$

We claim that

$$Si_* [\nu'] = [\varphi] \in HC_\alpha^*(C, SA)$$

To prove this calculate

$$Si_* \nu' = -Si_* \circ Sj_* \circ \psi + Si_* \circ h \circ \varphi' = Si_* \circ h \circ \varphi'$$

as $i \circ j = 0$. The commutative diagram

$$\begin{array}{ccccc}
SA & \xrightarrow{j'} & CA & \xrightarrow{eval} & A \\
\| & & \downarrow (eval, Cf) & & \| \\
SA & \xrightarrow{j \circ Sf} & C_f & \xrightarrow{i} & A
\end{array}$$

and the fact that the canonical nullhomotopy of $j \circ Sf$ is given by the factorization over the contractible cone CA in the diagram above show that

$$Si_* \circ h = h'$$

where

$$h' \in X_\alpha^1(S^2 A, SA)$$

is the Cartan homotopy operator associated to the canonical nullhomotopy of the composition

$$S^2 A \xrightarrow{Sj'} CSA \xrightarrow{eval \otimes id} SA$$

Note that h' is in fact a cocycle because the two evaluations at the endpoints coincide. Its cohomology class is well known

Lemma 9.7:

$$[h'] = (\tau_1 \times id_*^{SA}) \circ k_*^{-1} \circ Sj_*' = (j'^* \tau_1 \times id_*^{SA}) \circ k_*^{-1} = \alpha_{SA} \in HC_\alpha^1(S^2 A, SA)$$

Proof:

The lemma asserts that

$$h' = \tau_1 \times id_*^A \in HC_\epsilon^1(SA, A)$$

Let $\psi \in X^0(RSA, RA)$ be defined in even degrees by $\frac{1}{2}$ times the composition

$$X_0R(SA) \xrightarrow{\partial x_*} X_1R(SA) \xrightarrow{\int_0^1 h \frac{d}{dt}} X_0(RA)$$

i.e. by

$$\psi : \varrho\omega^n \to \frac{1}{2} \int_0^1 (\omega^n \dot\varrho + \sum_0^{n-1} \omega^{n-1-j} \varrho\omega^j \dot\varrho)\, dt$$

$$- \frac{1}{2} \int_0^1 \sum_0^{n-1} (\varrho\omega^{n-1-j} \varrho\omega^j \dot\varrho + \omega^{n-1-j} \varrho\omega^j \varrho\dot\varrho)\, dt$$

and in odd degrees by

$$\psi : \varrho\omega d\varrho \to \frac{1}{2} \int_0^1 (\varrho\omega^n d\dot\varrho)\, dt$$

Then $\psi \in X_\epsilon^0(SA, A)$ and

$$h' - \tau_1 \times id_*^A = \partial \circ \psi - \psi \circ \partial$$

as a lengthy but elementary calculation shows.

$\square$

Thus finally

$$[Si_* \nu'] = Si_* \circ h \circ [\varphi'] = h' \circ [\varphi'] = \alpha_{SA} \circ [\varphi'] = \alpha_{SA} \circ \beta_{SA} \circ [\varphi] = [\varphi]$$

by the periodicity theorem again.

The proof of exactness of the two six term exact sequences being achieved, we go on to define the desired quasiisomorphisms.

The map

$$\sigma : \quad X_\alpha^*(C, SC_f) \quad \to \quad\quad\quad\quad Cone(Sf_*)[-1]$$

$$\varphi \quad\quad \to \quad ((Si)_* \varphi, -(-1)^{deg\varphi} h_{S(f \circ i)} \circ \varphi)$$

is easily seen to be a map of complexes. (Here again, $h_{S(f \circ i)} \in X_\alpha^1(SC_f, SB)$ is the Cartan homotopy operator associated to the canonical nullhomotopy of $S(f \circ i)$.)

Moreover, this map of complexes fits into a diagram which commutes up to signs and up to homotopy:

$$X_\alpha^*(C, SA)[-1] \quad \to \quad X_\alpha^*(C, SB)[-1] \quad \overset{\delta}{\to} \quad X_\alpha^*(C, SC_f)$$

$$\downarrow\| \qquad\qquad\qquad \downarrow\| \qquad\qquad\qquad \downarrow \sigma$$

$$X_\alpha^*(C, SA)[-1] \quad \to \quad X_\alpha^*(C, SB)[-1] \quad \underset{i}{\to} \quad Cone(Sf_*)[-1]$$

$$X_\alpha^*(C, SC_f) \quad \to \quad X_\alpha^*(C, SA) \quad \to \quad X_\alpha^*(C, SB)$$

$$\downarrow \sigma \qquad\qquad\qquad \downarrow\| \qquad\qquad\qquad \downarrow\|$$

$$Cone(Sf_*)[-1] \quad \to \quad X_\alpha^*(C, SA) \quad \to \quad X_\alpha^*(C, SB)$$

Therefore σ is a quasiisomorphism by the exactness of the induced cohomology sequences and the five lemma.

In the case of the cohomological X-complex we put

$$\sigma': \quad Cone(Sf^*, X_\alpha^*(SB, D), X_\alpha^*(SA, D)) \quad \to \quad X_\alpha^*(C_f, D)$$

$$(\varphi, \psi) \qquad\qquad \to \quad \varphi \circ h_{S(f \circ i)} + (Si)^* \psi$$

The same reasoning is then valid in this case, too, and is left to the reader.

□

9-3 Stable cohomology of C^*-algebras and the second excision theorem

When does a short exact sequence of admissible Fréchet algebras give rise to a six term exact sequence on (stable) asymptotic cohomology ?

A look at the commutative diagram

$$0 \quad \to \quad J \quad \to \quad A \quad \to \quad B \quad \to \quad 0$$

$$\downarrow \qquad\quad \downarrow\sim \qquad\quad \|$$

$$0 \quad \to \quad C_f \quad \to \quad Cyl_f \quad \to \quad B \quad \to \quad 0$$

and the first excision theorem show that a necessary and sufficient condition is given by

Definition 9.8:

A unital epimorphism $f : A \to B$ of admissible Fréchet algebras satisfies **stable excision** iff

$$j : \quad J \quad \to \quad C_f$$

$$x \quad \to \quad (x, 0)$$

is a stable asymptotic HC-equivalence.

□

In general one would expect excision to hold if f would turn out to be a cofibration on some stable asymptotic homotopy category. This condition would then force j to be a stable homotopy equivalence. We have however not developed this point of view far enough to get any reasonable conclusions. Anyway, an arbitrary epimorphism of admissible Fréchet algebras should be far from satisfying excision. To obtain a sufficient criterion for excision note, that a stable asymptotic morphism strictly inverse to j would carry any positive, quasicentral, bounded approximate unit of C_f to a similar approximate unit of J. This leads one to restrict attention to separable C^*-algebras, which form essentially the largest category of Fréchet algebras, for which the kernel of any epimorphism possesses a poitive, quasicentral, bounded, approximate unit.

Theorem 9.9: (Second Excision Theorem)

Let

$$0 \to J \to A \xrightarrow{f} B \to 0$$

be a short exact sequence of separable C^*-algebras with f unital. Suppose that f admits a bounded, linear section.

Then f satisfies stable excision.

Consequently, there are natural six term exact sequences

$$HC^0_\alpha(C, SA) \xrightarrow{\ sf_* \ } HC^0_\alpha(C, SB)$$

$$\nearrow \qquad\qquad\qquad\qquad\qquad \searrow \partial$$

$$HC^0_\alpha(C, SJ) \qquad\qquad\qquad\qquad HC^1_\alpha(C, SJ)$$

$$\partial \nwarrow \qquad\qquad\qquad\qquad\qquad \swarrow$$

$$HC^1_\alpha(C, SB) \xleftarrow[\ sf_* \ ] HC^1_\alpha(C, SA)$$

$$HC_\alpha^0(SA, D) \xleftarrow{Sf^*} HC_\alpha^0(SB, D)$$

$$HC_\alpha^0(SJ, D) \qquad\qquad HC_\alpha^1(SJ, D)$$

$$HC_\alpha^1(SB, D) \xrightarrow{Sf^*} HC_\alpha^1(SA, D)$$

for any admissible Fréchet algebras C, D. They are natural in C, D under asymptotic morphisms and under maps of extensions

$$
\begin{array}{ccccccccc}
0 & \to & J & \to & A & \to & B & \to & 0 \\
  &   & \downarrow & & \downarrow & & \downarrow & & \\
0 & \to & J' & \to & A' & \to & B' & \to & 0
\end{array}
$$

Proof:

In Connes-Higson [CH] it has been shown that j is a stable asymptotic homotopy equivalence. As their notion of asymptotic morphism differs from ours, we repeat their argument with the necessary modifications.

We have to show that

$$Sj_* \in HC_\alpha^0(SJ, SC_f)$$

is an asymptotic HC-equivalence. We will construct an HC-inverse $[\Theta]$ of Sj_* explicitly.

To do this choose a positive, quasicentral, bounded approximate unit

$$(u_t) \in J, \ 0 \le u_t \le 1 \ \forall t \in \mathbb{R}_+$$

and fix a bounded, linear section of f

$$s : B \to A \ f \circ s = id_B$$

Let

$$(v_t) \in C_0(]0, 1[), \ 0 \le v_t \le 1 \ \forall t \in \mathbb{R}_+$$

be a bounded, positive, approximate unit for $C_0(]0, 1[)$ consisting of functions that differ from 1 only near the endpoints of the unit interval.

To define the bivariant cocycle Θ the asymptotic connecting map associated to the extension

$$0 \to SJ \to CA \xrightarrow{p} C_f \to 0$$

will be used. p also admits a bounded, linear section, for example

$$s' : \quad C_f \quad \to \quad\quad\quad\quad CA$$

$$(a, h(t)) \quad \to \quad \chi(t)(a - s(h(0))) + s(h(t))$$

where $\chi \in S\mathbb{C}$ equals 1 near 0 and vanishes near 1.

We put then

$$\theta_t : \quad S\mathbb{C} \otimes_\pi C_f \quad \to \quad\quad\quad SJ$$

$$g \otimes x \quad\quad \to \quad g(v_t \otimes u_t)\, s'(x)$$

It is easily verified that θ is an asymptotic morphism so that we may finally define

$$\Theta := \theta_* \circ (k_*^{C_f})^{-1} \in HC_\alpha^0(SC_f, SJ)$$

with

$$k : S\mathbb{C} \otimes_\pi C_f \to SC_f$$

given by the obvious inclusion.

First step:

$$\Theta \circ [Sj_*] = [id_*^{SJ}] \in HC_\alpha^0(SJ, SJ)$$

$$\Theta \circ [Sj_*] = [\theta_*] \circ [k_*^{C_f}]^{-1} \circ [Sj_*] = [\theta_*] \circ [(j \otimes_\pi id^{S\mathbb{C}})_*] \circ [k_*]^{-1}$$

by the naturality of k.

However

$$\theta \circ (j \otimes_\pi id^{S\mathbb{C}}) \sim k : S\mathbb{C} \otimes_\pi J \to SJ$$

are homotopic, as one can see by considering the family of asymptotic morphisms

$$F_\lambda : \quad S\mathbb{C} \otimes_\pi J \quad \to \quad\quad\quad SJ$$

$$g \otimes x \quad\quad \to \quad g(v_t \otimes ((1 - \lambda)\, u_t + \lambda\, 1_A))\, s'(j(x))$$

where $F_0 = \theta \circ (j \otimes_\pi id^{S\mathbb{C}})$ and F_1 is explicitly homotopic to k^J. So we conclude

$$\Theta \circ [Sj_*] = [(\theta \circ (j \otimes_\pi id))_*] \circ [k_*^J]^{-1} = [k_*^J] \circ [k_*^J]^{-1} = [id_*^{SJ}]$$

Second step:

$$[Sj_*] \circ \Theta = [id_*^{SC_f}] \in HC_\alpha^0(SC_f, SC_f)$$

Again

$$[Sj_*] \circ \Theta = [(Sj \circ \theta)_*] \circ [k_*^{C_f}]^{-1}$$

where $Sj \circ \theta$ is homotopic to k^{C_f}: If

$$(w_t) \in C_f \ 0 \le w_t \le 1 \ \forall t \in \mathbb{R}_+$$

is a bounded, approximate unit for C_f consisting of real valued functions that vanish around the vertex and are equal to 1 near the base of the mapping cone, then

$$G_\lambda : \quad S\mathbb{C} \otimes_\pi C_f \quad \to \quad\quad\quad SC_f$$

$$g \otimes y \quad\quad \to \quad g(v_t \otimes ((1 - \lambda)\, j(u_t) + \lambda\, w_t))\, y$$

is a continuous family of asymptotic morphisms with $G_0 = Sj \circ \theta$ and G_1 explicitly homotopic to k^{C_f}. Thus

$$[Sj_*] \circ \Theta = [(Sj \circ \theta)_*] \circ [k_*^{C_f}]^{-1} = [k_*^{C_f}] \circ [k_*^{C_f}]^{-1} = [id_*^{C_f}]$$

As $[Sj_*]$ is now known to be an asymptotic HC-equivalence, the assertion follows from the six term Puppe exact sequence of the first excision theorem.

□

Chapter 10: KK-theory and asymptotic cohomology

In this chapter we show that asymptotic cohomology is the target of a Chern character on Kasparovs bivariant K-theory. (The cyclic theories known so far do not achieve this goal due to their pathological behaviour for C^*-algebras.) Our construction generalizes the well known cases of Chern characters for finitely summable and Θ-summable Fredholm modules treated in $[CO]$. $[CO_2]$ and is the second main result of the paper.

Morally, the Chern character is given by the composition

$$ch: KK^*(A, B) \to E^*(A, B)" \to " HC_\alpha^*(SA, SB)$$

of the Connes-Higson map [CH] and the evident "map" associating to a homotopy class of asymptotic morphisms its bivariant cohomology class.

Due to the fact that E-theory is defined using nonlinear asymptotic morphisms the second "map" cannot be constructed rigorously. The Chern character can
nevertheless be obtained with the help of the derivation lemma because the image of bivariant K-theory in E-theory can be described by asymptotic morphisms whose restriction to suitable "smooth" subalgebras is linear.

The most fundamental property of the bivariant Chern character is its
compatibility with the Kasparov- resp. composition product, which can be viewed as a generalized Grothendieck-Riemann-Roch theorem.

To derive the compatibility of the bivariant Chern character with the Kasparov-resp. composition product (up to a period factor) we use Cuntzs description of bivariant K-theory which provides clean and simple formulas and give only afterwards a description in terms of linear asymptotic morphisms. The period factor $2\pi i$ has the same origin as the corresponding factor showing up for the ordinary Chern character (See Chapter 8).

After comparing the bivariant with the ordinary Chern character of the previous chapters we show finally that the complexified bivariant Chern character yields an isomorphism

$$ch: KK^*(A, B) \otimes_{\mathbb{Z}} \mathbb{C} \xrightarrow{\simeq} HC_\alpha^*(SA, SB)$$

on a class of separable C^*-algebras containing $\mathbb{C}$ and being closed under extensions with completely positive splitting and KK-equivalences. An explicit treatment of the most interesting special cases has still to be worked out.

10-1 The bivariant Chern character and the Riemann-Roch theorem

Theorem 10.1: (Grothendieck-Riemann-Roch)

a) There exists a natural transformation of bifunctors on the category of separable C^*-algebras

$$ch : KK^*(-,-) \to HC^*_\alpha(S-,S-)$$

called the **bivariant Chern character**

b) For any separable, unital $C*$-algebras A, B, C the diagram

$$
\begin{array}{ccc}
KK^j(A,B) \otimes KK^l(B,C) & \xrightarrow{\;\circ\;} & KK^{j+l}(A,C) \\
{\scriptstyle ch\otimes ch}\downarrow & & \downarrow{\scriptstyle ch} \\
HC^j_\alpha(SA,SB) \otimes HC^l_\alpha(SB,SC) & \xrightarrow[\frac{1}{(2\pi i)^{jl}}\circ]{} & HC^{j+l}_\alpha(SA,SC)
\end{array}
$$

commutes, where the upper horizontal map is the Kasparov product and the lower horizontal map is given by $\frac{1}{(2\pi i)^{jl}}$ times the composition product. (See Theorem 8.22 for an explanation of the factor $2\pi i$.)

c) (Grothendieck-Riemann-Roch Theorem)

Let $ch' := \frac{1}{(2\pi i)^j} ch : K_j \to HC^\alpha_j$ be the normalised Chern character K-theory. Let $\phi \in KK^l(A,B)$ and denote by $S\phi \in KK^l(SA,SB)$ the corresponding stabilised element. Then the diagram

$$
\begin{array}{ccc}
K_j(SA) & \xrightarrow{\;-\otimes S\phi\;} & K_{j+l}(SB) \\
{\scriptstyle ch'}\downarrow & & \downarrow{\scriptstyle ch'} \\
HC^\alpha_j(SA) & \xrightarrow[\frac{1}{(2\pi i)^{jl}}ch_{biv}(\phi)\circ-]{} & HC^\alpha_{j+l}(SB)
\end{array}
$$

commutes.

d) If

$$0 \to J \to A \to B \to 0$$

is an extension that admits a completely positive lifting, ch is compatible with long exact sequences, i.e.

$$
\begin{array}{ccccc}
\to & KK^j(C,B) & \xrightarrow{\;\partial\;} & KK^{j-1}(C,J) & \to \\
& \downarrow ch & & \downarrow ch & \\
\to & HC^j_\alpha(SC,SB) & \xrightarrow{(2\pi i)^j \partial} & HC^{j-1}_\alpha(SC,SJ) & \to
\end{array}
$$

and

$$\leftarrow \quad KK^{j+1}(B,D) \quad \overset{\partial}{\leftarrow} \quad KK^j(J,D) \quad \leftarrow$$

$$\downarrow ch \qquad\qquad\qquad \downarrow ch$$

$$\leftarrow \quad HC^{j+1}_\alpha(SB,SD) \quad \overset{(2\pi i)^j \partial}{\longleftarrow} \quad HC^j_\alpha(SJ,SD) \quad \leftarrow$$

commute.

$\square$

Theorem 10.2: (Explicit description of the bivariant Chern character)[CH]

Let A, B be separable C^*-algebras.

a) Let $x \in KK^0(A,B)$ be represented by a Kasparov bimodule of the form

$$x = \left(\mathcal{H}_B \oplus \mathcal{H}_B, \rho = (\rho_0, \rho_1) : A \to \mathcal{L}(\mathcal{H}_B) = M(\mathcal{K} \otimes_{\mathbb{C}^.} B), F = \begin{pmatrix} 0 & 1 \\ 1 & 0 \end{pmatrix} \right)$$

Choose a bounded, approximate unit $(u_t)_{t \in \mathbb{R}_+}, 0 \le u_t \le 1$ in $\mathcal{K} \otimes_{\mathbb{C}^.} B$ which is quasicentral with respect to the C^*-subalgebra of $\mathcal{L}(\mathcal{H}_B)$ generated by $\rho_0(A) \cup \rho_1(A)$.

Then the linear map

$$\Phi_{ev}: \quad SA \quad \to \quad M_2 S(\mathcal{K} \otimes_{\mathbb{C}^.} B) \simeq S(\mathcal{K} \otimes_{\mathbb{C}^.} B)$$

$$f \otimes a \quad \to \quad f(D_t) \begin{pmatrix} \rho_0(a) & 0 \\ 0 & \rho_1(a) \end{pmatrix}$$

$$D_t = \begin{pmatrix} sin\frac{\pi}{2}x & cos\frac{\pi}{2}x(1 - u_t) \\ cos\frac{\pi}{2}x(1 - u_t) & -sin\frac{\pi}{2}x \end{pmatrix} \quad x \text{ the coordinate function on }]-1,1[$$

defines an asymptotic morphism and the bivariant Chern character of x may be represented by the cocycle

$$ch(x) = [i_{SB*}]^{-1} \circ [\Phi_{ev*}] \circ [k^A_*]^{-1}$$

where

$$i_{SB}: \quad SB \quad \to \quad S(\mathcal{K} \otimes_{\mathbb{C}^.} B)$$

$$k^A: \quad SA \quad \to \quad SA$$

denote the inclusions.

b) Let $y \in KK^1(A, B)$ be represented by the extension

$$0 \to \mathcal{K} \otimes_{C^*} B \to E \to A \to 0$$

admitting a completely positive splitting s. Choose a bounded, approximate unit $(v_t)_{t \in \mathbb{R}_+}, 0 \leq v_t \leq 1$ of $\mathcal{K} \otimes_{C^*} B$ which is quasicentral in E. Then the linear map

$$\Phi_{odd} : \quad SSA \quad \to \quad S(\mathcal{K} \otimes_{C^*} B)$$

$$f \otimes a \quad \to \quad f(v_t)s(a)$$

defines an asymptotic morphism and the bivariant Chern character of y may be represented by the cocycle

$$ch(y) = [i_{SB*}]^{-1} \circ [\Phi_{odd*}] \circ [k_*^{SA}]^{-1} \circ \beta_{SA}$$

(Here suspensions correspond to functions on $]0, 1[$.)   □

Before coming to a proof of Theorems 10.1.,10.2 we want to recall Cuntz's description of KK-theory [CU] and the modifications in odd dimensions due to Zekri [Z].

For any C^*-algebra A, the free product $A * A$ of A with itself carries a natural C^*-norm. If the completion is denoted by QA there is a short exact sequence of C^*-algebras

$$0 \to qA \to QA \xrightarrow{id*id} A \to 0$$

splitting naturally in two ways. The morphisms

$$id * 0, 0 * id : QA \to A$$

restrict to maps

$$\pi_0, \pi_1 : qA \to A$$

These maps are KK-equivalences in fact. The interest in these algebras stems from the result of Cuntz that there is a natural isomorphism

$$KK^0(A, B) \xleftarrow{\simeq} [qA \otimes_{C^*} \mathcal{K}, qB \otimes_{C^*} \mathcal{K}]$$

between the group of homotopy classes of homomorphisms of C^*-algebras from $qA \otimes_{C^*} \mathcal{K}$ to $qB \otimes_{C^*} \mathcal{K}$ and $KK^0(A, B)$.

The group $\mathbb{Z}/2\mathbb{Z}$ acts on QA and its ideal qA by switching the two factors of A. The crossed product by this action is denoted by

$$\epsilon A := qA \times \mathbb{Z}/2\mathbb{Z}$$

It fits into a short exact sequence with a completely positive splitting

$$0 \to \epsilon A \to C^*(A[P]) \to A \oplus A \to 0$$

where

$$C^*(A[P]) \simeq QA \times \mathbb{Z}/2\mathbb{Z}$$

is the universal C^*-algebra generated by A and a selfadjoint involution P. For separable A the algebras $C^*(A[P])$ (resp.ϵA) are KK-equivalent to A by an even (resp. odd) KK-equivalence. Zekri proved that the natural map

$$KK^1(A, B) \xleftarrow{\cong} [\epsilon A \otimes_{C^*} \mathcal{K}, qB \otimes_{C^*} \mathcal{K}]$$

is an isomorphism providing thus a description of KK-theory in odd dimensions similar to that of Cuntz in the even case.

Proposition 10.3:

Let A be a separable C^*-algebra.

a) The maps

$$\pi_0, \pi_1 : qA \to A$$

induce asymptotic HC-equivalences

$$[S\pi_{0*}], [S\pi_{1*}] \in HC_\alpha^0(SqA, SA)$$

b) There exist natural asymptotic HC-equivalences

$$\alpha_\epsilon^A : \in HC_\alpha^1(S\epsilon A, SA), \quad \beta_\epsilon^A : \in HC_\alpha^1(SA, S\epsilon A)$$

inverse to each other. Naturality means that for any homomorphism of C^*-algebras $f : A \to B$ the equalities

$$f_* \circ \alpha_\epsilon^A = \alpha_\epsilon^B \circ (\epsilon f)_*, \quad \beta_\epsilon^B \circ f_* = (\epsilon f)_* \circ \beta_\epsilon^A$$

hold, i.e. the diagrams

$$
\begin{array}{ccc}
"S\epsilon A" & \xrightarrow{\alpha} & "SA" \\
{\scriptstyle S(\epsilon f)_*}\downarrow & & \downarrow{\scriptstyle Sf_*} \\
"S\epsilon B" & \xrightarrow{\alpha} & "SB"
\end{array}
\qquad
\begin{array}{ccc}
"SA" & \xrightarrow{\beta} & "S\epsilon A" \\
{\scriptstyle Sf_*}\downarrow & & \downarrow{\scriptstyle S(\epsilon f)_*} \\
"SB" & \xrightarrow{\beta} & "S\epsilon B"
\end{array}
$$

commute.

Proof:

We repeat the arguments of Cuntz and Zekri:

a) Consider the homomorphisms

$$(\pi_0, \pi_1) : A * A \to A \oplus A$$

and

$$A \oplus A \quad \to \quad M_2(A * A)$$

$$(x, y) \quad \to \quad \begin{pmatrix} i(x) & 0 \\ 0 & \bar{i}(y) \end{pmatrix}$$

with

$$i, \bar{i} : A \to A * A$$

the two inclusions. The compositions of them are homotopic to the canonical inclusions

$$A * A \to M_2(A * A)$$

$$A \oplus A \to M_2(A \oplus A)$$

which yield asymptotic HC-equivalences by (8.17). So both morphisms and their suspensions are asymptotic HC-equivalences and we find

$$HC_\alpha^*(SQA, -) \simeq HC_\alpha^*(SA \oplus SA, -) \simeq HC_\alpha^*(SA, -) \oplus HC_\alpha^*(SA, -)$$

$$\simeq S(id * 0)^* HC_\alpha^*(SA, -) \oplus S(0 * id)^* HC_\alpha^*(SA, -)$$

If A is separable, the second excision theorem may be applied to the splitting extension

$$0 \to qA \to QA \xrightarrow{id*id} A \to 0$$

and yields an isomorphism

$$HC_\alpha^*(SqA, -) \oplus HC_\alpha^*(SA, -) \simeq HC_\alpha^*(SQA, -) \simeq$$

$$\simeq S(id * 0)^* HC_\alpha^*(SA, -) \oplus S(0 * id)^* HC_\alpha^*(SA, -)$$

which can be provided either by π_0 or by π_1:

$$S\pi_0^*, S\pi_1^* : HC_\alpha^*(SA, -) \xrightarrow{\simeq} HC_\alpha^*(SqA, -)$$

Treating the case $HC_\alpha^*(-, SA)$ similarly allows finally to establish the claim.

b) The extension defining qA is equivariant under the involution switching the two factors of A and gives rise to an extension of crossed product algebras

$$0 \to \epsilon A \to QA \times \mathbb{Z}/2\mathbb{Z} \to A \times \mathbb{Z}/2\mathbb{Z} \to 0$$

$$\| \qquad\qquad \downarrow\simeq \qquad\qquad \downarrow\simeq$$

$$0 \to \epsilon A \to C^*(A[P]) \to A \oplus A \to 0$$

The crossed product $QA \times_{\mathbb{Z}/2\mathbb{Z}}$ is easily seen to be isomorphic to the universal C^*-algebra $C^*(A[P])$ with relation

$$(P^2 - P)A = 0$$

where

$$P = \frac{1 + F}{2}$$

and the inner automorphism defined by

$$F = F^*, \quad (F^2 - 1)A = 0$$

corresponds to the action of the nontrivial element of $\mathbb{Z}/2\mathbb{Z}$ on QA. Finally the epimorphism

$$(p_0, p_1): C^*(A[P]) \to A \times \mathbb{Z}/2\mathbb{Z} \simeq A \otimes \mathbb{C}[\mathbb{Z}/2\mathbb{Z}] \simeq A \otimes (\mathbb{C} \oplus \mathbb{C}) = A \oplus A$$

is given by putting $P = 0$ and $P = 1$, respectively. The homotopy

$$C^*(A[P]) \quad \to \quad M_2(C^*(A[P]))[0,1]$$

$$a \quad \to \quad \begin{pmatrix} a & 0 \\ 0 & 0 \end{pmatrix}$$

$$P \quad \to \quad u_t \begin{pmatrix} P & 0 \\ 0 & 0 \end{pmatrix} u_t^{-1}$$

with

$$u_t = \begin{pmatrix} \cos\frac{\pi}{2}t & -\sin\frac{\pi}{2}t \\ \sin\frac{\pi}{2}t & \cos\frac{\pi}{2}t \end{pmatrix} \in M_2\mathbb{C}$$

allows to conclude that the natural inclusion

$$A \to C^*(A[P])$$

as well as its suspension is an asymptotic HC-equivalence inverse to

$$p_0: C^*(A[P]) \to A \oplus A \xrightarrow{\pi_1} A$$

(resp. Sp_0). As the projection $C^*(A[P]) \to A \oplus A$ possesses an obvious completely positive lifting the second excision theorem may be applied in the case that A is separable and yields the exact sequences

$$0 \to HC_\alpha^*(S\epsilon A, -) \xrightarrow{\partial} HC_\alpha^{*+1}(SA, -)^2 \xrightarrow{(Sp_0, Sp_1)^*} HC_\alpha^{*+1}(SA, -) \to 0$$

$$0 \to HC_\alpha^*(-, SA) \xrightarrow{(Sp_0, Sp_1)^*} HC_\alpha^*(-, SA)^2 \xrightarrow{\partial} HC_\alpha^{*+1}(-, S\epsilon A) \to 0$$

from which the isomorphisms

$$HC_\alpha^*(S\epsilon A, -) \xrightarrow[\simeq]{\partial} HC_\alpha^{*+1}(SA, -)$$

and

$$HC_\alpha^*(-, SA) \xrightarrow[\simeq]{\partial} HC_\alpha^{*+1}(-, S\epsilon A)$$

are obtained. They are natural in A under morphisms of algebras (which give rise to a map of extensions defining ϵA) and under asymptotic morphisms in the free variable $-$. The Yoneda lemma shows then that these isomorphisms are induced from the composition product with canonical classes

$$\alpha_\epsilon^A \in HC_\alpha^1(S\epsilon A, SA), \quad \beta_\epsilon^A \in HC_\alpha^1(SA, S\epsilon A)$$

and the claimed equalities follow from the naturality of the boundary maps above.

$\square$

Proof of Theorem 10.1:

Construction of the Chern character

Define the Chern character in even dimensions by the composition

$$KK^0(A, B) \xrightarrow{\simeq} [S(qA \otimes_{C^.} \mathcal{K}), S(qB \otimes_{C^.} \mathcal{K})] \to HC_\alpha^0(S(qA) \otimes_{C^.} \mathcal{K}, S(qB) \otimes_{C^.} \mathcal{K})$$

$$\xrightarrow{\simeq} HC_\alpha^0(SqA, SqB) \xrightarrow{\simeq} HC_\alpha^0(SA, SB)$$

where the two isomorphisms in the lower line are given by

$$[i_*^{SqB}]^{-1} \circ - \circ [i_*^{SqA}] : HC_\alpha^0(SqA \otimes_{C^.} \mathcal{K}, SqB \otimes_{C^.} \mathcal{K}) \to HC_\alpha^0(SqA, SqB)$$

and

$$[S\pi_{0*}^B] \circ - \circ [S\pi_{0*}^A]^{-1} : HC_\alpha^0(SqA, SqB) \to HC_\alpha^0(SA, SB)$$

respectively.

Define the Chern character in odd dimensions to be the composition

$$KK^1(A, B) \xrightarrow{\simeq} [S(\epsilon A \otimes_{C^.} \mathcal{K}), S(qB \otimes_{C^.} \mathcal{K})] \to HC_\alpha^0(S(\epsilon A) \otimes_{C^.} \mathcal{K}, S(qB) \otimes_{C^.} \mathcal{K})$$

$$\xrightarrow{\simeq} HC_\alpha^0(S\epsilon A, SqB) \xrightarrow{\simeq} HC_\alpha^1(SA, SB)$$

where the two isomorphisms in the lower line are given by

$$[i_*^{SqB}]^{-1} \circ - \circ [i_*^{S\epsilon A}] : HC_\alpha^0((S\epsilon A) \otimes_{C^.} \mathcal{K}, (SqB) \otimes_{C^.} \mathcal{K}) \to HC_\alpha^0(S\epsilon A, SqB)$$

and

$$[S\pi_{0*}^B] \circ - \circ \beta_\epsilon^A : HC_\alpha^0(S\epsilon A, SqB) \to HC_\alpha^0(SA, SB)$$

respectively.

The Chern character is natural under morphisms of algebras and

$$ch(id_*^{KK}) = id_*^{HC_\alpha(S, S)}$$

Naturality of the Chern character

First case:

$$\begin{array}{ccc} KK^0 \otimes KK^0 & \longrightarrow & KK^0 \\ ch \otimes ch \downarrow & & \downarrow ch \\ HC_\alpha^0 \otimes HC_\alpha^0 & \longrightarrow & HC_\alpha^0 \end{array}$$

Obvious from the definitions.

Second case:

$$\begin{array}{ccc} KK^1 \otimes KK^0 & \longrightarrow & KK^1 \\ ch \otimes ch \downarrow & & \downarrow ch \\ HC_\alpha^1 \otimes HC_\alpha^0 & \longrightarrow & HC_\alpha^1 \end{array}$$

Obvious from the definitions, as the Kasparov product is just given by the composition of the corresponding morphisms of universal algebras.

Third case:

$$\begin{array}{ccc} KK^0 \otimes KK^1 & \longrightarrow & KK^1 \\ ch \otimes ch \downarrow & & \downarrow ch \\ HC_\alpha^0 \otimes HC_\alpha^1 & \longrightarrow & HC_\alpha^1 \end{array}$$

Let

$$x \in [qA \otimes_{C^.} \mathcal{K}, qB \otimes_{C^.} \mathcal{K}], \quad y \in [\epsilon B \otimes_{C^.} \mathcal{K}, qC \otimes_{C^.} \mathcal{K}]$$

The Kasparov product $x \otimes y$ of x and y is given by the composition

$$x \otimes y : \epsilon A \otimes_{C^.} \mathcal{K} \xrightarrow{\mu_A} \epsilon(qA) \otimes_{C^.} \mathcal{K} \xrightarrow{\epsilon(i) \otimes id} \epsilon(qA \otimes_{C^.} \mathcal{K}) \otimes_{C^.} \mathcal{K} \to$$

$$\xrightarrow{\epsilon(x) \otimes id} \epsilon(qB \otimes_{C^.} \mathcal{K}) \otimes_{C^.} \mathcal{K} \xrightarrow{k} \epsilon(qB) \otimes_{C^.} \mathcal{K} \otimes_{C^.} \mathcal{K} \xrightarrow{\epsilon(\pi_0)}$$

$$\epsilon(B) \otimes_{C^.} \mathcal{K} \otimes_{C^.} \mathcal{K} \xrightarrow{y \otimes id} qC \otimes_{C^.} \mathcal{K} \otimes_{C^.} \mathcal{K} \simeq qC \otimes_{C^.} \mathcal{K}$$

(See [Z]). The morphism

$$\mu_A : \epsilon A \otimes_{C^.} \mathcal{K} \to \epsilon(qA) \otimes_{C^.} \mathcal{K}$$

is a homotopy inverse of $\epsilon(\pi_0)$ up to stabilization by matrices: The composed morphisms

$$\epsilon A \otimes_{C^.} \mathcal{K} \xrightarrow{\mu_A} \epsilon(qA) \otimes_{C^.} \mathcal{K} \xrightarrow{\epsilon(\pi_0) \otimes id} \epsilon A \otimes_{C^.} \mathcal{K}$$

$$\epsilon(qA) \otimes_{C^.} \mathcal{K} \xrightarrow{\epsilon(\pi_0) \otimes id} \epsilon A \otimes_{C^.} \mathcal{K} \xrightarrow{\mu_A} \epsilon(qA) \otimes_{C^.} \mathcal{K}$$

are homotopic to the identity so that we may conclude

$$[S\mu_*^A] = [i_*] \circ [S\epsilon(\pi_0)_*]^{-1} \circ [i_*]^{-1} \in HC_\alpha^0(S\epsilon A \otimes_{C^.} \mathcal{K}, S\epsilon(qA) \otimes_{C^.} \mathcal{K})$$

The natural morphism

$$k : \epsilon(C \otimes_{C^.} \mathcal{K}) \to \epsilon C \otimes_{C^.} \mathcal{K}$$

is defined by the map of extensions

$$\epsilon(C \otimes_{C^.} \mathcal{K}) \rightarrow Q(C \otimes_{C^.} \mathcal{K}) \times \mathbb{Z}/2\mathbb{Z} \rightarrow C \otimes_{C^.} \mathcal{K} \oplus C \otimes_{C^.} \mathcal{K}$$

$$k \downarrow \qquad\qquad \downarrow \qquad\qquad\qquad \downarrow$$

$$\epsilon C \otimes_{C^.} \mathcal{K} \rightarrow QC \times \mathbb{Z}/2\mathbb{Z} \otimes_{C^.} \mathcal{K} \rightarrow C \otimes_{C^.} \mathcal{K} \oplus C \otimes_{C^.} \mathcal{K}$$

which shows also, after repeating the argument proving the existence of the elements α_ϵ, β_ϵ that

$$[Sk_*] \circ \beta_\epsilon^{C \otimes_{C^.} \mathcal{K}} \circ i_C = i_{\epsilon C} \circ \beta_\epsilon^C \in HC_\alpha^1(SC, S\epsilon(C) \otimes_{C^.} \mathcal{K})$$

Let us calculate the Chern character of $x \otimes y$: We find after suspending the sequence of morphisms defining $x \otimes y$

$$ch(x \otimes y) = [S\pi_{0*}^C] \circ [i_*^{SqC}]^{-1} \circ [i_*^{SqC \otimes \mathcal{K}}]^{-1} \circ [S(x \otimes y)_*] \circ [i_*^{S\epsilon A}] \circ \beta_\epsilon^A$$

$$= [S\pi_{0*}^C] \circ [i_*^{SqC}]^{-1} \circ [i_*^{SqC \otimes \mathcal{K}}]^{-1} \circ [S((y \circ \epsilon(\pi_0) \circ k \circ \epsilon(x) \circ \epsilon(i)) \otimes id)_*] \circ [S\mu_*^A] \circ [i_*^{S\epsilon A}] \circ \beta_\epsilon^A$$

$$= [S\pi_{0*}^C] \circ [i_*^{SqC}]^{-1} \circ [i_*^{SqC \otimes \mathcal{K}}]^{-1} \circ [S((y \circ \epsilon(\pi_0) \circ k \circ \epsilon(x \circ i)) \otimes id)_*] \circ$$
$$\circ [i_*] \circ [S\epsilon(\pi_0)_*]^{-1} \circ [i_*]^{-1} \circ [i_*] \circ \beta_\epsilon^A$$

$$= [S\pi_{0*}^C] \circ [i_*^{SqC}]^{-1} \circ [S(y \circ \epsilon(\pi_0))_*] \circ [Sk_*] \circ [S\epsilon(x \circ i)_*] \circ [S\epsilon(\pi_0)_*]^{-1} \circ \beta_\epsilon^A$$

$$= [S\pi_{0*}^C] \circ [i_*^{SqC}]^{-1} \circ [S(y \circ \epsilon(\pi_0))_*] \circ ([Sk_*] \circ \beta_\epsilon^{qB \otimes \mathcal{K}} \circ [i_*]) \circ ([i_*]^{-1} \circ [Sx] \circ [i_*] \circ [S\pi_{0*}]^{-1})$$

$$= [S\pi_{0*}^C] \circ [i_*^{SqC}]^{-1} \circ [S(y \circ i)_*] \circ ([S\epsilon(\pi_0)_*] \circ \beta_\epsilon^{qB}) \circ ([i_*]^{-1} \circ [Sx] \circ [i_*] \circ [S\pi_{0*}]^{-1})$$

$$= ([S\pi_{0*}^C] \circ [i_*^{SqC}]^{-1} \circ [Sy] \circ [i_*] \circ \beta_\epsilon^B) \circ ([S\pi_{0*}] \circ [i_*]^{-1} \circ [Sx] \circ [i_*] \circ [S\pi_{0*}]^{-1})$$

$$= ch(y) \circ ch(x) = ch(x) \otimes ch(y)$$

Before proceeding further it is necessary to investigate the behaviour of the Chern character with respect to boundary maps in the long exact sequences in KK-theory and asymptotic cohomology. We treat the homological case, the cohomological one being similar.

Let

$$0 \rightarrow J \rightarrow A \rightarrow B \rightarrow 0$$

be an extension of separable C^*-algebras admitting a completely positive splitting. If one considers the diagram

$$SA \xrightarrow{Sf} SB \xrightarrow{j} C_f \xrightarrow{i} A \xrightarrow{f} B$$
$$\uparrow$$
$$J$$

the KK-theoretic connecting map $\delta : KK^*(C, B) \rightarrow KK^{*+1}(C, J)$ is given by the composition

$$\delta : KK^*(C, B) \xrightarrow{\otimes \beta_{KK}} KK^{*+1}(C, SB) \xrightarrow{j_*} KK^{*+1}(C, C_f) \xleftarrow{\simeq} KK^{*+1}(C, J)$$

(The existence of a completely positive splitting is essential in showing that the canonical inclusion $J \to C_f$ is a KK-equivalence). It is also a stable asymptotic HC-equivalence by the second excision theorem (9.9) and the cohomological connecting map is defined similarly by the composition

$$\delta : HC_\alpha^*(SC, SB) \xrightarrow{\otimes \beta_{HC}} HC_\alpha^{*+1}(SC, S^2B) \xrightarrow{Sj_*}$$

$$\xrightarrow{Sj_*} HC_\alpha^{*+1}(SC, SC_f) \overset{\cong}{\leftarrow} HC_\alpha^{*+1}(SC, SJ)$$

So one obtains a diagram

$$
\begin{array}{ccccc}
\delta: & KK^*(C,B) & \xrightarrow{\otimes\beta_{KK}} & KK^{*+1}(C,SB) & \to & KK^{*+1}(C,J) \\
& ch \downarrow & & ch \downarrow & & ch \downarrow \\
\delta: & HC_\alpha^*(SC,SB) & \xrightarrow{\otimes\beta_{HC}} & HC_\alpha^{*+1}(SC,S^2B) & \to & HC_\alpha^{*+1}(SC,SJ)
\end{array}
$$

where the square on the right side is commutative by the naturality of the Chern character under algebra homomorphisms. So it remains to investigate the commutativity of the square on the left. The compatibility of the Chern character with the Kasparov- resp. composition-product already being established in the case where at least one factor is even-dimensional we see that the following diagrams commute

$$
\begin{array}{ccc}
KK^0(C,B) & \xrightarrow{\otimes\beta_{KK}} & KK^1(C,SB) \\
ch \downarrow & & ch \downarrow \\
HC_\alpha^0(SC,SB) & \xrightarrow{\otimes ch(\beta_{KK})} & HC_\alpha^1(SC,S^2B)
\end{array}
$$

$$
\begin{array}{ccc}
KK^1(C,B) & \xleftarrow{\otimes\alpha_{KK}} & KK^0(C,SB) \\
ch \downarrow & & ch \downarrow \\
HC_\alpha^1(SC,SB) & \xleftarrow{\otimes ch(\alpha_{KK})} & HC_\alpha^0(SC,S^2B)
\end{array}
$$

and consequently also the square

$$
\begin{array}{ccc}
KK^1(C,B) & \xrightarrow{\otimes\beta_{KK}} & KK^0(C,SB) \\
ch \downarrow & & ch \downarrow \\
HC_\alpha^1(SC,SB) & \xrightarrow{\otimes ch(\alpha_{KK})^{-1}} & HC_\alpha^0(SC,S^2B)
\end{array}
$$

The statement of Theorem 10.1.d) follows then from the

Lemma 10.4:

Let

$$\alpha_{KK} \in KK^1(SB,B), \quad \beta_{KK} \in KK^1(B,SB)$$

be the K-theoretical and

$$\alpha_{HC} \in HC_\alpha^1(S^2B,SB), \quad \beta_{HC} \in HC_\alpha^1(SB,S^2B)$$

the cohomological Dirac- resp. Bott-elements. Then

$$ch(\beta_{KK}) = \beta_{HC} \quad ch(\alpha_{KK}) = \frac{1}{2\pi i}\alpha_{HC}$$

$\square$

Assuming the lemma for the moment we go on to establish the compatibility of the bivariant Chern character with products in the remaining fourth case:

Fourth case:

$$KK^1 \otimes KK^1 \longrightarrow KK^0$$

$$ch \otimes ch \downarrow \qquad\qquad \downarrow ch$$

$$HC_\alpha^1 \otimes HC_\alpha^1 \longrightarrow HC_\alpha^0$$

$$[x] \in [\epsilon A, B \otimes_{C^\cdot} \mathcal{K}] \xleftarrow[\simeq]{(\pi_0 \otimes id)\circ -} [\epsilon A, qB \otimes_{C^\cdot} \mathcal{K}] \simeq KK^1(A,B)$$

$$[y] \in [\epsilon B, C \otimes_{C^\cdot} \mathcal{K}] \xleftarrow[\simeq]{(\pi_0 \otimes id)\circ -} [\epsilon B, qC \otimes_{C^\cdot} \mathcal{K}] \simeq KK^1(B,C)$$

Their Kasparov product $x \otimes y \in KK^0(A,C)$ can be represented by the composition (see [Z])

$$(x \otimes y) : qA \xrightarrow{\nu} \epsilon(\epsilon A) \otimes_{C^\cdot} \mathcal{K} \xrightarrow{\epsilon(x)} \epsilon(B \otimes_{C^\cdot} \mathcal{K}) \otimes_{C^\cdot} \mathcal{K} \to$$

$$\xrightarrow{k \otimes id} \epsilon B \otimes_{C^\cdot} \mathcal{K} \otimes_{C^\cdot} \mathcal{K} \xrightarrow{y \otimes id \otimes id} C \otimes_{C^\cdot} \mathcal{K} \otimes_{C^\cdot} \mathcal{K} \otimes_{C^\cdot} \mathcal{K} \simeq C \otimes_{C^\cdot} \mathcal{K}$$

The element $\nu \in KK^0(A, \epsilon(\epsilon A))$ corresponding to $\nu : qA \to \epsilon(\epsilon A) \otimes_{C^\cdot} \mathcal{K}$ is defined as follows: From the extension

$$0 \to \epsilon A \to QA \times \mathbb{Z}/2\mathbb{Z} \to A \oplus A \to 0$$

one obtains the connecting map

$$KK^0(A, A \oplus A) \xrightarrow{\partial} KK^1(A, \epsilon A)$$

$$(i_0)_* \qquad \to \qquad \beta_\epsilon^A$$

For a general separable C^*-algebra D, the connecting map

$$KK^*(D, A) \xrightarrow{\partial \circ i_0 *} KK^{*+1}(D, \epsilon A)$$

is induced by the Kasparov product with the KK-equivalence β_ϵ^A.

The KK-equivalence $[\nu]$ is then given by the square of the connecting map:

$$[\nu]: \quad KK^0(A, A) \xrightarrow{\partial \circ i_0 *} KK^1(A, \epsilon A) \xrightarrow{\partial \circ i_0 *} KK^0(A, \epsilon(\epsilon A))$$

$$id_* \qquad \to \qquad \beta_\epsilon^A \qquad \to \qquad \beta_\epsilon^A \otimes \beta_\epsilon^{\epsilon A}$$

The connecting maps being compatible with the Chern character, it is easy to calculate $ch(\nu)$ by having a look at the commutative diagram

$$id_* \in \quad KK^0(A,A) \xrightarrow{\partial \circ i_{0*}} KK^1(A, \epsilon A) \xrightarrow{\partial \circ i_{0*}} KK^0(A, \epsilon^2 A)$$

$$\downarrow \qquad\qquad ch \downarrow \qquad\qquad\qquad ch \downarrow \qquad\qquad\qquad\qquad ch \downarrow$$

$$id_* \in \quad HC^0_\alpha(SA, SA) \xrightarrow{\partial \circ i_{0*}} HC^1_\alpha(SA, S\epsilon A) \xrightarrow{(2\pi i)\partial \circ i_{0*}} HC^0_\alpha(SA, S\epsilon^2 A)$$

$$id_* \qquad\qquad \to \qquad\qquad \beta^A_\epsilon \qquad\qquad \to \qquad\qquad (2\pi i)\beta^{\epsilon A}_\epsilon \circ \beta^A_\epsilon$$

So
$$ch(\nu) = (2\pi i)\beta^{\epsilon A}_\epsilon \circ \beta^A_\epsilon \in HC^0_\alpha(SA, S\epsilon^2 A)$$

With this we find

$$ch(x \otimes y) = [i^{SC}_*]^{-1} \circ [i^{SC \otimes c \cdot \mathcal{K}}_*]^{-1} \circ [i^{SC \otimes c \cdot \mathcal{K} \otimes c \cdot \mathcal{K}}_*]^{-1} \circ$$

$$\circ [S((y \otimes id) \circ k \circ \epsilon(x)) \otimes id_*] \circ [S\nu_*] \circ [S\pi_{0*}]^{-1}$$

$$= [i^{SC}_*]^{-1} \circ [i^{SC \otimes c \cdot \mathcal{K}}_*]^{-1} \circ [S((y \otimes id) \circ k \circ \epsilon(x))_*] \circ ([i^{S\epsilon^2 A}_*]^{-1} \circ [S\nu_*] \circ [S\pi_{0*}]^{-1})$$

$$= (2\pi i)[i^{SC}_*]^{-1} \circ [Sy] \circ [i^{S\epsilon B}_*]^{-1} \circ [Sk_*] \circ [S\epsilon(x)_*] \circ \beta^{\epsilon A}_\epsilon \circ \beta^A_\epsilon$$

because

$$[i^{S\epsilon^2 A}_*]^{-1} \circ [S\nu_*] \circ [S\pi_{0*}]^{-1} = ch(\nu) = (2\pi i)\beta^{\epsilon A}_\epsilon \circ \beta^A_\epsilon$$

Thus $ch(x \otimes y)$ equals

$$(2\pi i)([i^{SC}_*]^{-1} \circ [Sy]) \circ ([i^{S\epsilon B}_*]^{-1} \circ [Sk_*] \circ \beta^{B \otimes c \cdot \mathcal{K}}_\epsilon) \circ ([Sx] \circ \beta^A_\epsilon)$$

$$= (2\pi i)([i^{SC}_*]^{-1} \circ [Sy]) \circ ([i^{S\epsilon B}_*]^{-1} \circ [i^{S\epsilon B}_*] \circ \beta^B_\epsilon) \circ ([i^{SB}_*]^{-1} \circ [Sx] \circ \beta^A_\epsilon)$$

because of

$$[Sk_*] \circ \beta^{B \otimes c \cdot \mathcal{K}}_\epsilon \circ [i^{SB}_*] = [i^{S\epsilon B}_*] \circ \beta^B_\epsilon$$

$$= (2\pi i)([i^{SC}_*]^{-1} \circ [Sy] \circ \beta^B_\epsilon) \circ ([i^{SB}_*]^{-1} \circ [Sx] \circ \beta^A_\epsilon)$$

$$= (2\pi i)ch(y) \circ ch(x) = (2\pi i)ch(x) \otimes ch(y)$$

The demonstration of Theorem 10.1.a) is thus achieved. Theorem 10.1.c) follows from Theorem 10.1.b), Lemma 10.4. and 10.5. and the stable periodicity theorem.

$\square$

Proof of Lemma 10.4:

Recall the extension

$$0 \to \epsilon B \to E_1 B \to B \to 0$$

defined as the pullback

$$
\begin{array}{ccccccccc}
0 & \to & \epsilon B & \to & E_1 B & \to & B & \to & 0 \\
& & \| & & \downarrow & & \downarrow i_0 & & \\
0 & \to & \epsilon B & \to & C^* B[P] & \to & B \oplus B & \to & 0
\end{array}
$$

of the extension defining ϵB in 9.3.(see [Z]). The inclusion $E_1 B \to M_2(E_1 B)$ is nullhomotopic which gives rise to maps of extensions

$$
\begin{array}{ccccccccc}
0 & \to & SB & \to & CB & \to & B & \to & 0 \\
& & \downarrow & & \downarrow & & \downarrow & & \\
0 & \to & M_2(SB) & \to & M_2(CB) & \to & M_2(B) & \to & 0 \\
& & s \uparrow & & \uparrow & & \uparrow & & \\
0 & \to & \epsilon B & \to & E_1 B & \to & B & \to & 0
\end{array}
$$

The KK-theoretic connecting map of the upper extension being the K-theoretic Bott element $\beta_{KK} \in KK^1(B, SB)$ we conclude that under the isomorphism

$$KK^1(B, SB) \xleftarrow{\sim} [\epsilon B, SB \otimes_{C^-} \mathcal{K}]$$

β_{KK} corresponds to the homomorphisms s. The Chern character $ch(\beta_{KK})$ can therefore be computed by composing Ss_* with the cohomological connecting map $\beta_\epsilon^B \in HC^1_\alpha(SB, S\epsilon B)$ which equals the cohomological connecting map of the extension

$$0 \to SB \to CB \to B \to 0$$

by naturality. A look at the definition of the cohomological connecting map shows that it coincides for the extension under discussion with the cohomological Bott element so that

$$ch(\beta_{KK}) = \beta_{HC}$$

which proves the first part of the lemma. To calculate the Chern character of the Dirac element α_{KK} we use the equation

$$\beta_{HC} = ch(\beta_{KK}) = ch(\alpha_{KK}\beta_{KK}^2) = ch(\alpha_{KK})ch(\beta_{KK}^2)$$

which is true because the compatibility of the bivariant Chern character holds already if at least one class of even degree is involved. By the proof of Theorem 9.4 and Lemma 10.5. we know however that

$$\langle ch(\beta_{KK}^2), \alpha_{HC}^2 \rangle = \langle ch([" Bott" \otimes id^B]), [(\tau_1 \times \tau_1) \times id_*^{SB}] \rangle = (2\pi i)[id_*^{SB}]$$

and consequently

$$ch(\beta_{KK}^2) = (2\pi i)\beta_{HC}^2$$

so that

$$\beta_{HC} = ch(\alpha_{KK})ch(\beta_{KK}^2) = (2\pi i)ch(\alpha_{KK})\beta_{HC}^2$$

and finally

$$ch(\alpha_{KK}) = \frac{1}{2\pi i}\alpha_{HC}$$

$\square$

Proof of Theorem 10.2:

We divide the demonstration into several steps.

1) Φ_{ev} and Φ_{odd} are asymptotic morphisms.

It can be readily verified that Φ_{ev} and Φ_{odd} define bounded families of continuous linear maps. To calculate their curvature note that

$$\omega_{\Phi_{ev}} = f(D_t)[g(D_t), \begin{pmatrix} \rho_0(a) & 0 \\ 0 & \rho_1(a) \end{pmatrix}]\begin{pmatrix} \rho_0(b) & 0 \\ 0 & \rho_1(b) \end{pmatrix}$$

$$\omega_{\Phi_{odd}} = fg(v_t)(s(ab) - s(a)s(b)) + f(v_t)[g(v_t), s(a)]s(b)$$

We may suppose in both cases that f,g are holomorphic near the intervals $[-1, 1]$ resp. $[0, 1]$. In the even case one finds

$$[g(D_t), \begin{pmatrix} \rho_0(a) & 0 \\ 0 & \rho_1(a) \end{pmatrix}] =$$

$$= \frac{1}{2\pi i}\int_\Gamma g(\lambda)(\lambda - D_t)^{-1}[(\lambda - D_t), \begin{pmatrix} \rho_0(a) & 0 \\ 0 & \rho_1(a) \end{pmatrix}](\lambda - D_t)^{-1}d\lambda$$

and

$$[(\lambda - D_t), \begin{pmatrix} \rho_0(a) & 0 \\ 0 & \rho_1(a) \end{pmatrix}] =$$

$$= \begin{pmatrix} 0 & \cos\frac{\pi}{2}x((1-u_t)\rho_1(a) - \rho_0(a)(1-u_t)) \\ \cos\frac{\pi}{2}x((1-u_t)\rho_0(a) - \rho_1(a)(1-u_t)) & 0 \end{pmatrix}$$

so that the estimate

$$\| (1 - u_t)\rho_1(a) - \rho_0(a)(1 - u_t) \| \leq \| (1 - u_t)(\rho_1(a) - \rho_0(a)) \| + \| [u_t, \rho_0(a)] \|$$

shows that the curvature of Φ_{ev} becomes arbitrarily small near infinity (note that $\rho_1(a) - \rho_0(a) \in \mathcal{K} \otimes_{C^*} B$). The same argument keeps track of the component $f(v_t)[g(v_t), s(a)]s(b)$ of the curvature of Φ_{odd} whereas the fact that $s(ab) - s(a)s(b) \in \mathcal{K} \otimes_{C^*} B$, the identity

$$f(v_t)j = f(v_t)j - f(1)j =$$

$$= \frac{1}{2\pi i}\int_\Gamma f(\lambda)((\lambda - v_t)^{-1} - (\lambda - 1)^{-1})jd\lambda$$

and the estimate

$$\| ((\lambda - v_t)^{-1} - (\lambda - 1)^{-1})j \| \leq \| (\lambda - v_t)^{-1}(\lambda - 1)^{-1} \| \| (v_t - 1)j \|$$

take care of the second component $fg(v_t)(s(ab) - s(a)s(b))$ of the curvature of Φ_{odd}.

2) The evendimensional case

Let

$$0 \to J \to A \xrightarrow{p} B \to 0$$

be a splitting (under s) extension of separable C^*-algebras and let (u_t) be a bounded, quasicentral, approximate unit in J. Then the linear map

$$\chi: \quad SA \quad \to \quad M_2 SJ$$

$$f \otimes a \quad \to \quad f(D_t) \begin{pmatrix} a & 0 \\ 0 & s(p(a)) \end{pmatrix}$$

is an asymptotic morphism by what we just proved. Moreover, following the arguments in the proof of 9.2 shows that the cohomology class

$$\xi := [Tr \times id_*^{SJ}] \circ [\chi_*] \circ [k_*^A]^{-1} \in HC_\alpha^0(SA, SJ)$$

satisfies

$$(Si)_* \xi = [Sid_*^A] - [S(s \circ p)_*] \in HC_\alpha^0(SA, SA)$$

If we apply this to the extension

$$0 \to qA \to QA \xrightarrow{id*id} A \to 0$$

with splitting $i_1 : A \to QA$, then $\xi \circ [Si_{0*}] \in HC_\alpha^0(SA, SqA)$ satisfies

$$[S\pi_{0*}] \circ \xi \circ [Si_{0*}] = [S(id*0)_*] \circ ([Sid_*] - [S(s \circ p)_*]) \circ [Si_{0*}] = [id_*] \in HC_\alpha^0(SA, SA)$$

so that we conclude

$$\xi \circ [Si_{0*}] = [S\pi_{0*}]^{-1} \in HC_\alpha^0(SA, SqA)$$

Let now $x \in KK^0(A, B)$ be represented by the Kasparov bimodule

$$x = \left(\mathcal{H}_B \oplus \mathcal{H}_B, \rho = (\rho_0, \rho_1) : A \to \mathcal{L}(\mathcal{H}_B) = M(\mathcal{K} \otimes_{C^*} B), F = \begin{pmatrix} 0 & 1 \\ 1 & 0 \end{pmatrix} \right)$$

Then there is a corresponding diagram of extensions

$$\begin{array}{ccccccccc}
0 & \to & \mathcal{K} \otimes_{C^*} B & \to & M(\mathcal{K} \otimes_{C^*} B) & \to & \mathcal{Q}(\mathcal{K} \otimes_{C^*} B) & \to & 0 \\
& & f \uparrow & & \uparrow \rho_0 * \rho_1 & & \uparrow & & \\
0 & \to & qA & \to & QA & \to & A & \to & 0
\end{array}$$

and the homomorphism

$$f : qA \to \mathcal{K} \otimes_{C^*} B$$

represents the class x under the Cuntz isomorphism

$$KK^0(A, B) \xleftarrow{\cong} [qA, B \otimes_{C^*} \mathcal{K}]$$

By definition

$$ch(x) = [i_{SB*}]^{-1} \circ [Sf_*] \circ [S\pi_{0*}]^{-1} = [i_{SB*}]^{-1} \circ [Sf_*] \circ \xi \circ [Si_{0*}]$$

$$= [i_{SB*}]^{-1} \circ [(Sf \circ \chi \circ Si_0)_*] \circ [k_*^A]^{-1} \in HC_\alpha^0(SA, SB)$$

Moreover, the fact that the bounded, quasicentral approximate units in an ideal of a separable C^*-algebra form a convex cone shows that the diagram

$$
\begin{array}{ccc}
S(\mathcal{K} \otimes_{C^*} B) & \xleftarrow{\ \Phi_{ev}\ } & SA \\
{\scriptstyle sf}\uparrow & & \downarrow{\scriptstyle Si_0} \\
S(qA) & \xleftarrow[\ \chi\ ] & S(QA)
\end{array}
$$

commutes up to homotopy which proves

$$[(Sf \circ \chi \circ Si_0)_*] = [\Phi_{ev*}] \in HC_\alpha^0(SA, S(\mathcal{K} \otimes_{C^*} B))$$

and establishes the claim.

3) The oddimensional case

If

$$0 \to \mathcal{K} \otimes_{C^*} B \to E \to A \to 0$$

is an extension admitting a completely positive splitting then it can be shown using Kasparov's generalized Stinespring theorem that there is a map of extensions

$$
\begin{array}{ccccccccc}
0 & \to & \mathcal{K} \otimes_{C^*} B & \to & E & \to & A & \to & 0 \\
 & & {\scriptstyle g}\uparrow & & \uparrow & & \| & & \\
0 & \to & \epsilon A & \to & E_1 A & \to & A & \to & 0
\end{array}
$$

where $E_1 A$ is the universal C^*-algebra of 10.4 (see [Z]) and g is the homomorphism corresponding to $y \in KK^1(A, B)$ under the Zekri isomorphism

$$KK^1(A, B) \xleftarrow{\ \cong\ } [\epsilon A, \mathcal{K} \otimes_{C^*} B]$$

By definition of the bivariant Chern character

$$ch(y) = [i_{SB*}]^{-1} \circ [Sg_*] \circ \beta_\epsilon^A$$

where $\beta_\epsilon^A \in KK^1(SA, S\epsilon A)$ is the cohomological connecting map of the extension

$$0 \to \epsilon A \to E_1 A \to A \to 0$$

The connecting map being natural we conclude

$$ch(y) = [i_{SB*}]^{-1} \circ \delta \circ [id_*^{SA}]$$

where δ is the cohomological connecting map of the extension

$$0 \to \mathcal{K} \otimes_{C^*} B \to E \to A \to 0$$

defining y.

Recall that the connecting map was defined as follows:

$$\delta = \Theta \circ [Sj_*] \circ \beta_{SA}$$

in terms of the morphisms in the diagram

$$
\begin{array}{ccccccc}
SA & \xrightarrow{j} & C_f & \rightarrow & E & \xrightarrow{f} & A \\
 & & {\scriptstyle i}\uparrow & & \| & & \| \\
0 & \rightarrow & K \otimes_{C} \cdot B & \rightarrow & E & \rightarrow & A & \rightarrow & 0
\end{array}
$$

where

$$\Theta = [\theta_*] \circ [k_*^{C_f}]^{-1} = [Si_*]^{-1}$$

(see the two excision theorems 9.6,9.9). Thus

$$\delta = [\theta_*] \circ [k_*^{C_f}]^{-1} \circ [Sj_*] \circ \beta_{SA} =$$

$$= [\theta_*] \circ [Sj_*] \circ [k_*^{SA}]^{-1} \circ \beta_{SA}$$

As the asymptotic morphism $\theta \circ Sj$ is readily seen to be homotopic to Φ_{odd} we conclude

$$ch(y) = [i_{SB*}]^{-1} \circ \delta = [i_{SB*}]^{-1} \circ [\Phi_{odd*}] \circ [k_*^{SA}]^{-1} \circ \beta_{SA}$$

as claimed.

□

Lemma 10.5:

The normalised ordinary Chern character ch' (10.1.b)) and the bivariant Chern character ch_{biv} (10.1.a)) coincide, i.e. for a separable C^*-algebra A the diagram

$$
\begin{array}{ccc}
K_j(A) & \xrightarrow{\simeq} & KK^j(\mathbb{C}, A) \\
{\scriptstyle ch'}\Big\downarrow & & \\
HC_*^\epsilon(A) & & \Big\downarrow \quad ch_{biv} \\
{\scriptstyle \simeq}\Big\downarrow & & \\
HC_\alpha^j(\mathbb{C}, A) & \xleftarrow[\alpha_A \circ - \circ \beta_{\mathbb{C}}]{} & HC_\alpha^j(S\mathbb{C}, SA)
\end{array}
$$

commutes.

Proof:

The diagram being natural under algebra homomorphisms we may suppose $A = \mathbb{C}$ in the even and $A = S\mathbb{C}$ in the odd case. In the even case the commutativity of the diagram is obvious. In the odd case, the fundamental class u of $KK^1(\mathbb{C}, S\mathbb{C}) \simeq \mathbb{Z}$ is represented by the extension $0 \rightarrow S\mathbb{C} \rightarrow C\mathbb{C} \rightarrow \mathbb{C} \rightarrow 0$ whose bivariant Chern character is given by $\beta_{S\mathbb{C}} \in HC_\alpha^1(S\mathbb{C}, S^2\mathbb{C})$. Then

$$\alpha_{S\mathbb{C}} \circ ch_{biv}(u) \circ \beta_{\mathbb{C}} = \alpha_{S\mathbb{C}} \circ \beta_{S\mathbb{C}} \circ \beta_{\mathbb{C}} = \beta_{\mathbb{C}}$$

in $HC_\alpha^1(\mathbb{C}, S\mathbb{C}) \simeq \mathbb{C}$.

On the other hand the ordinary Chern character of the fundamental class $u \in K_1(S\mathbb{C})$ satisfies

$$\langle ch(u), \alpha_\mathbb{C} \rangle = 2\pi i$$

from which the claim

$$ch(u) = 2\pi i \, ch_{biv}(u)$$

immediately follows.

$\square$

Corollary 10.6:

Let A, B be separable C^*-algebras. Let

$$\phi \in KK^*(A, B)$$

be a KK-equivalence. Then

$$ch(\phi) \in HC^*_\alpha(SA, SB)$$

is an asymptotic HC-equivalence. Consequently

$$HC^*_\alpha(SA, C) \simeq HC^{*+|\varphi|}_\alpha(SB, C) \quad HC^*_\alpha(D, SA) \simeq HC^{*+|\varphi|}_\alpha(D, SB)$$

$\square$

Theorem 10.7:

Let $\mathcal{C}$ be the smallest class of separable C^*-algebras satisfying

1) $\mathbb{C} \in \mathcal{C}$

2) If in an extension

$$0 \to J \to A \to B \to 0$$

with completely positive splitting two algebras belong to $\mathcal{C}$, then also the third one.

3) $\mathcal{C}$ is closed under KK-equivalence.

Then for any $A, B \in \mathcal{C}$ the bivariant Chern character yields an isomorphism

$$ch : KK^*(A, B) \otimes_\mathbb{Z} \mathbb{C} \xrightarrow{\simeq} HC^*_\alpha(SA, SB)$$

Proof:

Consider the class $\mathcal{C}'$ of separable C^*-algebras A such that

$$ch : KK^*(A, \mathbb{C}) \otimes_{\mathbb{Z}} \mathbb{C} \to HC^*_\alpha(SA, S\mathbb{C})$$

is an isomorphism. By our calculation of $HC^*_\alpha(S\mathbb{C}, S\mathbb{C})$ (9.5) we know that $\mathbb{C} \in \mathcal{C}'$.

An extension of separable C^*-algebras with completely positive lifting yields six term exact sequences in KK-theory and bivariant cohomology, compatible (up to multiplication by $2\pi i$) under the Chern character:

$$\overset{\partial}{\leftarrow} \quad KK^*(J, \mathbb{C}) \quad \leftarrow \quad KK^*(A, \mathbb{C}) \quad \leftarrow \quad KK^*(B, \mathbb{C}) \quad \overset{\partial}{\leftarrow}$$

$$\downarrow ch \qquad\qquad \downarrow ch \qquad\qquad \downarrow ch$$

$$\overset{\partial}{\leftarrow} \quad HC^*_\alpha(SJ, S\mathbb{C}) \quad \leftarrow \quad HC^*_\alpha(SA, S\mathbb{C}) \quad \leftarrow \quad HC^*_\alpha(SB, S\mathbb{C}) \quad \overset{\partial}{\leftarrow}$$

The five lemma shows then that $\mathcal{C}'$ satisfies also condition 2) above. It is also clear that $\mathcal{C}'$ is closed under KK-equivalence as the diagram (commuting up to multiplication by constants)

$$KK^*(B, \mathbb{C}) \xrightarrow{\varphi \otimes -} KK^{*+|\varphi|}(A, \mathbb{C})$$

$$ch \downarrow \qquad\qquad\qquad \downarrow ch$$

$$HC^*_\alpha(SB, S\mathbb{C}) \xrightarrow{ch(\varphi) \otimes -} HC^{*+|\varphi|}_\alpha(SA, S\mathbb{C})$$

shows. Thus

$$ch : KK^*(A, \mathbb{C}) \otimes_{\mathbb{Z}} \mathbb{C} \overset{\cong}{\to} HC^*_\alpha(SA, S\mathbb{C})$$

for any algebra A in $\mathcal{C}$.

Running the same argument for the class $\mathcal{C}''_A$ of separable C^*-algebras B such that

$$ch : KK^*(A, B) \otimes_{\mathbb{Z}} \mathbb{C} \to HC^*_\alpha(SA, SB)$$

is an isomorphism completes the proof of the theorem.

$\square$

Corollary 10.8:

Let A be a separable C^*-algebra belonging to the class $\mathcal{C}$. Then the Chern character defines isomorphisms

$$ch : K_*(A) \otimes_{\mathbb{Z}} \mathbb{C} \overset{\cong}{\to} HC^*_\alpha(\mathbb{C}, S^2 A)$$

$$ch : K^*(A) \otimes_{\mathbb{Z}} \mathbb{C} \overset{\cong}{\to} HC^*_\alpha(S^2 A, \mathbb{C})$$

between the complexified K-theory (K-homology) of A and the asymptotic cyclic homology (cohomology) of $S^2 A$.

$\square$

11 Examples

Finally two explicit calculations of asymptotic cyclic (co)homology groups are presented. The two examples are of a very different nature.

In the first, the stable bivariant asymptotic cyclic homology of separable, commutative C^*-algebras is computed. The arguments are exclusively based on the functorial, homotopy- and excision-properties of asymptotic cohomology developed hitherto. If A is a separable, commutative C^*-algebra with associated locally compact Hausdorff space X, the asymptotic cyclic homology of A equals the ($\mathbb{Z}/2\mathbb{Z}$-periodic) sheaf cohomology of X with compact supports and coefficients in the constant sheaf $\mathbb{C}$:

$$HC_*^\alpha(S^2 A) \simeq \bigoplus_{n=-\infty}^{\infty} H_c^{*+2n}(X, \mathbb{C})$$

This is in some sense the most natural answer one could hope for and again provides evidence that asymptotic cohomology yields a reasonable cohomology theory for Banach algebras.

The second example illustrates, how asymptotic cyclic groups can be calculated by methods of homological algebra. We treat the case of the Banach group algebra $l^1(F_n)$ of a free group on n generators. One obtains an isomorphism between asymptotic homology and group homology

$$HC_*^\alpha(l^1(F_n)) \simeq H_*(F_n, \mathbb{C})$$

as in the case of the algebraic cyclic homology of the ordinary group algebra. The result coincides with that for the (stable) asymptotic homology of the reduced group C^*-algebra:

$$HC_*^\alpha(S^2 C_r^*(F_n)) \simeq H_*(F_n, \mathbb{C})$$

(This follows from the fact that the group C^*-algebra is KK-equivalent to a commutative C^*-algebra (whose homology is known by the first example) and from the existence and properties of the bivariant Chern character of chapter 10.)

We emphasize however that it is not the result but rather the way to obtain it, which might be of some interest. The case treated here is particularly simple, but the calculation as such applies (in principle) to a larger class of algebras.

Finally it should be mentioned that the calculations in the cohomological case are more involved. They do not yield the full bivariant asymptotic cohomology but closely related groups which will be studied elsewhere.

11-1 Asymptotic cyclic cohomology of commutative C^*-algebras

In this section the stable asymptotic (co)homology of separable, commutative C^*-algebras will be computed.

Recall that every commutative C^*-algebra coincides with the C^*-algebra of continuous functions on a compact Hausdorff space in the unital case and with the C^*-algebra of continuous functions vanishing at infinity on a locally compact Hausdorff space in the nonunital case. The algebra is separable if and only if the corresponding compact space (the one point compactification of the corresponding locally compact space) is metrisable.

First of all it is shown that stable asymptotic (co)homology defines a (co)homology theory in the sense of Eilenberg Steenrod on the category of separable, commutative C^*-algebras (i.e. compact Hausdorff spaces).

Theorem 11.1:

Let A, B be admissible Fréchet algebras. For any pair $X \supset X'$ of compact, metrisable spaces denote by C_p the mapping cone of the natural surjection $p : C(X) \to C(X')$. Then the functors

$$H_A^*(X, X') := HC_\alpha^*(SA, SC_p)$$

$$H_*^B(X, X') := HC_\alpha^*(SC_p, SB)$$

define generalised, $\mathbb{Z}/2\mathbb{Z}$-periodic cohomology (homology) theories on the category of pairs of compact, metrisable spaces.

Proof:

By definition, the functors H_A^*, H_*^B are $\mathbb{Z}/2\mathbb{Z}$-graded. It follows from the homotopy invariance of bivariant, asymptotic cyclic cohomology (Theorem 6.15) that H_A^*, H_*^B are homotopy functors. If (X, X') is a pair of compact, metrisable spaces, then the first excision theorem (9.6.), applied to the homomorphism $p : C(X) \to C(X')$, yields the six term exact sequenes

$$
\begin{array}{ccccc}
H_A^*(X') & \leftarrow & H_A^*(X) & \leftarrow & H_A^*(X, X') \\
\partial \downarrow & & & & \uparrow \partial \\
H_A^{*+1}(X, X') & \to & H_A^{*+1}(X) & \to & H_A^{*+1}(X')
\end{array}
$$

$$
\begin{array}{ccccc}
H_*^B(X') & \to & H_*^B(X) & \to & H_*^B(X, X') \\
\partial \uparrow & & & & \downarrow \partial \\
H_{*+1}^B(X, X') & \leftarrow & H_{*+1}^B(X) & \leftarrow & H_{*+1}^B(X')
\end{array}
$$

The same holds if one starts with a triad of spaces.

Consider the extension of C^*-algebras

$$0 \to C(X, X') \to C(X) \xrightarrow{p} C(X') \to 0$$

where $C(X, X')$ is the algebra of continuous functions on X vanishing along X'. The second excision theorem in asymptotic cohomology (9.9) implies

$$H_A^*(X, X') = HC_\alpha^*(SA, SC_p) \simeq HC_\alpha^*(SA, SC(X, X'))$$

$$H_*^B(X, X') = HC_\alpha^*(SC_p, SB) \simeq HC_\alpha^*(SC(X, X'), SB)$$

From this it is clear that the functors H_A^*, H_*^B satisfy the following strong version of the excision axiom:

Strong excision axiom:

If $f : (X, X') \to (Y, Y')$ is a map of pairs of compact, metrisable spaces which induces a homeomorphism of $X - X'$ onto $Y - Y'$ then

$$f_* : H_*(X, X') \xrightarrow{\simeq} H_*(Y, Y') \quad f^* : H^*(X, X') \xleftarrow{\simeq} H^*(Y, Y')$$

are isomorphisms.

$\square$

To identify the (co)homology theories occuring in this way the following special case has to be considered first.

Theorem 11.2:

Let X, Y be finite CW-complexes. Then (in the notations of 11.1)

$$H_A^*(X) \simeq \bigoplus_{n=0}^{\infty} H^n(X, \mathbb{C}) \otimes H_A^{*-n}(pt)$$

$$H_*^B(Y) \simeq \prod_{n=0}^{\infty} Hom(H^n(Y, \mathbb{C}), H_{*-n}^B(pt))$$

Especially

$$HC_\alpha^*(SC(X), SC(Y)) \simeq Hom^* \left(\bigoplus_{n=0}^{\infty} H^n(X, \mathbb{C}), \bigoplus_{m=0}^{\infty} H^m(Y, \mathbb{C}) \right)$$

where the grading is such that the components Φ_{nm} of an element Φ of the right hand side vanish if $n - m \neq *$ mod (2).

Proof:

The groups above are well defined because every finite CW-complex is compact and metrisable. It is clear that the last statement follows from the two previous ones, applied to the cases $A = \mathbb{C}$ and $B = C(Y)$ respectively.

Consider the contravariant functor H_A^* on the category of finite CW-complexes. By Theorem 11.1, it is a homotopy functor taking values in abelian groups, taking cofibration sequences of spaces into exact sequences of abelian groups and satisfying the weak wedge axiom. The Brown representation theorem tells then that there exist CW-complexes E_n and a natural equivalence of functors

$$[-, E_n] \xrightarrow{\simeq} H_A^n(-)$$

Moreover, as the functors under consideration are group valued, the complexes E_n are actually H-spaces and consequently nilpotent spaces. The homotopy groups

$$\pi_k(E_n) \xrightarrow{\simeq} H_A^n(S^k)$$

being Q-vector spaces and E_n being nilpotent, the complexes E_n are Q-local, i.e. coincide with their Q-localisations. As the k-invariants of H-spaces vanish rationally, every Q-local H-space is a product of Eilenberg-MacLane spaces and a check of homotopy groups shows

$$E_n \simeq \prod_{k=0}^{\infty} K(H_A^n(S^k), k)$$

Thus

$$H_A^n(X) \simeq [X, E_n] \simeq \prod_{k=0}^{\infty} H^k(X, H_A^n(S^k))$$

$$\simeq \bigoplus_{k=0}^{\infty} H^k(X, \mathbb{C}) \otimes H_A^{n-k}(pt))$$

Due to the stable periodicity theorem (9.4) the functor H_*^B is not only defined on the homotopy category of finite CW-complexes but extends to a functor on the homotopy category of finite spectra. For a finite spectrum Y let Y^* be its Spanier-Whitehead dual. Y^* is again a finite spectrum and $(Y^*)^* \simeq Y$. For any finite spectrum put

$$F_B^m(Y) := H_{-m}^B(Y^*)$$

This defines a contravariant homotopy functor on the category of finite spectra. As Spanier-Whitehead duality turns cofibration sequences into cofibration sequences, the Brown representation theorem can be applied again. Repeating the arguments above one finds

$$H_m^B(Y) \simeq F_B^{-m}(Y^*) \simeq \bigoplus_{l=-\infty}^{\infty} H^l(Y^*, \mathbb{C}) \otimes H_m^B(S^{-l})$$

$$\simeq \prod_{l=0}^{\infty} Hom(H^l(Y, \mathbb{C}), \mathbb{C}) \otimes H_{m-l}^B(pt)$$

$$\simeq \prod_{l=0}^{\infty} Hom(H^l(Y, \mathbb{C}), H^B_{m-l}(pt))$$

$\square$

The following well known lemma is the key to extend the results obtained so far to general compact spaces. The presentation follows the article [M] of Milnor.

Lemma 11.3:

Every compact, metrisable space is homeomorphic to an inverse limit (indexed by $\mathbb{Z}_+$) of finite, simplicial complexes.

Proof (Sketch) [M]:

Let X be a compact, metric space. Let $(\mathcal{U}_n, n \in \mathbb{Z}_+)$ be a sequence of finite open covers of X satisfying the following conditions.

a) The diameters of the open sets of the cover $\mathcal{U}_n$ tend to zero as n approaches ∞.

b) For $n < n'$ the cover $\mathcal{U}_{n'}$ is a refinement of $\mathcal{U}_n$

To every cover $\mathcal{U}_n$ there is an associated simplicial set X_n, its nerve, which captures the combinatorial data of the cover. As all covers $\mathcal{U}_n$ are finite, the geometric realisations $|X_n|$ of its nerves are finite simplicial complexes. A refinement $\mathcal{U}'$ of a cover $\mathcal{U}$ gives rise to a map of geometric realisations of the nerves $|X_{\mathcal{U}'}| \to |X_{\mathcal{U}}|$ which is well defined up to homotopy. Choosing representatives of these maps one can form $\mathcal{X} := \lim_{\leftarrow} |X_n|$. Because the diameters of the open sets in the covers $\mathcal{U}_n$ tend to zero it is possible to identify the inverse limit over the nerves with the original space homeomorphically.

$\square$

This allows to extend the calculation of the (co)homology from finite CW-complexes to arbitrary compact, metrisable spaces, provided the considered cohomology theory behaves well with respect to inverse limits. The relevant conditions are as follows.

Lemma 11.4: [M]

Let H^* (resp. H_*) be a cohomology (homology) theory on the category of compact, metrisable spaces. Assume that H^* (resp. H_*) satisfies the strong excision property (11.1).

1) In the cohomological case the following assertions are equivalent.

a) If $(X_n, n \in \mathbb{Z}_+)$ is an inverse system of compact, metrisable spaces, then

$$H^*(\varprojlim X_n) \xleftarrow{\cong} \varinjlim H^*(X_n)$$

b) If $Y = \vee_{i=1}^\infty Y_i$ is an infinite union of compact, metric spaces with diameters tending to zero which intersect pairwise in a single point y_0, then

$$H^*(Y, y_0) \xrightarrow{\cong} \bigoplus_{i=1}^\infty H^*(Y_i, y_0)$$

2) In the homological case the following assertions are equivalent.

a)

$$H_*(\varprojlim X_n) \xrightarrow{\cong} R\varprojlim H^*(X_n)$$

where $R\varprojlim$ denotes the total right derived functor of the inverse limit functor.

b)

$$H_*(Y, y_0) \xrightarrow{\cong} \prod_{i=1}^\infty H_*(Y_i, y_0)$$

Proof: [M]

The implications $a) \Rightarrow b)$ are clear because the infinite wedge sum Y is the inverse limit $Y \simeq \vee_{i=1}^n Y_i$ under the obvious contractions.

Implication $b) \Rightarrow a)$. To begin with, we add a one point space $X_0 = pt$ to the given inverse system which does not change the projective limit. Consider the mapping telescopes Z_n of the finite sequences

$$X_0 \leftarrow X_1 \leftarrow \cdots \leftarrow X_n$$

and let $Z := \varprojlim Z_n$ be their projective limit. The space Z contains $X = \varprojlim X_n$ as compact subspace and the complement can be identified with the union of the finite mapping telescopes

$$Z - X \simeq \bigcup_{n=0}^\infty Z_n$$

The point $Z_0 = X_0 = pt$ is taken as base point of Z. It does not belong to $X \subset Z$.

Claim: Z is contractible:

A contraction $c : Z \times [0,1] \to Z$ is defined as follows. Let $c(z,0) := z$ and let $c(z, \frac{1}{n})$ denote the image of z under the projection map $Z \to Z_{n-1} \subset Z$. The deformation retraction of Z_n onto Z_{n-1} is used to define $c(z,t)$ for $\frac{1}{n+1} < t < \frac{1}{n}$. As $Z_0 = pt$ the map $c(-,1)$ is constant.

The complement $Z - X$ of X in Z can be decomposed as union of two subspaces made up by the even (resp. odd) parts of the mapping telescopes.

$$Z - X = Y' \cup Y''$$

$$Y' \sim \amalg X_{2i} \quad Y'' \sim \amalg X_{2i+1}$$

$$Y' \cap Y'' = \amalg X_i$$

($\sim$ denotes "homotopy equivalent").

For any locally compact space U denote by $\overline{U}$ its one point compactification. The long exact cohomology sequence

$$0 = H^*(Z, pt) \to H^*(X \cup pt, pt) \to H^{*+1}(Z, X \cup pt) \to H^{*+1}(Z, pt) = 0$$

for the pair (Z, X) shows that

$$H^*(X) \simeq H^*(X \cup pt, pt) \simeq H^{*+1}(Z, X \cup pt) \simeq H^{*+1}(\overline{Z - X}, \infty \cup pt)$$

The cohomology sequence

$$\to H^{*+1}(\overline{Z - X}, \infty \cup pt) \to H^{*+1}(\overline{Y'}, \infty \cup pt) \oplus H^{*+1}(\overline{Y''}, \infty) \to H^{*+1}(\overline{Y' \cap Y''}) \to$$

for the triad $(\overline{Z - X}, \overline{Y'}, \overline{Y''})$ provides a long exact sequence

$$\to H^*(\vee_{n=1}^\infty (X_n \cup \infty), \infty) \to H^*(\vee_{n=1}^\infty (X_n \cup \infty), \infty) \to H^*(X) \to$$

A word about the metrics of the spaces involved. If X_n is an inverse system of compact, metric spaces, then $\prod_{n=1}^\infty X_n$ is compact and metrisable again and a metric on the product space is obtained in an evident way from metrics on the individual factors, provided that the diameters of the factors tend to zero: $\lim_{n \to \infty} diam(X_n) = 0$. The restriction of this metric to the inverse limit $\lim_{\leftarrow} X_n \subset \prod_n X_n$ defines the inverse limit topology. From this remark it is clear that the diameters of the wedge summands $X_n \cup \infty$ above tend to zero. Therefore the assumed continuity property b) leads to the exact sequence

$$\to \bigoplus_{n=1}^\infty H^*(X_n) \xrightarrow{j} \bigoplus_{n=1}^\infty H^*(X_n) \to H^*(X) \to$$

It is not difficult to identify the map j in the above sequence as

$$j(\ldots, a_n, \ldots) = (\ldots, a_n - f_n^*(a_{n-1}), \ldots)$$

which allows finally to deduce

$$H^*(\lim_{\leftarrow} X_n) = H^*(X) \simeq \lim_{\rightarrow} H^*(X_n)$$

The reasoning in the homological case is analogous.

$\square$

Some of the considered cohomology theories possess the continuity property described above.

Lemma 11.5:

Let $X = \vee_{n=1}^{\infty} X_n$ be an infinite wedge sum of compact, metrisable spaces with diameters tending to zero. Then

$$HC_\alpha^*(\mathbb{C}, SC(\vee_{n=1}^{\infty} X_n)) \simeq \bigoplus_{n=1}^{\infty} HC_\alpha^*(\mathbb{C}, SC(X_n))$$

Proof:

Let $p_n : \vee_{i=1}^{\infty} X_i \to \vee_{i=1}^{n} X_i$ be the natural contraction. The induced homomorphism $p_n^* : C(\vee_{i=1}^{n} X_i) \to C(\vee_{i=1}^{\infty} X_i)$ maps $C(\vee_{i=1}^{n} X_i)$ isomorphically onto the algebra of continuous functions on $\vee_{i=1}^{\infty} X_i$ which are constant on the subspace $\vee_{i=n+1}^{\infty} X_i$.

Let $\mathcal{K} := \{(K, N)\}$ be as in 5.5, 5.6 where K runs through the family of compact subsets of the open unit ball in $C(\vee_{i=1}^{\infty} X_i)$ and let $\mathcal{K}_n$ be the corresponding families for the algebras $p_n^*(C(\vee_{i=1}^{n} X_i)) \subset C(\vee_{i=1}^{\infty} X_i)$. Then by definition of the asymptotic, resp. analytic cyclic homology one has

$$HC_*^\alpha(SC(\vee_{i=1}^{\infty} X_i)) \simeq HC_*^\epsilon(SC(\vee_{i=1}^{\infty} X_i)) \simeq HC_*^\epsilon(C(\vee_{i=1}^{\infty} SX_i)) =$$

$$= H_*(\lim_{\to \mathcal{K}} X_*(RC(\vee_{i=1}^{\infty} SX_i)_{(K,N)}) \simeq \lim_{\to \mathcal{K}} H_*(X_*(RC(\vee_{i=1}^{\infty} SX_i)_{(K,N)}))$$

We claim that the natural map

$$\lim_{\to n} \lim_{\to \mathcal{K}_n} H_*(X_*(RC(\vee_{i=1}^{\infty} SX_i)_{(K,N)})) \to \lim_{\to \mathcal{K}} H_*(X_*(RC(\vee_{i=1}^{\infty} SX_i)_{(K,N)}))$$

is an isomorphism.

In fact it is immediately clear that the selfmaps $f_n : X \xrightarrow{p_n} X_n \xrightarrow{i_n} X$ give rise to an asymptotic morphism $\varphi_t : C(X) \to C(X)$ such that $\varphi_n := f_n^*$ and φ_t for noninteger values of t is defined by linear interpolation. This asymptotic morphism is naturally homotopic to the identity. Let now $(K, N) \in \mathcal{K}$. If n is choosen large enough (so that the curvature of φ_t becomes very small on the multiplicative closure K^∞ of K for $t \geq n$) one can find $(K', N') \in \mathcal{K}$ by (5.12) so that the composition

$$X_*(RC(SX)_{(K,N)}) \xrightarrow{(\varphi_n)_* = f_n^*} X_*(RC(SX)_{(p_n^* K, N)}) \to X_*(RC(SX)_{(K',N')})$$

is chain homotopic to the identity. This shows that the map considered in the claim is surjective and the injectivity follows from a similar argument.

From this one obtains by definition

$$HC_*^\epsilon(SC(X)) \simeq \lim_{\to \mathcal{K}} H_*(X_*(RC(SX))) \simeq \lim_{\to n} \lim_{\to \mathcal{K}_n} H_*(X_*(RC(SX))) \simeq$$

$$\simeq \lim_{\to n} HC_*^\epsilon(SC(\vee_{i=1}^{n} X_i)) \simeq \lim_{\to n} HC_*^\alpha(SC(\vee_{i=1}^{n} X_i)$$

which, by the second excision theorem (9.9), equals

$$\lim_{\to n} \left(\bigoplus_{i=1}^{n} HC_*^\alpha(SC(X_i)) \right) = \bigoplus_{n=1}^{\infty} HC_*^\alpha(SC(X_n))$$

$\square$

Lemma 11.6:

Let $X = \vee_{n=1}^{\infty} X_n$ be an infinite wedge sum of compact, metric spaces with diameters tending to zero. Then the sheaf cohomology with coefficients in the constant sheaf $\mathbb{C}$ satisfies

$$H^*(X, \mathbb{C}) \simeq \bigoplus_{n=1}^{\infty} H^*(X_n, \mathbb{C})$$

Proof:

Let X be a compact Hausdorff space and let $(\mathcal{F}_n, n \in \mathbb{Z}_+)$ be an inductive system of sheaves of abelian groups on X. The sheaf associated to the presheaf $U \to \lim_{\to n} \mathcal{F}_n(U)$ is called the direct limit $\lim_{\to n} \mathcal{F}_n$ of the sheaves $\mathcal{F}_n$. The stalks of a direct limit are the direct limits of the stalks of the individual sheaves:

$$\left(\lim_{\to n} \mathcal{F}_n\right)_x = \lim_{\to n} (\mathcal{F}_n)_x$$

Consequently the functor $\lim_{\to} : Sh_X^{\mathbb{Z}_+} \to Sh_X$ is exact. The obvious homomorphisms of sheaves $\mathcal{F}_k \to \lim_{\to n} \mathcal{F}_n$ give rise to a natural transformation of left exact functors

$$\lim_{\to n} \Gamma(\mathcal{F}_n) \to \Gamma(\lim_{\to n} \mathcal{F}_n)$$

If X is compact and Hausdorff, this is actually a natural equivalence. In this case the derived functors of these functors also coincide. Therefore

$$\lim_{\to n} H^*(X, \mathcal{F}_n) \xrightarrow{\sim} H^*(X, \lim_{\to n} \mathcal{F}_n)$$

if X is compact and Hausdorff.

Let now $X = \vee_{n=1}^{\infty} X_i$ be as above and denote by $i_n : \vee_{i=1}^{n} X_i \to X$ the natural inclusion which maps $\vee_{i=1}^{n} X_i$ homeomorphically onto a compact subspace of X. The direct limit $\lim_{\to n} (i_{n*} \mathbb{C})$ over the direct image sheaves $i_{n*} \mathbb{C}$ is then easily identified with the constant sheaf $\mathbb{C}$ over X:

$$\mathbb{C}_X \simeq \lim_{\to n} i_{n*} \mathbb{C}_{\vee_{i=1}^{n} X_i}$$

For the cohomology one finds therefore

$$H^*(X, \mathbb{C}) \simeq H^*(X, \lim_{\to n} i_{n*} \mathbb{C}_{\vee_{i=1}^{n} X_i}) \simeq \lim_{\to n} H^*(X, i_{n*} \mathbb{C}_{\vee_{i=1}^{n} X_i}) \simeq$$

$$\simeq \lim_{\to n} H^*(\vee_{i=1}^{n} X_i, \mathbb{C}) \simeq \bigoplus_{n=1}^{\infty} H^*(X_i, \mathbb{C})$$

□

What has been obtained so far can be summarised in the

Theorem 11.7:

Let A be a separable, commutative C^*-algebra and let X be the associated locally compact space. Then there is a natural isomorphism

$$HC_*^\epsilon(S^2A) \simeq HC_*^\alpha(S^2A) \simeq \bigoplus_{n=-\infty}^{\infty} H_c^{*+2n}(X, \mathbb{C})$$

where H_c^* denotes sheaf cohomology with compact supports.

Proof:

If A is unital, then X is compact and the compact support condition is empty for the cohomology of X: $H_c^*(X, -) \simeq H^*(X, -)$. If A is not unital, the corresponding space X is only locally compact. Denote by $\overline{X}$ its one point compactification. Then there are exact sequences

$$0 \to HC_*^\alpha(S^2C_0(X)) \to HC_*^\alpha(S^2C(\overline{X})) \to HC_*^\alpha(S^2\mathbb{C}) = HC_*^\alpha(\mathbb{C}) \to 0$$

$$0 \to H_c^*(X, \mathbb{C}) \to H^*(\overline{X}, \mathbb{C}) \to H^*(pt, \mathbb{C}) \to 0$$

which shows that one can assume $A = C(X)$ unital and X compact. the separability of A implies that X (resp. $\overline{X}$) is metrisable. By Lemma 11.3 X can be identified with the inverse limit of the nerves of finer and finer open finite covers: $X = \lim_{\leftarrow} X_n$, X_n finite simplicial complexes. Thus

$$HC_*^\alpha(S^2C(X)) = HC_*^\alpha(S^2C(\lim_{\leftarrow} X_n)) \simeq \lim_{\to n} HC_*^\alpha(S^2C(X_n))$$

by Lemma 11.4 and 11.5

$$\simeq \lim_{\to n} \bigoplus_{k=-\infty}^{\infty} H^{*+2k}(X_n, \mathbb{C})$$

by Theorem 11.2

$$\simeq \bigoplus_{k=-\infty}^{\infty} H^{*+2k}(\lim_{\leftarrow} X_n, \mathbb{C}) = \bigoplus_{k=-\infty}^{\infty} H^{*+2k}(X, \mathbb{C})$$

by Lemma 11.6. The naturality needs some arguments but is not difficult to show.

□

The calculation of the asymptotic cyclic cohomology of a commutative C^*-algebra turns out to be more complicated however. As the cohomology of an inverse limit of complexes is not related to the cohomology of the individual complexes in general one cannot hope to get a closed expression for the asymptotic (bivariant) cohomology groups. It is however possible to introduce closely related groups which will in fact turn out to be computable. As this should be treated elsewhere we will be brief and content ourselves with some remarks.

Local cyclic cohomology with compact supports

Definition 11.8:

Let A, B be admissible Fréchet algebras and let $\mathcal{K}_A, \mathcal{K}_B$ be the families of compact subsets of the open unit balls as in 5.5, 5.6. The bivariant local cyclic cohomology with compact supports of the pair (A, B) is defined as

$$HC_{lc}^*(A, B) := R \varprojlim_{\leftarrow \mathcal{K}_A} \varinjlim_{\rightarrow \mathcal{K}_B} Hom_{cont}^*(X_* RA_{(K,N)}, X_* RB_{(K',N')})$$

where $R\varprojlim$ denotes the total derived functor of the inverse limit functor and both sides are viewed as objects in the derived category of the category of complex vector spaces.

□

Remark 11.9:

a) There exist natural transformations of functors

$$HP^*(-) \to HC_\epsilon^*(-) \to HC_\alpha^*(-) \to HC_{lc}^*(-)$$

$$HP_*(-) \leftarrow HC_*^\epsilon(-) \xrightarrow{\simeq} HC_*^\alpha(-) \xrightarrow{\simeq} HC_*^{lc}(-)$$

$$HC_\epsilon^*(-,-) \to HC_\alpha^*(-,-) \to HC_{lc}^*(-,-)$$

b) There exists a composition product in bivariant local cyclic cohomology with compact supports such that the diagram

$$\begin{array}{ccccc}
HC_\alpha^*(-,-) & \otimes & HC_\alpha^*(-,-) & \to & HC_\alpha^*(-,-) \\
\downarrow & & \downarrow & & \downarrow \\
HC_{lc}^*(-,-) & \otimes & HC_{lc}^*(-,-) & \to & HC_{lc}^*(-,-)
\end{array}$$

commutes.

□

Contrary to the asymptotic case, the bivariant local cyclic cohomology with compact supports of separable, commutative C^*-algebras can be computed.

Theorem 11.10:

Let A, B be separable, commutative C^*-algebras with corresponding locally compact spaces X, Y. Then

$$HC_{lc}^*(SA, SB) \simeq Hom^* \left(\bigoplus_{n=0}^\infty H_c^n(X, \mathbb{C}), \bigoplus_{m=0}^\infty H_c^m(Y, \mathbb{C}) \right)$$

where the grading is such that the components Φ_{nm} of an element Φ of the right hand side vanish if $n - m \neq * \mod(2)$.

□

Before we come to the proof a few more properties of local cyclic cohomology are needed.

Remark 11.11:

a) The first and second excision theorems (9.6,9.9) hold for $HC_{lc}^*(-,-)$.

b) The natural transformation $HC_\alpha^*(-,-) \to HC_{lc}^*(-,-)$ commutes with Puppe sequences and boundary maps in long exact cohomology sequences.

□

Corollary 11:12:

a) The bifunctor $(X,Y) \to HC_{lc}^*(SC(X),SC(Y))$ defines a generalised, bivariant cohomology theory on the category of pairs of compact, metrisable spaces.

b) If X,Y are finite CW-complexes the natural map

$$HC_\alpha^*(SC(X),SC(Y)) \xrightarrow{\cong} HC_{lc}^*(SC(X),SC(Y))$$

is an isomorphism.

Proof:

This follows from Theorems 11.1, 11.2 and the preceding remark.

□

The generalised homology theories obtained from the local cyclic theory satisfy the same continuity property as the homology considered before.

Lemma 11.13:

Let $X = \vee_{n=0}^\infty X_n$ be an infinite wedge sum of compact metric spaces whose diameters tend to zero. Then for any admissible Fréchet algebra B the natural map

$$HC_{lc}^*(SC(\vee_{n=0}^\infty X_n), SB) \xrightarrow{\cong} \prod_{n=0}^\infty HC_{lc}^*(SC(X_n), SB)$$

is an isomorphism.

Proof:

Let $\Phi_t : C(X) \to C(X)$ be the asymptotic morphism defined by the retraction of X onto successively larger finite wedge sums (see the proof of 11.5). Let, in the notations of 11.5, be $\mathcal{K}_n = p_n^* \mathcal{K} \subset \mathcal{K}$ be the family of compact sets of continuous functions of norm smaller than one on X which are constant on $\vee_{i=n+1}^\infty X_i$. It is then not difficult to establish the following facts:

a) (Φ_t) gives rise to a bivariant cocycle

$$\Phi_* \in R\lim_{\leftarrow \mathcal{K}} \lim_{\rightarrow \cup \mathcal{K}_n} Hom^0(X_*(RA_\mathcal{K}), X_*(RA_{\mathcal{K}_n}))$$

(For n-tuples $((K_1, N_1), \ldots, (K_k, N_k))$ the necessary higher chain homotopies between the individual chain maps are provided by the evident linear homotopies between $p_{n_1}, \ldots, p_{n_k} : C(X) \to C(X)$.

b) Let

$$i_* \in R\lim_{\leftarrow \cup \mathcal{K}_n} \lim_{\rightarrow \mathcal{K}} Hom^0(X_*(RA_{\mathcal{K}_n}), X_*(RA_\mathcal{K}))$$

be the obvious inclusion. Then

$$\Phi_* \circ i_* = Id \in R\lim_{\leftarrow \cup \mathcal{K}_n} \lim_{\rightarrow \mathcal{K}_n} Hom^0(X_*(RA_{\mathcal{K}_n}), X_*(RA_{\mathcal{K}_n}))$$

(even on the "chain level")

$$i_* \circ \Phi_* = Id \in HC^0_{lc}(C(X), C(X))$$

because the asymptotic morphism $\Phi : C(X) \to C(X)$ is naturally homotopic to the identity.

This implies (by using the composition product) that the natural map

$$HC^*_{lc}(C(X), B) = R\lim_{\leftarrow \mathcal{K}} \lim_{\rightarrow \mathcal{K}_B} Hom^*(X_*(RC(X)_\mathcal{K}), X_*(RB_{\mathcal{K}_B})) \to$$

$$\to R\lim_{\leftarrow \cup \mathcal{K}_n} \lim_{\rightarrow \mathcal{K}_B} Hom^*(X_*(RC(X)_{\mathcal{K}_n}), X_*(RB_{\mathcal{K}_B}))$$

is a quasiisomorphism. The latter complex can however be identified as

$$R\lim_{\leftarrow \cup \mathcal{K}_n} \lim_{\rightarrow \mathcal{K}_B} Hom^*(X_*(RC(X)_{\mathcal{K}_n}), X_*(RB_{\mathcal{K}_B})) \xrightarrow{qis}$$

$$\xrightarrow{qis} R(\lim_{\leftarrow n} \circ \lim_{\leftarrow \mathcal{K}_n}) \lim_{\rightarrow \mathcal{K}_B} Hom^*(X_*(RC(X)_{\mathcal{K}_n}), X_*(RB_{\mathcal{K}_B})) \xrightarrow{qis}$$

$$\xrightarrow{qis} R\lim_{\leftarrow n} \left(R\lim_{\leftarrow \mathcal{K}_n} \lim_{\rightarrow \mathcal{K}_B} Hom^*(X_*(RC(X)_{\mathcal{K}_n}), X_*(RB_{\mathcal{K}_B})) \right) =$$

$$= R\lim_{\leftarrow n} HC^*_{lc}(C(\vee^n_{i=1} X_i), B)$$

A similar sequence holds after suspension. Then the second excision theorem may be applied by Remark 11.11 and yields

$$HC^*_{lc}(SC(X), SB) \simeq R\lim_{\leftarrow n} HC^*_{lc}(SC(\vee^n_{i=1} X_i), SB) \simeq \prod_{n=1}^{\infty} HC^*_{lc}(SC(X_n), SB)$$

$\square$

Note that as a consequence of the foregoing lemma and 11.4, one obtains for an inverse limit of compact, metrisable spaces a quasiisomorphism

$$HC_{lc}^*(SC(\varprojlim_n X_n), SB) \xrightarrow{qis} R\varprojlim_n HC_{lc}^*(SC(X_n), SB)$$

Proof of Theorem 11.10:

Because excision holds for both sheaf cohomology and local cyclic cohomology, one may assume A and B to be unital: $A = C(X), B = C(Y)$, X, Y compact, metrisable. Realise X as inverse limit of finite, simplicial complexes (Lemma 11.3): $X \simeq \varprojlim_k X_k$. Then

$$HC_{lc}^*(SA, SB) \simeq HC_{lc}^*(SC(\varprojlim_k X_k), SB) \simeq R\varprojlim_k HC_{lc}^*(SC(X_k), SB)$$

Now for a finite, simplicial complex the arguments in the proof of Theorem 11.2 carry over from asymptotic to local cyclic cohomology with compact supports and show

$$HC_{lc}^*(SC(X_k), SB) \simeq \prod_{n=0}^{\infty} Hom(H^n(X_k, \mathbb{C}), HC_{*-n}^{lc}(S^2 B)$$

As the three homology theories $HC_*^\epsilon, HC_*^\alpha, HC_*^{lc}$ coincide by 11.9, Theorem 11.7 shows further that

$$HC_{lc}^*(SC(X_k), SC(Y)) \simeq Hom^* \left(\bigoplus_{n=0}^{\infty} H^n(X_k, \mathbb{C}), \bigoplus_{m=0}^{\infty} H^m(Y, \mathbb{C}) \right)$$

naturally. Consequently

$$HC_{lc}^*(SC(X), SC(Y)) \simeq R\varprojlim_k Hom^* \left(\bigoplus_{n=0}^{\infty} H^n(X_k, \mathbb{C}), \bigoplus_{m=0}^{\infty} H^m(Y, \mathbb{C}) \right)$$

On the category $Vect^{\mathbb{Z}_+}$ of inductive systems of complex vector spaces the two left exact functors

$$(C_n) \rightarrow \varprojlim_n Hom(C, D)$$

$$(C_n) \rightarrow Hom(\varinjlim_n C_n, D)$$

(D a fixed complex vector space) are naturally equivalent. Therefore their right derived functors are also naturally equivalent:

$$R\varprojlim_n Hom(C_*^n, D_*) \xleftarrow{\simeq} Hom(\varinjlim_n C_*^n, D_*)$$

Thus

$$HC_{lc}^*(SC(X), SC(Y)) \simeq R\varprojlim_k Hom^* \left(\bigoplus_{n=0}^{\infty} H^n(X_k, \mathbb{C}), \bigoplus_{m=0}^{\infty} H^m(Y, \mathbb{C}) \right) \simeq$$

$$\simeq Hom^* \left(\varinjlim_k \bigoplus_{n=0}^{\infty} H^n(X_k, \mathbb{C}), \bigoplus_{m=0}^{\infty} H^m(Y, \mathbb{C}) \right)$$

$$\simeq Hom^* \left(\bigoplus_{n=0}^{\infty} \varinjlim_k H^n(X_k, \mathbb{C}), \bigoplus_{m=0}^{\infty} H^m(Y, \mathbb{C}) \right)$$

$$\simeq Hom^* \left(\bigoplus_{n=0}^{\infty} H^n(\varprojlim_k X_k, \mathbb{C}), \bigoplus_{m=0}^{\infty} H^m(Y, \mathbb{C}) \right)$$

$$\simeq Hom^* \left(\bigoplus_{n=0}^{\infty} H^n(X, \mathbb{C}), \bigoplus_{m=0}^{\infty} H^m(Y, \mathbb{C}) \right)$$

$\square$

11-2 Explicit calculation of asymptotic cohomology groups

Contrary to K-theory, cyclic homology theories are defined by natural chain complexes. This should enable one, at least in principle, to calculate cyclic (co)homology groups with the tools of homological algebra. In this paragraph we will illustrate a rather general scheme for the calculation of asymptotic (local) cyclic groups by an example. We will treat the case of the convolution Banach algebra of summable functions on a free group. Although the cohomology groups are known (stably) by the general excision properties of the cyclic theories their determination will be quite different now as it is based on a purely homological calculation.

Local cyclic cohomology and the approximation property

From its definition it is clear that analytic cyclic homology is a dirct limit of periodic cyclic homology groups

$$HC_*^{lc}(A) = \lim_{\to \mathcal{K}} HP_*(RA_{(K,N)})$$

whereas local cyclic cohomology is the limit of a convergent spectral sequence

$$E_2^{pq} = R^p \lim_{\leftarrow \mathcal{K}}(HP^q(RA_{(K,N)})) \Rightarrow Gr^p HC_{lc}^{p+q}(A)$$

To be able to use this for computations one has to find criteria for cutting down the the direct limit, resp. the spectral sequence to a controllable size.

Definition 11.14:

Let $\mathcal{A}$ be a countably generated, normed algebra and let A be its completion. (If $\mathcal{A}$ is infinite dimensioal it will be of countable dimension and thus never complete.) Let $\{x_1, \ldots, x_k, \ldots\}$ be a sequence of generators and let $K_N \subset \mathcal{A} \subset A$ be the set of elements $x \in A$ satisfying

a) x belongs to the (finite dimensional) span of all monomials of length at most n in $x_1, \ldots, x_n$.

b) $\| x \| \leq \frac{n}{n+1}$

$\square$

Then K_n is a compact subset of the unit ball of A.

Definition 11.15:

Let $\mathcal{A}$ be a countably generated normed algebra and let A be its completion. The local cyclic (co)homology with finite supports of the pair $(A, \mathcal{A})$ is defined as

$$HC_*^{lf}(A, \mathcal{A}) := H_*(\lim_{\to n} X_*(RA_{(K_n,n)}))$$

$$HC_{lf}^*(A, \mathcal{A}) := H^*(R \lim_{\leftarrow n} X^*(RA_{(K_n,n)}))$$

This definition does not depend on the choice of generators of $\mathcal{A}$.

$\square$

It is clear from the definition that

$$HC_*^{lf}(A, \mathcal{A}) \simeq \varinjlim_{n} HP_*(RA_{(K_n, n)})$$

and that the corresponding spectral sequence in the cohomological case collapses to a short exact sequence

$$0 \to \varprojlim_{n}{}^{1} HP^{*-1}(RA_{(K_n, n)}) \to HC_{lf}^*(A, \mathcal{A}) \to \varprojlim_{n} HP^*(RA_{(K_n, n)}) \to 0$$

There is an important class of algebras for which the local cohomologies with finite respectively compact supports coincide. To describe the class recall the

Definition 11.16:

Let E be a Banach space. Then E has the Grothendieck approximation property if, given any compact subset $K \subset E$ and any $\epsilon > 0$ there exists a (bounded) linear selfmap $\phi : E \to E$ of finite rank such that $\sup_{x \in K} \| x - \phi(x) \| < \epsilon$ In other words, the identity belongs to the closure of the finite rank operators in the topology of compact convergence.

□

Typical examples of Banach spaces satisfying the Grothendieck approximation property are all kinds of L^p-spaces. Examples of C^*-algebras having the approximation property are all nucear C^*-algebras and also the reduced group C^*-algebras of the free groups and of discrete, cocompact subgroups of simple Lie groups of real rank one.

For the class of algebras with approximation property one has

Remark 11.17:

Let A be a Banach algebra and suppose that the underlying Banach space satisfies the Grothendieck approximation property. Let $\mathcal{A} \subset A$ be any dense, countably generated subalgebra. Then the natural maps between the local cyclic (co)homology groups with compact, resp. finite supports

$$HC_*^{lc}(A) \xleftarrow{\simeq} HC_{lf}^*(A, \mathcal{A})$$

$$HC_{lc}^*(A) \xrightarrow{\simeq} HC_*^{lf}(A, \mathcal{A})$$

are isomorphisms.

□

Therefore the calculation of the asymptotic homology (local cyclic cohomology) of a Banach algebra is reduced to the problem of calculating the (co)homology of the complexes $X_*(RA_{(K_n, n)})$, i.e. of the periodic cyclic (co)homology of some completions of finitely generated subalgebras of the tensor algebra over $\mathcal{A} \subset A$. This can be done (in principle) if A possesses a dense subalgebra $\mathcal{A}$ of finite cohomological dimension.

Algebras of finite cohomological dimension and connections

In this paragraph we will collect some facts (taken from Cuntz,Quillen [CQ] and Khalkhali [K]) about algebras of finite Hochschild cohomological dimension. All algebras are viewed as abstract algebras (not equipped with any topology).

Let A be a complex algebra. The category of A-bimodules is an abelian category with enough injective and enough projective objects. The Hochschild (co)homology groups of the pair (M, N) of A-bimodules are defined as

$$HH_*^A(M, N) := Tor_*^{A \otimes_{\mathbb{C}} A^{op}}(M, N) \quad HH_A^*(M, N) := Ext_{A \otimes_{\mathbb{C}} A^{op}}^*(M, N)$$

Definition and Proposition 11.18: ([CQ])

For an algebra A the following conditions are equivalent.

a) The A bimodule A possesses a resolution by projective A bimodules of length $\leq n$.

b) The A bimodule $\Omega^n A$ of formal differential forms of degree n is projective.

c) There exists a connection

$$\nabla : \Omega^n A \rightarrow \Omega^{n+1} A$$

i.e. a linear map ∇ satisfying

$$\nabla(a\omega) = a\nabla(\omega) \quad \nabla(\omega a) = \nabla(\omega)a + (-1)^{|\omega|}\omega da \quad \forall a \in A, \, \omega \in \Omega^n A$$

For a given algebra A, its cohomological dimension is the smallest integer satisfying the conditions above. If the conditions are not satisfied for any integer, the cohomological dimension of A is defined to be infinity.

Proof:

a) $\Leftrightarrow$ b): A possesses a standard resolution by free A-bimodules given by.

$$A \xleftarrow{m} P_0 := A \otimes A \leftarrow \ldots \xleftarrow{b'} P_k := A \otimes \overline{A}^{\otimes k} \otimes A \xleftarrow{b'} \ldots$$

$$m(a \otimes a') := aa'$$

$$b'(a^0 \otimes \cdots \otimes \overline{a}^j \otimes \cdots \otimes a^k) = \sum_{j=0}^{k-1}(-1)^j a^0 \otimes \cdots \otimes \overline{a^j a^{j+1}} \otimes \cdots \otimes a^k$$

This resolution can be written as

$$A \leftarrow \Omega^0 A \otimes A \leftarrow \ldots \xleftarrow{\partial} \Omega^k A \otimes A \xleftarrow{\partial} \ldots$$

with differential

$$\partial := j \circ m$$

where
$$m : \quad \Omega^n A \otimes A \quad \to \quad \Omega^n A$$

$$\omega \otimes a \quad \to \quad \omega a$$

is the (right module) multiplication and

$$j : \quad \Omega^n A \quad \to \quad \Omega^{n-1} A \otimes A$$

$$\omega^{n-1} da \quad \to \quad (-1)^{n-1}(\omega^{n-1} a \otimes 1 - \omega^{n-1} \otimes a)$$

identifies $\Omega^n A$ with the kernel of $\partial : \Omega^{n-1} A \otimes A \to \Omega^{n-2} A \otimes A$. As A has a projective resolution of length n iff the kernel K in any projective resolution

$$A \leftarrow P_0 \leftarrow \cdots \leftarrow P_{n-1} \leftarrow K \leftarrow 0$$

is projective itself, we are done.

b) $\Leftrightarrow$ c): If $\Omega^n A$ is projective as A-bimodule there exists an A-bimodule splitting s of the multiplication
$$m : \Omega^n A \otimes A \to \Omega^n A$$

Then

$$\nabla : \quad \Omega^n A \quad \xrightarrow{s} \quad \Omega^n A \otimes A \quad \to \quad \Omega^{n+1} A$$

$$\omega \otimes a \quad \to \quad (-1)^n \omega da$$

is a connection on $\Omega^n A$.

Conversely, let ∇ be a connection on $\Omega^n A$. Then

$$s : \quad \Omega^n A \quad \to \quad \Omega^n A \otimes A$$

$$\omega \quad \to \quad -j(\nabla \omega) + \omega \otimes 1$$

is a bimodule splitting of the multiplication which shows that $\Omega^n A$ is projective.

$\square$

Note that if ∇ is a connection on $\Omega^n A$, then

$$\nabla' : \quad \Omega^{n+k} A \quad \to \quad \Omega^{n+k+1} A$$

$$a^0 da^1 \ldots da^{n+k} \quad \to \quad \nabla(a^0 da^1 \ldots da^n) da^{n+1} \ldots da^{n+k}$$

is a connection on $\Omega^{n+k} A$. This induced connection will be denoted in the sequel by the same letter ∇. From this one easily obtains contracting homotopies of the standard projective resolution of A in degree larger than the cohomological dimension of A.

Recall (3.10, 3.11) that the periodic cyclic homology $HP_*(A)$ of a unital algebra A can be calculated by any of the three following complexes:

a) The X-complex of the I-adic completion $\widehat{RA}$ of RA.

b) The periodic de-Rham complex or normalised (b, B)-bicomplex of A

c) The full (b, B)-bicomplex of A.

The three complexes are related by quasiisomorphisms

$$X_*(\widehat{RA}) \xrightarrow{\text{qis}} \widehat{\Omega}_*^{pdR}(A) \xleftarrow{\text{qis}} \widehat{CC}_*^{per}(A)$$

which are actually deformation retractions. All three complexes can be identified with complexes of differential forms. After this identification homotopy inverses of both quasiisomorphisms in the diagram above are provided by the Cuntz-Quillen projection (see Chapter 3). Similarly, these chain maps yield (according to 5.25-5.27) quasiisomorphisms (homotopy inverse to each other) in the topological context

$$X_*(RA_{(K,N)}) \xrightarrow{\text{qis}} CC_*(A)_{(K,N)} \xrightarrow{\text{qis}} X_*(RA_{(K,N)})$$

Recall that the norm of a homogeneous differential form in $\Omega^{2n}(A)(\Omega^{2n+1}) \subset CC_*(A)_{(K,N)}$ is given by

$$\| \omega^{2n} \|_{N,m} = \inf_{\omega = \sum_\beta \lambda_\beta a_\beta^0 da_\beta^1 \dots da_\beta^{2n}} \sum_\beta |\lambda_\beta| (1+n)^m N^{-n} (n!)^{-1}$$

$$\| \omega^{2n+1} \|_{N,m} = \inf_{\omega = \sum_\beta \lambda_\beta a_\beta^0 da_\beta^1 \dots da_\beta^{2n+1}} \sum_\beta |\lambda_\beta| (1+n)^m N^{-n} (n!)^{-1}$$

Denote by $A_K \subset A$ the quotient algebra

$$A_K := RA_{(K,N)} / IA_{(K,N)}$$

It is an admissible Fréchet algebra.

The subcomplexes F^n generated by differential forms of degree at least n provide a natural filtration (Hodge filtration) of these complexes.

If A is an algebra of Hochschild cohomological dimension at most n, the subcomplex $F^{n+1}\widehat{CC}_*^{per}(A)$ of $\widehat{CC}_*^{per}(A)$ becomes contractible and allows to calculate the periodic cyclic homology of A by a much smaller complex. Using the quasiisomorphisms above, the corresponding X-complexes can be simplified in a similar manner. As we want to do this in a topological setting explicit formulas for the contracting homotopies are needed.

Let A be an algebra of cohomological dimension n. Let ∇ be a connection on $\Omega^n A$, denote the associated connections on $\Omega^k A$, $k > n$ by the same letter. Then there is a particularly simple contracting homotopy in degrees above n for the standard complex calculating the Hochschild homology of A, discovered by M.Khalkhali:

Lemma 11.19:([K])

Let
$$C_*A : 0 \leftarrow A \overset{b}{\leftarrow} \Omega^1 A \leftarrow \ldots \overset{b}{\leftarrow} \Omega^k A \leftarrow \ldots$$

be the standard complex calculating $HH_*(A, A)$.

Let ∇ be a connection on $\Omega^k A$, $k \geq n$. Then
$$\nabla \circ b + b \circ \nabla = Id \text{ on } \Omega^j A \quad j > n$$

Proof:

$$(\nabla \circ b + b \circ \nabla)(\omega da) = \nabla((-1)^{|\omega|}[\omega, a]) + b(\nabla(\omega da))$$
$$= (-1)^{|\omega|}[\nabla \omega, a]) + \omega da + (-1)^{|\omega|+1}[\nabla \omega, a])$$

$\square$

From the contracting homotopy of the Hochschild complex one can derive contracting homotopies of the cyclic bicomplexes.

Lemma 11.20:

Let A be of cohomological dimension n. Let ∇ be a connection on $\Omega^j A$, $j \geq n$. Consider the exact sequence of complexes

$$0 \to F^{n+1}CC_*^{per} A \to CC_*^{per} A \overset{p}{\to} CC_*^{per} A/F^{n+1}CC_*^{per} A \to 0$$

a) The operator

$$h := \sum_{k=0}^{\infty}(-\nabla B)^k \nabla : \quad F^{n+1}CC_*^{per} \to F^{n+1}CC_{*+1}^{per}$$

defines a nullhomotopy of F^{n+1}:

$$h \circ (b + B) + (b + B) \circ h = Id_{F^{n+1}CC_*^{per} A}$$

b) The map
$$s' : CC_*^{per} A/F^{n+1}CC_*^{per} A \to CC_*^{per} A$$
$$s' := \begin{cases} Id - b\nabla & \text{on } \Omega^n A/[\Omega^n A, A] \\ Id & \text{on } \Omega^{<n} A \end{cases}$$

defines a linear section of p:
$$p \circ s' = Id$$

c) The map
$$s : CC_*^{per} A/F^{n+1}CC_*^{per} A \to CC_*^{per} A$$
$$s := \begin{cases} Id & \text{on } \Omega^{<n-1} A \\ Id + \sum_{k=0}^{\infty}(-\nabla B)^k(\nabla b)(-\nabla B) & \text{on } \Omega^{n-1} A \\ \sum_{k=0}^{\infty}(-\nabla B)^k(Id - b\nabla) & \text{on } \Omega^n A/[\Omega^n A, A] \end{cases}$$

defines a chain map splitting p:

$$p \circ s = Id_{CC_*^{per} A / F^{n+1} CC_*^{per} A}$$

d) The chain map $s \circ p : CC_*^{per} A \to CC_*^{per} A$ is chain homotopic to the identity. An explicit homotopy is provided by

$$h' := h \circ (Id - s \circ p)$$

$$h' = \begin{cases} 0 & \text{on } \Omega^{<n-1} A \\ \sum_{k,l=0}^{\infty}(-\nabla B)^k \nabla(-\nabla B)^l (\nabla b)(\nabla B) & \text{on } \Omega^{n-1} A \\ \sum_{k=0}^{\infty}(-\nabla B)^k \nabla(Id - \sum_{l=0}^{\infty}(-\nabla B)^l (Id - b\nabla)) & \text{on } \Omega^n A \\ \sum_{k=0}^{\infty}(-\nabla B)^k \nabla & \text{on } \Omega^{>n} A \end{cases}$$

$$h' \circ (b+B) + (b+B) \circ h' = Id - s \circ p$$

$\square$

The calculation of the local cyclic (co)homology of a Banach algebra A can now be done as follows:

a) Check that A possesses the Grothendieck approximation property.

b) Find a dense, countably generated subalgebra $\mathcal{A} \subset A$ which is of finite cohomological dimension (as an abstract algebra).

$$(HC_*^{\alpha}(A) \simeq \lim_{\to n} H_*(X_*(R\mathcal{A}_{(K_n,n)})), HC_{lc}^*(A) \simeq R\lim_{\leftarrow n} H^*(X^*(R\mathcal{A}_{(K_n,n)})))$$

c) Construct a sufficiently well behaved connection ∇ on $\Omega\mathcal{A}$ which provides natural retractions of $X_*(R\mathcal{A}_{(K_n,n)})$.
(Then $HC_*^{\alpha} \simeq \lim_{\to n} HP_*(\mathcal{A}_{K_n})$ and $HC_{lc}^*(A) \simeq R\lim_{\leftarrow n} HP^*(\mathcal{A}_{K_n})$.)

d) Calculate $\lim_{\to n} HP_*(\mathcal{A}_{K_n})$ resp. $R\lim_{\leftarrow n} HP^*(\mathcal{A}_{K_n})$.

We will carry this out in a particularly simple case, namely for the completed group algebra $l^1(F_n)$ of the free group on n generators.

Example: Cohomology of $l^1(F_n)$

Let F_n be the free group on n generators. Let A be a Banach algebra which contains the group algebra $\mathbb{C}[F_n]$ as dense subalgebra and satisfies the Grothendieck approximation property. Examples of such algebras are the Banach algebra of summable functions $l^1(F_n)$ and the reduced group C^*-algebra $C_r^*(F_n)$, the norm completion of $\mathbb{C}[F_n]$ acting as linear operators on
$$\mathcal{H} = l^2(F_n).$$
In this situation the calculation scheme described before applies because the group algebra $\mathbb{C}[F_n]$ is of cohomological dimension one. This follows from the

Lemma 11.21:

Let F_n be the free group with generators $t_1, \ldots, t_n$ and let $t_1, \ldots, t_n, t_1^{-1}, \ldots, t_n^{-1}$ be the corresponding set of generators of the group algebra $\mathbb{C}[F_n]$. The associated word length function will be denoted by $|-|$.

a) There exists a connection

$$\nabla : \Omega^1 \mathbb{C}[F_n] \to \Omega^2 \mathbb{C}[F_n]$$

which is uniquely characterised by the property

$$\nabla(dt_i) = 0 \quad i = 1, \ldots, n$$

Therefore $\mathbb{C}[F_n]$ is of cohomological dimension at most one.

b) Let N be a positive integer and consider the compact set $K_N \subset \mathbb{C}[F_n]$ corresponding to the inclusion $\mathbb{C}[F_n] \subset A$ (11.14).

Then there exists a constant $C(N) > 1$ such that for all $a \in K_N$

$$\nabla(da) = \sum_\alpha b_\alpha^0 db_\alpha^1 db_\alpha^2$$

where $b_\alpha^j \in K_N$ and the number of summands is bounded by $C(N)$.

Proof:

a) It is easily seen that $\Omega^1 \mathbb{C}[F_n]$ is a free bimodule over $\mathbb{C}[F_n]$ with generators $dt_1, \ldots, dt_n$. Thus giving a connection is equivalent to specifying a set of differential forms $\nabla(dt_1), \ldots, \nabla(dt_n) \in \Omega^2$.

b) Each element $a \in K_N$ is contained in the linear span of the finite set of monomials $\{\prod_{i=1}^k t_i^{\pm 1}, k \leq N\}$.
For such a monomial

$$d\left(\prod_{i=1}^k t_i^{\pm 1}\right) = \sum_{i=1}^k \left(\prod_{i=1}^{j-1} t_i^{\pm 1}\right) dt_j^{\pm 1} \left(\prod_{l=j+1}^k t_l^{\pm 1}\right)$$

and

$$\nabla\left(\left(\prod_{i=1}^{j-1} t_i^{\pm 1}\right) dt_j \left(\prod_{l=j+1}^k t_l^{\pm 1}\right)\right) = -\left(\prod_{i=1}^{j-1} t_i^{\pm 1}\right) dt_j d\left(\prod_{l=j+1}^k t_l^{\pm 1}\right)$$

$$\nabla\left(\left(\prod_{i=1}^{j-1} t_i^{\pm 1}\right) dt_j^{-1} \left(\prod_{l=j+1}^k t_l^{\pm 1}\right)\right) = \left(\prod_{i=1}^{j-1} t_i^{\pm 1}\right) t_j^{-1} dt_j d\left(t_j^{-1} \prod_{l=j+1}^k t_l^{\pm 1}\right)$$

As connections are linear maps this shows that $\nabla(\prod_{i=1}^k t_i^{\pm 1})$ and also $\nabla(da)$ are of the form claimed in the assertion.

That there is a bound on the number of summands follows from the fact that every linear map between normed, finite dimensional vector spaces is bounded. Of course there is no control of the size of the constant $C(N)$ as it depends in general strongly on the choosen norm and connection on the considered algebra.

Lemma 11.22:

Let $N \geq 1$ be an integer. Then there exists $M > N$ such that the natural map

$$\sum_{k=0}^{\infty}(-\nabla \circ B)^k : \widehat{CC}_*(\mathbb{C}[F_l]) \to \widehat{CC}_*(\mathbb{C}[F_l])$$

extends to a bounded, linear operator

$$\sum_{k=0}^{\infty}(-\nabla \circ B)^k : CC_*(\mathbb{C}[F_l])_{(K_N,N)} \to CC_*(\mathbb{C}[F_l])_{(K_M,M)}$$

Proof:

Let $\omega \in \Omega^{2n}(\mathbb{C}[F_l]_N)$ and choose a presentation

$$\omega = \sum_{\beta} \lambda_\beta a_\beta^0 da_\beta^1 \ldots da_\beta^{2n}$$

with $a_\beta^i = \prod_{j=1}^{m_i} a_{j\beta}^i, a_{j\beta}^i \in K_N$ such that

$$\sum_{\beta} |\lambda_\beta| N^{-n} \frac{1}{n!} \leq \| \omega \|_{N,0} + \epsilon$$

Let $m := \sum_{i=1}^{2n} m_i$ which of course depends on the presentation of ω.

Now

$$-\nabla B(a^0 da^1 \ldots da^{2n}) = -\sum_{i=0}^{2n} \nabla(da^i)da^{i+1} \ldots da^{i-1}$$

and

$$\nabla(da^i) = \nabla(\sum_{j=1}^{m_i} a_1^i \ldots a_{j-1}^i d(a_j^i)a_{j+1}^i \ldots a_{m_i}^i) =$$

$$= \sum_{j=1}^{m_i} a_1^i \ldots a_{j-1}^i \nabla(d(a_j^i))a_{j+1}^i \ldots a_{m_i}^i - \sum_{j=1}^{m_i} a_1^i \ldots a_{j-1}^i d(a_j^i)d(a_{j+1}^i \ldots a_{m_i}^i)$$

Thus $-\nabla B(a_\beta^0 da_\beta^1 \ldots da_\beta^{2n})$ can be written (Lemma 11.21) as a sum of at most $4C(N)m$ terms of the form $a_0' da_1' \ldots da_{2n+2}'$ with $a_i' = \prod_{j=1}^{m_i'} a_{ij}', a_{ij}' \in K_N$ and such that $m \leq m' := \sum_{i=1}^{2n+2} m_i' \leq m + 2$.

Iterating the calculation shows that

$$(-\nabla \circ B)^k(a^0 da^1 \ldots da^{2n}) = \sum a_0'' da_1'' \ldots da_{2n+2k}''$$

with $a_i'' = \prod_{j=1}^{m_i''} a_{ij}''$, $a_{ij}'' \in K_N$ such that $m \le m'' := \sum_{i=1}^{2n+2k} m_i'' \le m + 2k$. and the number of summands is bounded by $(4C(N))^k m(m+2) \ldots (m+2k-2)$.

Choice of M: Choose first a number $M' \ge 2$ large enough that $2^{\frac{1}{M'}} < \left(\frac{N+1}{N}\right)^{\frac{1}{2}}$ and let M be an integer such that $M > 32C(N)M'$.

For the norms one finds

$$\| (-\nabla B)^k(a^0 da^1 \ldots da^{2n}) \|_{M,r} \le$$

$$\le (4C(N))^k m(m+2) \ldots (m+2k-2) max \| a_0'' da_1'' \ldots da_{2n+2k}'' \|_{M,r}$$

Now

$$m(m+2) \ldots (m+2k-2) = M'^k \frac{m}{M'} \ldots \frac{m+2k-2}{M'}$$

$$\le M'^k \frac{m}{M'} \left(\frac{m}{M'} + 1\right) \ldots \left(\frac{m}{M'} + k - 1\right)$$

$$\le M'^k \binom{\frac{m}{M'} + k}{k} k! \le M'^k 2^{\frac{m}{M'} + k} k! \le (2M')^k \left(\frac{N+1}{N}\right)^{\frac{m}{2}} k!$$

by the choice of M.

Considering the norm of $a_0'' da_1'' \ldots da_{2n+2k}''$ note that

$$a_i'' = \prod_{j=1}^{m_i''} a_{ij}'' = \left(\prod_{l=1}^{[\frac{m_i''}{2}]} a_{i(2l-1)}'' a_{i(2l)}''\right) a'''$$

and that

$$b_{il} := \frac{a_{i(2l-1)}'' a_{i(2l)}''}{\| a_{i(2l-1)}'' a_{i(2l)}'' \|_A} \frac{N}{N+1} \in K_M$$

because $M \ge 2N$ (the elements under consideration are linear combinations of monomials of word length at most $2N < M$). Thus

$$\| a_0'' da_1'' \ldots da_{2n+2k}'' \|_{M,r} =$$

$$\prod_{i=0}^{2n+2k} \prod_{l=1}^{[\frac{m_i''}{2}]} \| a_{i(2l-1)}'' a_{i(2l)}'' \|_A \frac{N+1}{N} \| b_0'' db_1'' \ldots db_{2n+2k}'' \|_{M,r}$$

with $b_i'' = \prod b_{il}''$, $b_{il}'' \in K_M$. Therefore

$$\| b_0'' db_1'' \ldots db_{2n+2k}'' \|_{M,r} \le (1+n+k)^r M^{-(n+k)} \frac{1}{(n+k)!}$$

and one concludes that

$$\| a_0'' da_1'' \ldots da_{2n+2k}'' \|_{M,r} \leq \left(\frac{N}{N+1} \right)^{\sum [\frac{m_i''}{2}]} (1+n+k)^r M^{-(n+k)} \frac{1}{(n+k)!}$$

Taking into account that

$$\sum_{i=0}^{2n+2k} [\frac{m_i''}{2}] \geq \sum \frac{m_i''}{2} - (2n+2k+1) \geq \frac{m}{2} - (2n+2k+1)$$

and putting together the estimates obtained so far yields

$$\| (-\nabla B)^k (a^0 da^1 \ldots da^{2n} \|_{M,r} \leq$$

$$\leq (4C(N))^k (2M')^k \left(\frac{N+1}{N} \right)^{(2n+2k+1)} (1+n+k)^r M^{-(n+k)} \frac{k!}{(n+k)!} =$$

$$= (1+n+k)^r \left(8C(N) (\frac{N+1}{N})^2 M' \frac{1}{M} \right)^k (\frac{N+1}{N})^{2n+1} M^{-n} \frac{k!}{(n+k)!} \leq$$

$$\leq C'(1+n+k)^r C^k 4^n M^{-n} \frac{k!}{(n+k)!}$$

for some constants $C, C', \ C < 1$.

Thus finally

$$\| (-\nabla B)^k \omega \|_{M,r} \leq \sum_\beta |\lambda_\beta| \| a_\beta^0 da_\beta^1 \ldots da_\beta^{2n} \|_{M,r} \leq$$

$$\leq (\sum_\beta |\lambda_\beta|) \sup_\beta \| a_\beta^0 da_\beta^1 \ldots da_\beta^{2n} \|_{M,r} \leq$$

$$N^n n! (\| \omega \|_{N,0} + \epsilon) C' (1+n+k)^r C^k 4^n M^{-n} \frac{k!}{(n+k)!} \leq$$

$$\leq C'(1+n+k)^r C^k \left(\frac{4N}{M} \right)^n (\| \omega \|_{N,0} + \epsilon)$$

and the norm of the total sum can be bounded by

$$\| \sum_{k=0}^\infty (-\nabla B)^k \omega \|_{M,r} \leq C' \left(\sum_{k=0}^\infty (1+n+k)^r C^k \right) \left(\frac{4N}{M} \right)^n (\| \omega \|_{N,0} + \epsilon)$$

$$\leq \left(\sum_{k=0}^\infty (k+1)^r C^k \right) C' n^r \left(\frac{4N}{M} \right)^n (\| \omega \|_{N,0} + \epsilon) \leq C''(\| \omega \|_{N,0} + \epsilon)$$

$\square$

Lemma 11.23:

The notations are those of 11.14 and the remarks after 11.18. The natural projections

$$\varinjlim_{N} X_*(R\mathbb{C}[F_n]_{(K_N,N)}) \xrightarrow{qis} \varinjlim_{N} X_*(\mathbb{C}[F_n]_{K_N})$$

$$R\varprojlim_{N} X^*(R\mathbb{C}[F_n]_{(K_N,N)}) \xleftarrow{qis} R\varprojlim_{N} X^*(\mathbb{C}[F_n]_{K_N})$$

are quasiisomorphisms.

Proof:

Consider the comutative diagram

$$
\begin{array}{ccc}
\varinjlim_{N} X_*(R\mathbb{C}[F_n]_{(K_N,N)}) & \xrightarrow{\;qis\;} & \varinjlim_{N} CC_*(\mathbb{C}[F_n])_{(K_N,N)} \\
\pi \downarrow & & \downarrow \pi' \\
\varinjlim_{N} X_*(\mathbb{C}[F_n]_{K_N}) & \xrightarrow[qis]{} & \varinjlim_{N} CC_*/F^2 CC_*(\mathbb{C}[F_n])_{K_N}
\end{array}
$$

The horizontal maps are quasiisomorphisms by the remarks after 11.18. It suffices to show that π' is a quasiisomorphism which follows from 11.20 and 11.22. The cohomological case is similar.

$\square$

Corollary 11.24:

Let A be one of the algebras $l^1(F_n), C_r^*(F_n)$. Let K_N be the corresponding families of compact subsets of $\mathbb{C}[F_n]$ (see 11.14). Then there are isomorphisms

$$HC_*^{lc}(l^1(F_n)) \simeq \varinjlim_{N} H_*(X_*(\mathbb{C}[F_n]_{K_N(l^1)}))$$

$$HC_*^{lc}(C_r^*(F_n)) \simeq \varinjlim_{N} H_*(X_*(\mathbb{C}[F_n]_{K_N(C^*)}))$$

and in the cohomological case short exact sequences

$$0 \to \varprojlim_{N}{}^1 H^{*-1}(X^*(\mathbb{C}[F_n]_{K_N(l^1)})) \to HC_{lc}^*(l^1(F_n)) \to \varprojlim_{N} H^*(X^*(\mathbb{C}[F_n]_{K_N(l^1)})) \to 0$$

$$0 \to \varprojlim_{N}{}^1 H^{*-1}(X^*(\mathbb{C}[F_n]_{K_N(C^*)})) \to HC_{lc}^*(C_r^*(F_n)) \to \varprojlim_{N} H^*(X^*(\mathbb{C}[F_n]_{K_N(C^*)})) \to 0$$

Proof:

This follows from 11.17 and 11.23.

Recall the cyclic cohomology of group rings. For a (torsion free) group Γ, the cyclic cohomology of $\mathbb{C}[\Gamma]$ decomposes as a direct product

$$HC^*(\mathbb{C}[\Gamma]) \simeq H^*(\Gamma, \mathbb{C}) \times \prod_{\langle x \rangle} HC^*(\mathbb{C}[\Gamma])_{\langle x \rangle}$$

of the group cohomology of Γ and of groups labeled by the (nontrivial) conjugacy classes $\langle x \rangle$ of Γ. For $x \in \Gamma$ let (x) be the infinite cyclic subgroup generated by x. Let N_x be the centraliser of (x) in Γ and let

$$1 \to (x) \to N_x \to S_x \to 1$$

be the associated central extension with corresponding class $\xi \in H^2(S_x, \mathbb{Z})$. Then the group $HC^*(\mathbb{C}[\Gamma])_{\langle x \rangle}$ is a module under $H^*(S_x, \mathbb{Z}) \otimes \mathbb{C}$ and the multiplication with ξ corresponds to the S-operation on cyclic cohomology under the isomorphism above.

If $\Gamma = F_n$ is a free group, the groups S_x associated to the nontrivial conjugacy classes are all finite cyclic, so that the action of the corresponding class ξ_x on $HC^*(\mathbb{C}[\Gamma])_{\langle x \rangle}$ is zero. Thus all contributions to the cyclic cohomology of $\mathbb{C}[F_n]$ from nontrivial conjugacy classes are annihilated by the S-operation and therefore

$$HP^*(\mathbb{C}[F_n]) \simeq H^*(X^*(\mathbb{C}[F_n])) \simeq H^*(F_n, \mathbb{C})$$

This will now be carried over to the topological setting.

Proposition 11.25:

The local cyclic (co)homology with compact supports of the Banach algebra $l^1(F_n)$ of the free group on n generators is given by

$$HC_*^{lc}(l^1(F_n)) \simeq \begin{cases} \mathbb{C} & * = 0 \\ \mathbb{C}^n & * = 1 \end{cases}$$

$$HC_{lc}^*(l^1(F_n)) \simeq \begin{cases} \mathbb{C} & * = 0 \\ \mathbb{C}^n & * = 1 \end{cases}$$

Proof:

For $g \in F_n$ denote by $|g|$ its word length with respect to the generators $t_1, \ldots, t_n, t_1^{-1}, \ldots, t_n^{-1}$. Consider the X-complex $X_*(\mathbb{C}[F_n])$. Here

$$X_0(\mathbb{C}[F_n]) = \mathbb{C}[F_n] \quad X_1(\mathbb{C}[F_n]) = \bigoplus_{i=1}^n \mathbb{C}[F_n]dt_i$$

This complex decomposes according to the different conjugacy classes of F_n:

$$X_0(\mathbb{C}[F_n]) = \bigoplus_{\langle x \rangle} X_0(\mathbb{C}[F_n])_{\langle x \rangle} = \bigoplus_{\langle x \rangle} \left(\bigoplus_{z \in \langle x \rangle} \mathbb{C}z \right)$$

$$X_1(\mathbb{C}[F_n]) = \bigoplus_{\langle x \rangle} X_1(\mathbb{C}[F_n])_{\langle x \rangle} = \bigoplus_{\langle x \rangle} \left(\bigoplus_{z \in \langle x \rangle} \bigoplus_{i=1}^n \mathbb{C}zt_i^{-1}dt_i \right)$$

For $z \in F_n, z \neq e$ let $x_1, \ldots, x_m$ be the (finite) set of elements of minimalword length in the conjugacy class of z. For fixed $i, 1 \leq i \leq m$ the set $Y_i(z) := \{y | y x_i y^{-1} = z\}$ contains a unique element y_i of minimal word length. Define an operator

$$\chi_* : X_*(\mathbb{C}[F_n]) \to X_{*+1}(\mathbb{C}[F_n])$$

by

$$\chi_0(z) := -\frac{1}{m} \sum_{i=1}^{m} x_i y_i^{-1} dy_i$$

$$\chi_1(adt_i) := \begin{cases} \frac{1}{|at_i|} at_i & |at_i| > |a| \\ -\frac{1}{|t_i a|} t_i a & |at_i| < |a| \\ 0 & a = t_i^{-1} \end{cases}$$

Then it is easily verified that

$$\chi \circ \partial_\chi + \partial_\chi \circ \chi = Id - P$$

where P is the projection onto the complex $X_*(\mathbb{C}[F_n])_{\langle e \rangle}$ associated to the trivial conjugacy class under the decomposition above. Thus the operator χ provides an explicit chain homotopy annihilating the contributions from the nontrivial conjugacy classes to the cohomology of $\mathbb{C}[F_n]$.

The complex $X_*(\mathbb{C}[F_n])_{\langle e \rangle}$ is easily identified:
$X_0(\mathbb{C}[F_n])_{\langle e \rangle} = \mathbb{C}e$ and $X_1(\mathbb{C}[F_n])_{\langle e \rangle} = \bigoplus_{i=1}^{n} \mathbb{C}t_i^{-1} dt_i$.
The differentials in this complex are zero.

To carry out the analogous calculation in the topological context, the algebras $\mathbb{C}[F_n]_{K_N(l^1)} \subset l^1(F_n)$ have to be identified.
Recalling that

$$K_N(l^1) = \{\sum_g a_g u_g \| |g| \leq N, \sum_g |a_g| \leq \frac{N}{N+1}\}$$

it is not difficult to show that

$$\mathbb{C}[F_n]_{K_N(l^1)} = \{\sum_g a_g u_g | \sum_g |a_g| \left(\frac{N+1}{N}\right)^{\frac{|g|}{N}} < \infty\}$$

and that the norm on this algebra is equivalent to

$$\| a \|_N := \| \sum_g a_g \left(\frac{N+1}{N}\right)^{\frac{|g|}{N}} u_g \|_{l^1}$$

To check the continuity of the homotopy operator χ it suffices, as one works with l^1-algebras, to consider elements of the form $a = u_g \in X_0, adb = u_{g'} du_{t_i} \in X_1$. As the operator χ_1 increases the word length of such an element by at most one and all its coefficients $\pm |g't|^{-1}$ are bounded, it extends to a bounded operator $\chi_1 : X_1(\mathbb{C}[F_n]_{K_N}) \to X_0(\mathbb{C}[F_n]_{K_N})$ for all N. Concerning the operator χ_0 let for $z \in F_n$ be $z = y_i x_i y_i^{-1}, i = 1, \ldots, m$ be presentations as considered before. Then one finds some $i, 1 \le i \le m$ such that $z = y_i x_i y_i^{-1}$ is a minimal presentation of z, i.e. $|z| = |X_i| + 2|y_i|$. Therefore $2|y_i| < |z|$ and for the other elements $y_j, j \ne i$ one easily verifies $|y_j| \le |y_i| + |x_i|$ so that $2|y_j| \le 3|z|$.

Thus

$$\| \chi_0(z) \|_M \le \sum_{i=1}^{m} \frac{1}{m} \| x_i y_i^{-1} dy_i \|_M \le max_i \| x_i y_i^{-1} dy_i \|_M \le$$

$$\le max_i |y_i| \left(\frac{M+1}{M} \right)^{|x_i|+2|y_i|} \le 2|z| \left(\frac{M+1}{M} \right)^{4|z|} \le C \left(\frac{N+1}{N} \right)^{|z|}$$

provided that $\left(\frac{M+1}{M} \right)^4 < \frac{N+1}{N}$. Therefore the operator

$$\chi : X_*(\mathbb{C}[F_n]_{K_N}) \to X_{*+1}(\mathbb{C}[F_n]_{K_M})$$

is continuous and the homotopy formula for χ implies that not the individual complexes $X_*(\mathbb{C}[F_n]_{K_N})$ but their direct limit is quasiisomorphic to the finite dimensional complex $X_*(\mathbb{C}[F_n])_{\langle e \rangle}$:

$$\lim_{\to N} X_*(\mathbb{C}[F_n]_{K_N}) \xrightarrow{qis} \lim_{\to N} X_*(\mathbb{C}[F_n]_{K_N})_{\langle e \rangle} = X_*(\mathbb{C}[F_n])_{\langle e \rangle}$$

From this the conclusion follows.

$\square$

Index of Symbols:

Bibliography:

[B]   B.Blackadar, K-Theory for Operator Algebras,
Springer, (1986)

[CO]   A.Connes, Noncommutative Geometry,
Academic Press, (1995)

[CO]   A.Connes, Noncommutative Differential Geometry,
Publ. Math. IHES 62, (1985), 41-144

[CO_2]   A.Connes, Entire Cyclic Cohomology of Banach Algebras,
and Characters of Theta-summable Fredholm-modules,
Journal of K-Theory 1, (1988), 519-548

[CGM]   A.Connes, M.Gromov, H.Moscovici, Conjecture de Novikov
et Fibres presque plats,
CRAS 310, (1990), 273-277

[CH]   A.Connes, N.Higson, Deformations, Morphismes asymptotiques
et K-Theorie bivariante,
CRAS 311, (1990), 101-106

[CM]   A.Connes, H.Moscovici, Cyclic cohomology, the Novikov Conjecture
and Hyperbolic Groups,
Topology 29, (1990), 345-388

[CQ]   J.Cuntz, D.Quillen, Algebra Extensions and Nonsingularity,
Journal of the AMS 8(2), (1995), 251-289

[CQ]   J.Cuntz, D.Quillen, Cyclic Homology and Nonsingularity,
Journal of the AMS 8(2), (1995), 373-442

[D]   J.Dixmier, C^*-Algebras,
        Gauthier-Villars, (1964)

[GO]   R.Godement, Theorie des Faisceaux,
        Hermann, (1964)

[K]   M.Khalkhali, Algebraic Connections, Universal Bimodules
        and Entire Cyclic Cohomology,
        Communications in Mathematical Physics 161, (1994), 433-446

[M]   J.Milnor, On the Steenrod Homology Theory,
        in Ferry, Ranicki, Rosenberg,
        Novikov Conjectures, Index Theorems and Rigidity,
        LMS Lecture Notes 226, (1995), 79-96

[P]   M.Puschnigg, Explicit Product Structures in Cyclic
        Homology Theories,
        Submitted to Journal of K-theory

[Z]   R.Zekri, Abstract Bott Periodicity in KK-Theory,
        Journal of K-Theory 3, (1990), 543-561

Vol. 1549: G. Vainikko, Multidimensional Weakly Singular Integral Equations. XI, 159 pages. 1993.

Vol. 1550: A. A. Gonchar, E. B. Saff (Eds.), Methods of Approximation Theory in Complex Analysis and Mathematical Physics IV, 222 pages, 1993.

Vol. 1551: L. Arkeryd, P. L. Lions, P.A. Markowich, S.R. S. Varadhan. Nonequilibrium Problems in Many-Particle Systems. Montecatini, 1992. Editors: C. Cercignani, M. Pulvirenti. VII, 158 pages 1993.

Vol. 1552: J. Hilgert, K.-H. Neeb, Lie Semigroups and their Applications. XII, 315 pages, 1993.

Vol. 1553: J.-L- Colliot-Thélène, J. Kato, P. Vojta. Arithmetic Algebraic Geometry. Trento, 1991. Editor: E. Ballico. VII, 223 pages. 1993.

Vol. 1554: A. K. Lenstra, H. W. Lenstra, Jr. (Eds.), The Development of the Number Field Sieve. VIII, 131 pages. 1993.

Vol. 1555: O. Liess, Conical Refraction and Higher Microlocalization. X, 389 pages. 1993.

Vol. 1556: S. B. Kuksin, Nearly Integrable Infinite-Dimensional Hamiltonian Systems. XXVII, 101 pages. 1993.

Vol. 1557: J. Azéma, P. A. Meyer, M. Yor (Eds.), Séminaire de Probabilités XXVII. VI, 327 pages. 1993.

Vol. 1558: T. J. Bridges, J. E. Furter, Singularity Theory and Equivariant Symplectic Maps. VI, 226 pages. 1993.

Vol. 1559: V. G. Sprindžuk, Classical Diophantine Equations. XII, 228 pages. 1993.

Vol. 1560: T. Bartsch, Topological Methods for Variational Problems with Symmetries. X, 152 pages. 1993.

Vol. 1561: I. S. Molchanov, Limit Theorems for Unions of Random Closed Sets. X, 157 pages. 1993.

Vol. 1562: G. Harder, Eisensteinkohomologie und die Konstruktion gemischter Motive. XX, 184 pages. 1993.

Vol. 1563: E. Fabes, M. Fukushima, L. Gross, C. Kenig, M. Röckner, D. W. Stroock, Dirichlet Forms. Varenna, 1992. Editors: G. Dell'Antonio, U. Mosco. VII, 245 pages. 1993.

Vol. 1564: J. Jorgenson, S. Lang, Basic Analysis of Regularized Series and Products. IX, 122 pages. 1993.

Vol. 1565: L. Boutet de Monvel, C. De Concini, C. Procesi, P. Schapira, M. Vergne. D-modules, Representation Theory, and Quantum Groups. Venezia, 1992. Editors: G. Zampieri, A. D'Agnolo. VII, 217 pages. 1993.

Vol. 1566: B. Edixhoven, J.-H. Evertse (Eds.), Diophantine Approximation and Abelian Varieties. XIII, 127 pages. 1993.

Vol. 1567: R. L. Dobrushin, S. Kusuoka, Statistical Mechanics and Fractals. VII, 98 pages. 1993.

Vol. 1568: F. Weisz, Martingale Hardy Spaces and their Application in Fourier Analysis. VIII, 217 pages. 1994.

Vol. 1569: V. Totik, Weighted Approximation with Varying Weight. VI, 117 pages. 1994.

Vol. 1570: R. deLaubenfels, Existence Families, Functional Calculi and Evolution Equations. XV, 234 pages. 1994.

Vol. 1571: S. Yu. Pilyugin, The Space of Dynamical Systems with the C^0-Topology. X, 188 pages. 1994.

Vol. 1572: L. Göttsche, Hilbert Schemes of Zero-Dimensional Subschemes of Smooth Varieties. IX, 196 pages. 1994.

Vol. 1573: V. P. Havin, N. K. Nikolski (Eds.), Linear and Complex Analysis – Problem Book 3 – Part I. XXII, 489 pages. 1994.

Vol. 1574: V. P. Havin, N. K. Nikolski (Eds.), Linear and Complex Analysis – Problem Book 3 – Part II. XXII, 507 pages. 1994.

Vol. 1575: M. Mitrea, Clifford Wavelets, Singular Integrals, and Hardy Spaces. XI, 116 pages. 1994.

Vol. 1576: K. Kitahara, Spaces of Approximating Functions with Haar-Like Conditions. X, 110 pages. 1994.

Vol. 1577: N. Obata, White Noise Calculus and Fock Space. X, 183 pages. 1994.

Vol. 1578: J. Bernstein, V. Lunts, Equivariant Sheaves and Functors. V, 139 pages. 1994.

Vol. 1579: N. Kazamaki, Continuous Exponential Martingales and BMO. VII, 91 pages. 1994.

Vol. 1580: M. Milman, Extrapolation and Optimal Decompositions with Applications to Analysis. XI, 161 pages. 1994.

Vol. 1581: D. Bakry, R. D. Gill, S. A. Molchanov, Lectures on Probability Theory. Editor: P. Bernard. VIII, 420 pages. 1994.

Vol. 1582: W. Balser, From Divergent Power Series to Analytic Functions. X, 108 pages. 1994.

Vol. 1583: J. Azéma, P. A. Meyer, M. Yor (Eds.), Séminaire de Probabilités XXVIII. VI, 334 pages. 1994.

Vol. 1584: M. Brokate, N. Kenmochi, I. Müller, J. F. Rodriguez, C. Verdi, Phase Transitions and Hysteresis. Montecatini Terme, 1993. Editor: A. Visintin. VII. 291 pages. 1994.

Vol. 1585: G. Frey (Ed.), On Artin's Conjecture for Odd 2-dimensional Representations. VIII, 148 pages. 1994.

Vol. 1586: R. Nillsen, Difference Spaces and Invariant Linear Forms. XII, 186 pages. 1994.

Vol. 1587: N. Xi, Representations of Affine Hecke Algebras. VIII, 137 pages. 1994.

Vol. 1588: C. Scheiderer, Real and Étale Cohomology. XXIV, 273 pages. 1994.

Vol. 1589: J. Bellissard, M. Degli Esposti, G. Forni, S. Graffi, S. Isola, J. N. Mather, Transition to Chaos in Classical and Quantum Mechanics. Montecatini Terme, 1991. Editor: S. Graffi. VII, 192 pages. 1994.

Vol. 1590: P. M. Soardi, Potential Theory on Infinite Networks. VIII, 187 pages. 1994.

Vol. 1591: M. Abate, G. Patrizio, Finsler Metrics – A Global Approach. IX, 180 pages. 1994.

Vol. 1592: K. W. Breitung, Asymptotic Approximations for Probability Integrals. IX, 146 pages. 1994.

Vol. 1593: J. Jorgenson & S. Lang, D. Goldfeld, Explicit Formulas for Regularized Products and Series. VIII, 154 pages. 1994.

Vol. 1594: M. Green, J. Murre, C. Voisin, Algebraic Cycles and Hodge Theory. Torino, 1993. Editors: A. Albano, F. Bardelli. VII, 275 pages. 1994.

Vol. 1595: R.D.M. Accola, Topics in the Theory of Riemann Surfaces. IX, 105 pages. 1994.

Vol. 1596: L. Heindorf, L. B. Shapiro, Nearly Projective Boolean Algebras. X, 202 pages. 1994.

Vol. 1597: B. Herzog, Kodaira-Spencer Maps in Local Algebra. XVII, 176 pages. 1994.

Vol. 1598: J. Berndt, F. Tricerri, L. Vanhecke, Generalized